Year 3A
A Guide to Teaching for Mastery

Series Editor: Tony Staneff
Lead author: Josh Lury

Contents

Introduction to the author team

Power Maths arises from the work of maths mastery experts who are committed to proving that, given the right mastery mindset and approach, **everyone can do maths**. Based on robust research and best practice from around the world, *Power Maths* was developed in partnership with a group of UK teachers to make sure that it not only meets our children's wide-ranging needs but also aligns with the National Curriculum in England.

Power Maths – White Rose Maths edition

This edition of *Power Maths* has been developed and updated by:

Tony Staneff, Series Editor and Author

Vice Principal at Trinity Academy, Halifax, Tony also leads a team of mastery experts who help schools across the UK to develop teaching for mastery via nationally recognised CPD courses, problem-solving and reasoning resources, schemes of work, assessment materials and other tools.

Josh Lury, Lead Author

Josh is a specialist maths teacher, author and maths consultant with a passion for innovative and effective maths education.

The first edition of *Power Maths* was developed by a team of experienced authors, including:

- **Tony Staneff and Josh Lury**

- **Trinity Academy Halifax** (Michael Gosling CEO, Emily Fox, Kate Henshall, Rebecca Holland, Stephanie Kirk, Stephen Monaghan and Rachel Webster)

- **David Board, Belle Cottingham, Jonathan East, Tim Handley, Derek Huby, Neil Jarrett, Stephen Monaghan, Beth Smith, Tim Weal, Paul Wrangles** – skilled maths teachers and mastery experts

- **Cherri Moseley** – a maths author, former teacher and professional development provider

- **Professors Liu Jian and Zhang Dan**, Series Consultants and authors, and their team of mastery expert authors: **Wei Huinv, Huang Lihua, Zhu Dejiang, Zhu Yuhong, Hou Huiying, Yin Lili, Zhang Jing, Zhou Da and Liu Qimeng**

  Used by over 20 million children, Professor Liu Jian's textbook programme is one of the most popular in China. He and his author team are highly experienced in intelligent practice and in embedding key maths concepts using a C-P-A approach.

- **A group of 15 teachers and maths co-ordinators**

  We consulted our teacher group throughout the development of *Power Maths* to ensure we are meeting their real needs in the classroom.

What is *Power Maths*?

Created especially for UK primary schools, and aligned with the new National Curriculum, *Power Maths* is a whole-class, textbook-based mastery resource that empowers every child to understand and succeed. *Power Maths* rejects the notion that some people simply 'can't do' maths. Instead, it develops growth mindsets and encourages hard work, practice and a willingness to see mistakes as learning tools.

Best practice consistently shows that mastery of small, cumulative steps builds a solid foundation of deep mathematical understanding. *Power Maths* combines interactive teaching tools, high-quality textbooks and continuing professional development (CPD) to help you equip children with a deep and long-lasting understanding. Based on extensive evidence, and developed in partnership with practising teachers, *Power Maths* ensures that it meets the needs of children in the UK.

Power Maths and Mastery

Power Maths makes mastery practical and achievable by providing the structures, pathways, content, tools and support you need to make it happen in your classroom.

To develop mastery in maths, children must be enabled to acquire a deep understanding of maths concepts, structures and procedures, step by step. Complex mathematical concepts are built on simpler conceptual components and when children understand every step in the learning sequence, maths becomes transparent and makes logical sense. Interactive lessons establish deep understanding in small steps, as well as effortless fluency in key facts such as tables and number bonds. The whole class works on the same content and no child is left behind.

Power Maths

⚡ Builds every concept in small, progressive steps

⚡ Is built with interactive, whole-class teaching in mind

⚡ Provides the tools you need to develop growth mindsets

⚡ Helps you check understanding and ensure that every child is keeping up

⚡ Establishes core elements such as intelligent practice and reflection

The *Power Maths* approach

Everyone can!

Founded on the conviction that every child can achieve, *Power Maths* enables children to build number fluency, confidence and understanding, step by step.

Child-centred learning

Children master concepts one step at a time in lessons that embrace a concrete-pictorial-abstract (C-P-A) approach, avoid overload, build on prior learning and help them see patterns and connections. Same-day intervention ensures sustained progress.

Continuing professional development

Embedded teacher support and development offer every teacher the opportunity to continually improve their subject knowledge and manage whole-class teaching for mastery.

Whole-class teaching

An interactive, whole-class teaching model encourages thinking and precise mathematical language and allows children to deepen their understanding as far as they can.

What's different in the new edition?

If you have previously used the first editions of *Power Maths*, you might be interested to know how this edition is different. All of the improvements described below are based on feedback from *Power Maths* customers.

Changes to units and the progression

⚡ The order of units has been slightly adjusted, creating closer alignment between adjacent year groups, which will be useful for mixed age teaching.

⚡ The flow of lessons has been improved within units to optimise the pace of the progression and build in more recap where needed. For key topics, the sequence of lessons gives more opportunities to build up a solid base of understanding. Other units have fewer lessons than before, where appropriate, making it possible to fit in all the content.

⚡ Overall, the lessons put more focus on the most essential content for that year, with less time given to non-statutory content.

⚡ The progression of lessons matches the steps in the new White Rose Maths schemes of learning.

Lesson resources

⚡ There is a Quick recap for each lesson in the Teacher Guide, which offers an alternative lesson starter to the Power Up for cases where you feel it would be more beneficial to surface prerequisite learning than general number fluency.

⚡ In the **Discover** and **Share** sections there is now more of a progression from 1 a) to 1 b). Whereas before, 1 b) was mainly designed as a separate question, now 1 a) leads directly into 1 b). This means that there is an improved whole-class flow, and also an opportunity to focus on the logic and skills in more detail. As a teacher, you will be using 1 a) to lead the class into the thinking, then 1 b) to mould that thinking into the core new learning of the lesson.

⚡ In the **Share** section, for KS1 in particular, the number of different models and representations has been reduced, to support the clarity of thinking prompted by the flow from 1 a) into 1 b).

⚡ More fluency questions have been built into the guided and independent practice.

⚡ Pupil pages are as easy as possible for children to access independently. The pages are less full where this supports greater focus on key ideas and instructions. Also, more freedom is offered around answer format, with fewer boxes scaffolding children's responses; squared paper backgrounds are used in the Practice Books where appropriate. Artwork has also been revisited to ensure the highest standards of accessibility.

New components

480 Individual Practice Games are available in *ActiveLearn* for practising key facts and skills in Years 1 to 6. These are designed in an arcade style, to feel like fun games that children would choose to play outside school. They can be accessed via the Pupil World for homework or additional practice in school – and children can earn rewards. There are Support, Core and Extend levels to allocate, with Activity Reporting available for the teacher. There is a Quick Guide on *ActiveLearn* and you can use the Help area for support in setting up child accounts.

There is also a new set of lesson video resources on the Professional Development tile, designed for in-school training in 10- to 20-minute bursts. For each part of the *Power Maths* lesson sequence, there is a slide deck with embedded video, which will facilitate discussions about how you can take your *Power Maths* teaching to the next level.

Your *Power Maths* resources

Pupil Textbooks

> **Discover**, **Share** and **Think together** sections promote discussion and introduce mathematical ideas logically, so that children understand more easily.

> Using a Concrete-Pictorial-Abstract approach, clear mathematical models help children to make connections and grasp concepts.

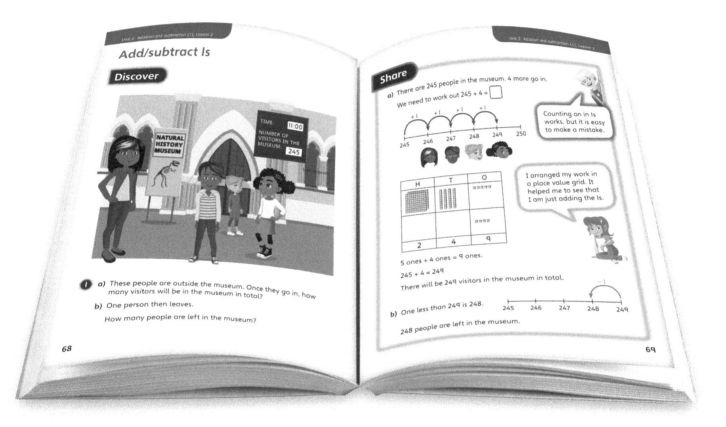

> Appealing scenarios stimulate curiosity, helping children to identify the maths problem and discover patterns and relationships for themselves.

> Friendly, supportive characters help children develop a growth mindset by prompting them to think, reason and reflect.

To help you teach for mastery, *Power Maths* comprises a variety of high-quality resources.

The coherent *Power Maths* lesson structure carries through into the vibrant, high-quality textbooks. Setting out the core learning objectives for each class, the lesson structure follows a carefully mapped journey through the curriculum and supports children on their journey to deeper understanding.

Pupil Practice Books

The Practice Books offer just the right amount of intelligent practice for children to complete independently in the final section of each lesson.

Practice questions are finely tuned to move children forward in their thinking and to reveal misconceptions.

The practice questions are for everyone – each question varies one small element to move children on in their thinking.

Calculations are connected so that children think about the underlying concept.

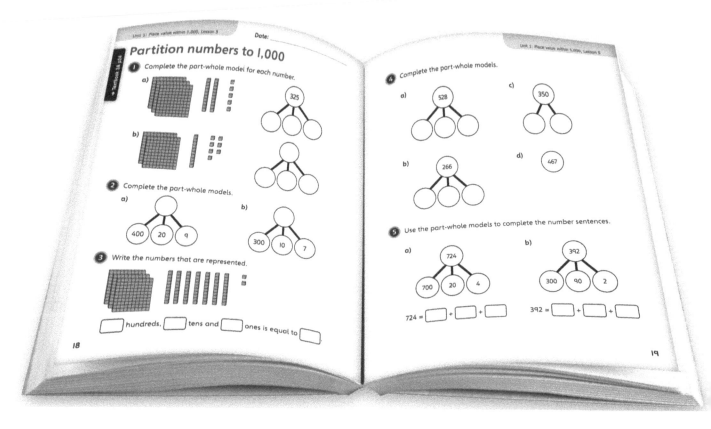

The *Power Maths* characters support and encourage children to think and work in different ways.

Challenge questions allow children to delve deeper into a concept.

Think differently questions encourage children to use reasoning as well as their mathematical knowledge to reach a solution.

Reflect questions reveal the depth of each child's understanding before they move on.

Online subscription

The online subscription will give you access to additional resources and answers from the Textbook and Practice Book.

eTextbooks

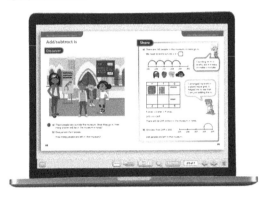

Digital versions of *Power Maths* Textbooks allow class groups to share and discuss questions, solutions and strategies. They allow you to project key structures and representations at the front of the class, to ensure all children are focusing on the same concept.

Teaching tools

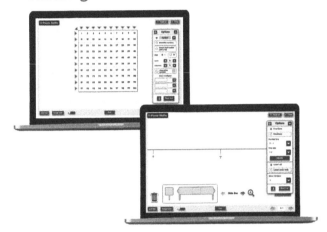

Here you will find interactive versions of key *Power Maths* structures and representations.

Power Ups

Use this series of daily activities to promote and check number fluency.

Online versions of Teacher Guide pages

PDF pages give support at both unit and lesson levels. You will also find help with key strategies and templates for tracking progress.

Unit videos

Watch the professional development videos at the start of each unit to help you teach with confidence. The videos explore common misconceptions in the unit, and include intervention suggestions as well as suggestions on what to look out for when assessing mastery in your students.

End of unit Strengthen and Deepen materials

The Strengthen activity at the end of every unit addresses a key misconception and can be used to support children who need it. The Deepen activities are designed to be low ceiling/high threshold and will challenge those children who can understand more deeply. These resources will help you ensure that every child understands and will help you keep the class moving forward together. These printable activities provide an optional resource bank for use after the assessment stage.

Individual Practice Games

These enjoyable games can be used at home or at school to embed key number skills (see page 6).

Professional Development videos and slides

These slides and videos of *Power Maths* lessons can be used for ongoing training in short bursts or to support new staff.

The *Power Maths* teaching model

At the heart of *Power Maths* is a clearly structured teaching and learning process that helps you make certain that every child masters each maths concept securely and deeply. For each year group, the curriculum is broken down into core concepts, taught in units. A unit divides into smaller learning steps – lessons. Step by step, strong foundations of cumulative knowledge and understanding are built.

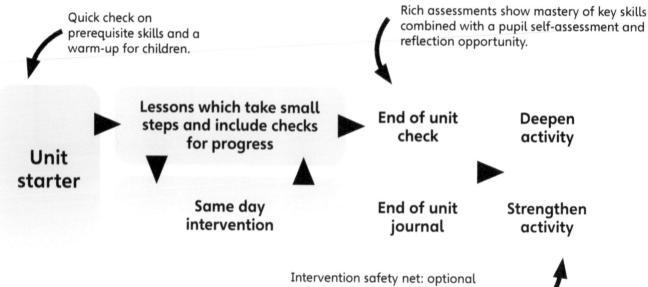

Quick check on prerequisite skills and a warm-up for children.

Rich assessments show mastery of key skills combined with a pupil self-assessment and reflection opportunity.

Unit starter

Lessons which take small steps and include checks for progress

Same day intervention

End of unit check

End of unit journal

Deepen activity

Strengthen activity

Intervention safety net: optional activities to use if assessment shows some children still have misconceptions.

Unit starter

Each unit begins with a unit starter, which introduces the learning context along with key mathematical vocabulary and structures and representations.

- The Textbooks include a check on readiness and a warm-up task for children to complete.

- Your Teacher Guide gives support right from the start on important structures and representations, mathematical language, common misconceptions and intervention strategies.

- Unit-specific videos develop your subject knowledge and insights so you feel confident and fully equipped to teach each new unit. These are available via the online subscription.

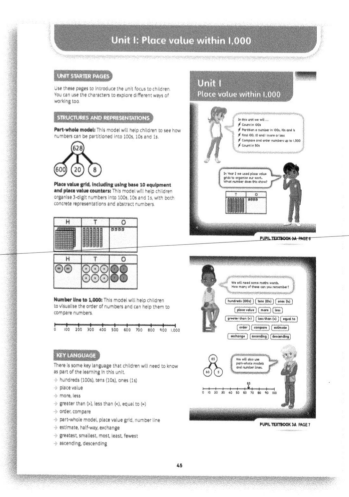

Lesson

Once a unit has been introduced, it is time to start teaching the series of lessons.

- Each lesson is scaffolded with Textbook and Practice Book activities and begins with a Power Up activity (available via online subscription) or the Quick recap activity in the Teacher Guide (see page 15).

- *Power Maths* identifies lesson by lesson what concepts are to be taught.

- Your Teacher Guide offers lots of support for you to get the most from every child in every lesson. As well as highlighting key points, tricky areas and how to handle them, you will also find question prompts to check on understanding and clarification on why particular activities and questions are used.

Same-day intervention

Same-day interventions are vital in order to keep the class progressing together. This can be during the lesson as well as afterwards (see page 29). Therefore, *Power Maths* provides plenty of support throughout the journey.

- Intervention is focused on keeping up now, not catching up later, so interventions should happen as soon as they are needed.

- Practice section questions are designed to bring misconceptions to the surface, allowing you to identify these easily as you circulate during independent practice time.

- Child-friendly assessment questions in the Teacher Guide help you identify easily which children need to strengthen their understanding.

End of unit check and journal

For each unit, the End of unit check in the Textbook lets you see which children have mastered the key concepts, which children have not and where their misconceptions lie. The Practice Books also include an End of unit journal in which children can reflect on what they have learned. Each unit also offers Strengthen and Deepen activities, available via the online subscription.

The Teacher Guide offers different ways of managing the End of unit assessments as well as giving support with handling misconceptions.

The End of unit check presents multiple-choice questions. Children think about their answer, decide on a solution and explain their choice.

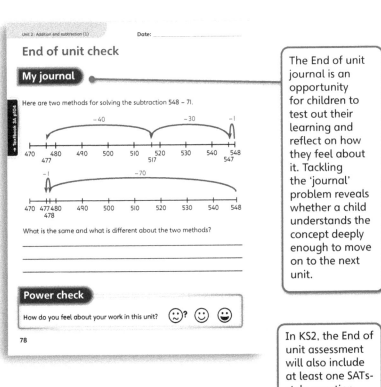

The End of unit journal is an opportunity for children to test out their learning and reflect on how they feel about it. Tackling the 'journal' problem reveals whether a child understands the concept deeply enough to move on to the next unit.

In KS2, the End of unit assessment will also include at least one SATs-style question.

The *Power Maths* lesson sequence

At the heart of *Power Maths* is a unique lesson sequence designed to empower children to understand core concepts and grow in confidence. Embracing the National Centre for Excellence in the Teaching of Mathematics' (NCETM's) definition of mastery, the sequence guides and shapes every *Power Maths* lesson you teach.

Flexibility is built into the *Power Maths* programme so there is no one-to-one mapping of lessons and concepts and you can pace your teaching according to your class. While some children will need to spend longer on a particular concept (through interventions or additional lessons), others will reach deeper levels of understanding. However, it is important that the class moves forward together through the termly schedules.

Power Up ⏱ 5 minutes

Each lesson begins with a Power Up activity (available via the online subscription) which supports fluency in key number facts.

The whole-class approach depends on fluency, so the Power Up is a powerful and essential activity.

The Quick recap is an alternative starter, for when you think some or all children would benefit more from revisiting pre-requisite work (see page 15).

TOP TIP
If the class is struggling with the task, revisit it later and check understanding.

Power Ups reinforce the two key things that are essential for success: times-tables and number bonds.

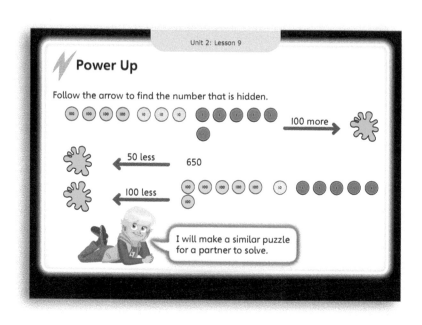

Discover ⏱ 10 minutes

A practical, real-life problem arouses curiosity. Children find the maths through story telling.

A real-life scenario is provided for the **Discover** section but feel free to build upon these with your own examples that are more relevant to your class, or get creative with the context.

TOP TIP
Discover works best when run at tables, in pairs with concrete objects.

Question ❶ a) tackles the key concept and question ❶ b) digs a little deeper. Children have time to explore, play and discuss possible strategies.

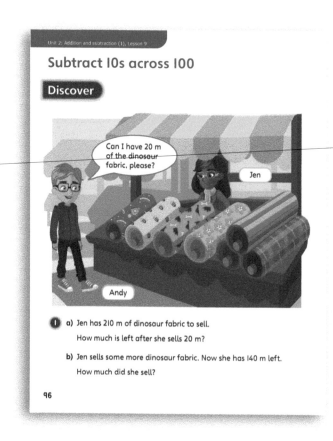

Subtract 10s across 100

Discover

❶ a) Jen has 210 m of dinosaur fabric to sell.
How much is left after she sells 20 m?

b) Jen sells some more dinosaur fabric. Now she has 140 m left.
How much did she sell?

96

Share ⏱ 10 minutes

Teacher-led, this interactive section follows the **Discover** activity and highlights the variety of methods that can be used to solve a single problem.

TOP TIP

Pairs sharing a textbook is a great format for **Share**!

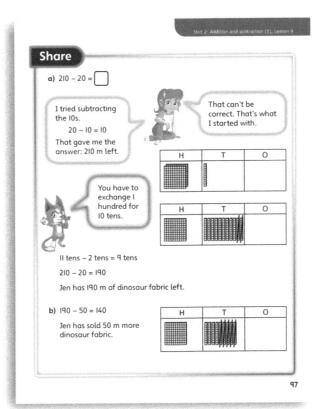

Your Teacher Guide gives target questions for children. The online toolkit provides interactive structures and representations to link concrete and pictorial to abstract concepts.

Bring children to the front to share and celebrate their solutions and strategies.

Think together

⏱ 10 minutes

Children work in groups on the carpet or at tables, using their textbooks or eBooks.

TOP TIP

Make sure children have mini whiteboards or pads to write on if they are not at their tables.

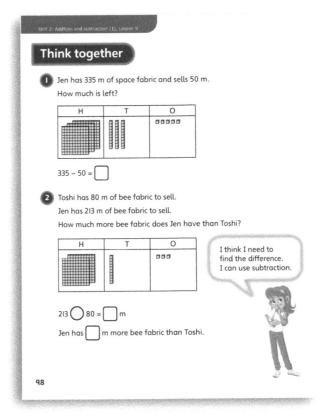

Using the Teacher Guide, model question ① for your class.

Question ② is less structured. Children will need to think together in their groups, then discuss their methods and solutions as a class.

In question ③ children try working out the answer independently. The openness of the **Challenge** question helps to check depth of understanding.

Using their Practice Books, children work independently while you circulate and check on progress.

Questions follow small steps of progression to deepen learning.

TOP TIP
Some children could work separately with a teacher or assistant.

Are some children struggling? If so, work with them as a group, using mathematical structures and representations to support understanding as necessary.

There are no set routines: for real understanding, children need to think about the problem in different ways.

Unit 2: Addition and subtraction (1), Lesson 9   Date: _____

Subtract 10s across 100

↑ Textbook 3A p96

1 Work out the subtractions.

a)

H	T	O

225 – 50 = ☐

I will use base 10 equipment to help me.

b)

H	T	O

231 – 60 = ☐

c)

H	T	O

315 – 80 = ☐

72

'Spot the mistake' questions are great for checking misconceptions.

The Reflect section is your opportunity to check how deeply children understand the target concept.

The Practice Books use various approaches to check that children have fully understood each concept.

Looking like they understand is not enough! It is essential that children can show they have grasped the concept.

Unit 2: Addition and subtraction (1), Lesson 9

5 Complete the part-whole model and missing base 10 equipment. Then complete the calculations.

a) 326 → (200) (120) (6)

326 – 60 = ☐

b) 632 → (◯) (130) (◯)

632 – 80 = ☐

6 Reena thinks of a number. She adds 40 and then adds 50. She ends up with 231. What number did she start with? Talk to a partner about your method. CHALLENGE

_____

_____

_____

Reflect

If you know that 5 + 9 = 14, what other facts can you write? Discuss with a partner.

74

14

Using the *Power Maths* Teacher Guide

Think of your Teacher Guides as *Power Maths* handbooks that will guide, support and inspire your day-to-day teaching. Clear and concise, and illustrated with helpful examples, your Teacher Guides will help you make the best possible use of every individual lesson. They also provide wrap-around professional development, enhancing your own subject knowledge and helping you to grow in confidence about moving your children forward together.

There is a Teacher Guide per year group for every term, with unit and lesson level guidance and support.

Never feel stuck! You will find ideas for introducing every unit and lesson and questions to encourage teacher reflection before and after each lesson.

Tips and advice on key elements such as C-P-A approaches, misconceptions, language, modelling growth mindsets and same day intervention.

Annotations for every Textbook and Practice Book page, providing prompts for key questions to ask to expose understanding and explanations as to why key questions have been chosen.

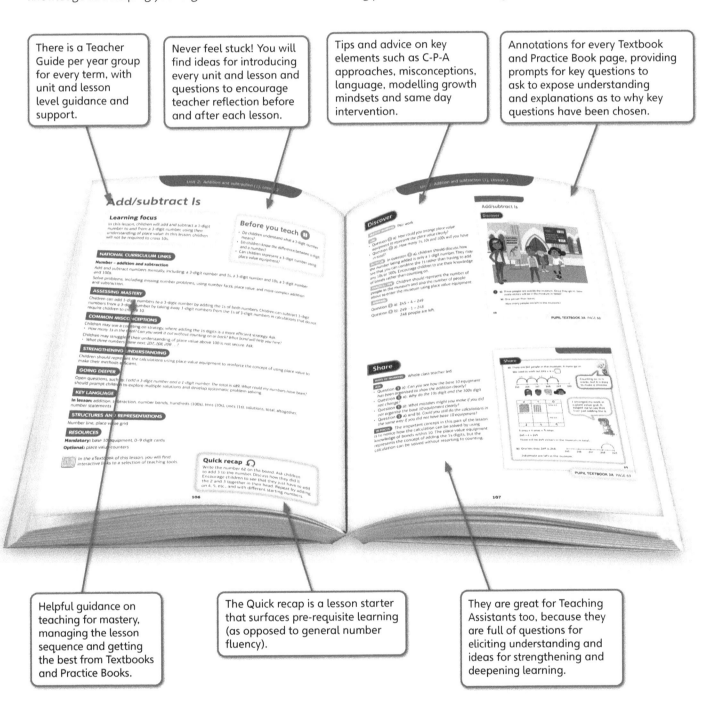

Helpful guidance on teaching for mastery, managing the lesson sequence and getting the best from Textbooks and Practice Books.

The Quick recap is a lesson starter that surfaces pre-requisite learning (as opposed to general number fluency).

They are great for Teaching Assistants too, because they are full of questions for eliciting understanding and ideas for strengthening and deepening learning.

At the end of each unit, your Teacher Guide helps you identify who has fully grasped the concept, who has not and how to move every child forward. This is covered later in the Assessment strategies section.

Power Maths Year 3, yearly overview

Textbook	Strand	Unit		Number of lessons
Textbook A / Practice Workbook A (Term 1)	Number – number and place value	1	Place value within 1,000	13
	Number – addition and subtraction	2	Addition and subtraction (1)	10
	Number – addition and subtraction	3	Addition and subtraction (2)	13
	Number – multiplication and division	4	Multiplication and division (1)	5
	Number – multiplication and division	5	Multiplication and division (2)	13
Textbook B / Practice Workbook B (Term 2)	Number – multiplication and division	6	Multiplication and division (3)	13
	Measurement	7	Length and perimeter	11
	Number – fractions	8	Fractions (1)	10
	Measurement	9	Mass	7
	Measurement	10	Capacity	6
Textbook C / Practice Workbook C (Term 3)	Number – fractions	11	Fractions (2)	8
	Measurement	12	Moneys	5
	Measurement	13	Time	12
	Geometry – properties of shapes	14	Angles and properties of shapes	9
	Statistics	15	Statistics	7

Power Maths Year 3, Textbook 3A (Term I) overview

Strand	Unit		Lesson number	Lesson title	NC Objective 1	NC Objective 2
Number – number and place value	Unit 1	Place value within 1,000	1	Represent and partition numbers to 100	Recognise the place value of each digit in a two-digit number (tens, ones) (Year 2)	Identify, represent and estimate numbers using different representations, including the number line
Number – number and place value	Unit 1	Place value within 1,000	2	Number line to 100	Compare and order numbers up to 1,000	Identify, represent and estimate numbers using different representations, including the number line
Number – number and place value	Unit 1	Place value within 1,000	3	100s	Count from 0 in multiples of 4, 8, 50 and 100; find 10 or 100 more or less than a given number	Recognise the place value of each digit in a three-digit number (hundreds, tens, ones)
Number – number and place value	Unit 1	Place value within 1,000	4	Represent numbers to 1,000	Identify, represent and estimate numbers using different representations, including the number line	Recognise the place value of each digit in a three-digit number (hundreds, tens, ones)
Number – number and place value	Unit 1	Place value within 1,000	5	Partition numbers to 1,000	Recognise the place value of each digit in a three-digit number (100s, 10s, 1s)	Identify, represent and estimate numbers using different representations, including the number line
Number – number and place value	Unit 1	Place value within 1,000	6	Partition numbers to 1,000 flexibly	Recognise the place value of each digit in a three-digit number (100s, 10s, 1s)	
Number – number and place value	Unit 1	Place value within 1,000	7	100s, 10s and 1s	Recognise the place value of each digit in a three-digit number (100s, 10s, 1s)	Identify, represent and estimate numbers using different representations, including the number line
Number – number and place value	Unit 1	Place value within 1,000	8	Use a number line to 1,000	Identify, represent and estimate numbers using different representations, including the number line	Recognise the place value of each digit in a three-digit number (100s, 10s, 1s)
Number – number and place value	Unit 1	Place value within 1,000	9	Estimate on a number line to 1,000	Identify, represent and estimate numbers using different representations, including the number line	
Number – number and place value	Unit 1	Place value within 1,000	10	Find 1, 10 and 100 more or less	Count from 0 in multiples of 4, 8, 50 and 100; find 10 or 100 more or less than a given number	Recognise the place value of each digit in a three-digit number (100s, 10s, 1s)
Number – number and place value	Unit 1	Place value within 1,000	11	Compare numbers to 1,000	Compare and order numbers up to 1,000	Recognise the place value of each digit in a three-digit number (100s, 10s, 1s)
Number – number and place value	Unit 1	Place value within 1,000	12	Order numbers to 1,000	Compare and order numbers up to 1,000	Recognise the place value of each digit in a three-digit number (100s, 10s, 1s)
Number – number and place value	Unit 1	Place value within 1,000	13	Count in 50s	Count from 0 in multiples of 4, 8, 50 and 100; find 10 or 100 more or less than a given number	
Number – addition and subtraction	Unit 2	Addition and subtraction (1)	1	Apply number bonds within 10	Recognise the place value of each digit in a two-digit number (10s, 1s) (Year 2)	Add and subtract numbers mentally, including: a three-digit number and ones, a three-digit number and tens, a three-digit number and hundreds
Number – addition and subtraction	Unit 2	Addition and subtraction (1)	2	Add/subtract 1s	Add and subtract numbers mentally, including: a three-digit number and ones, a three-digit number and tens, a three-digit number and hundreds	Solve problems, including missing number problems, using number facts, place value, and more complex addition and subtraction
Number – addition and subtraction	Unit 2	Addition and subtraction (1)	3	Add/subtract 10s	Add and subtract numbers mentally, including: a three-digit number and ones, a three-digit number and tens, a three-digit number and hundreds	
Number – addition and subtraction	Unit 2	Addition and subtraction (1)	4	Add/subtract 100s	Add and subtract numbers mentally, including: a three-digit number and ones, a three-digit number and tens, a three-digit number and hundreds	

Strand	Unit		Lesson number	Lesson title	NC Objective 1	NC Objective 2
Number – addition and subtraction	Unit 2	Addition and subtraction (1)	5	Spot the pattern	Add and subtract numbers with up to three digits, using formal written methods of columnar addition and subtraction	Add and subtract numbers mentally, including: a three-digit number and ones, a three-digit number and tens, a three-digit number and hundreds
Number – addition and subtraction	Unit 2	Addition and subtraction (1)	6	Add 1s across 10	Add and subtract numbers with up to three digits, using formal written methods of columnar addition and subtraction	Add and subtract numbers mentally, including: a three-digit number and ones, a three-digit number and tens, a three-digit number and hundreds
Number – addition and subtraction	Unit 2	Addition and subtraction (1)	7	Add 10s across 100	Add and subtract numbers with up to three digits, using formal written methods of columnar addition and subtraction	Add and subtract numbers mentally, including: a three-digit number and ones, a three-digit number and tens, a three-digit number and hundreds
Number – addition and subtraction	Unit 2	Addition and subtraction (1)	8	Subtract 1s across 10	Add and subtract numbers with up to three digits, using formal written methods of columnar addition and subtraction	Add and subtract numbers mentally, including: a three-digit number and ones, a three-digit number and tens, a three-digit number and hundreds
Number – addition and subtraction	Unit 2	Addition and subtraction (1)	9	Subtract 10s across 100	Add and subtract numbers with up to three digits, using formal written methods of columnar addition and subtraction	Add and subtract numbers mentally, including: a three-digit number and ones, a three-digit number and tens, a three-digit number and hundreds
Number – addition and subtraction	Unit 2	Addition and subtraction (1)	10	Making connections	Solve problems, including missing number problems, using number facts, place value, and more complex addition and subtraction	
Number – addition and subtraction	Unit 3	Addition and subtraction (2)	1	Add two numbers	Add and subtract numbers with up to three digits, using formal written methods of columnar addition and subtraction	Add and subtract numbers mentally, including: a three-digit number and ones, a three-digit number and tens, a three-digit number and hundreds
Number – addition and subtraction	Unit 3	Addition and subtraction (2)	2	Subtract two numbers	Add and subtract numbers with up to three digits, using formal written methods of columnar addition and subtraction	Add and subtract numbers mentally, including: a three-digit number and ones, a three-digit number and tens, a three-digit number and hundreds
Number – addition and subtraction	Unit 3	Addition and subtraction (2)	3	Add two numbers (across 10)	Add and subtract numbers with up to three digits, using formal written methods of columnar addition and subtraction	Add and subtract numbers mentally, including: a three-digit number and ones, a three-digit number and tens, a three-digit number and hundreds
Number – addition and subtraction	Unit 3	Addition and subtraction (2)	4	Add two numbers (across 100)	Add and subtract numbers with up to three digits, using formal written methods of columnar addition and subtraction	Add and subtract numbers mentally, including: a three-digit number and ones, a three-digit number and tens, a three-digit number and hundreds
Number – addition and subtraction	Unit 3	Addition and subtraction (2)	5	Subtract two numbers (across 10)	Add and subtract numbers with up to three digits, using formal written methods of columnar addition and subtraction	Add and subtract numbers mentally, including: a three-digit number and ones, a three-digit number and tens, a three-digit number and hundreds
Number – addition and subtraction	Unit 3	Addition and subtraction (2)	6	Subtract two numbers (across 100)	Add and subtract numbers with up to three digits, using formal written methods of columnar addition and subtraction	Add and subtract numbers mentally, including: a three-digit number and ones, a three-digit number and tens, a three-digit number and hundreds
Number – addition and subtraction	Unit 3	Addition and subtraction (2)	7	Add a 3-digit and 2-digit number	Add and subtract numbers with up to three digits, using formal written methods of columnar addition and subtraction	Add and subtract numbers mentally, including: a three-digit number and ones, a three-digit number and tens, a three-digit number and hundreds
Number – addition and subtraction	Unit 3	Addition and subtraction (2)	8	Subtract a 2-digit number from a 3-digit number	Add and subtract numbers with up to three digits, using formal written methods of columnar addition and subtraction	Add and subtract numbers mentally, including: a three-digit number and ones, a three-digit number and tens, a three-digit number and hundreds

Strand	Unit		Lesson number	Lesson title	NC Objective 1	NC Objective 2
Number – addition and subtraction	Unit 3	Addition and subtraction (2)	9	Complements to 100	Add and subtract numbers with up to three digits, using formal written methods of columnar addition and subtraction	
Number – addition and subtraction	Unit 3	Addition and subtraction (2)	10	Estimate answers	Estimate the answer to a calculation and use inverse operations to check answers	
Number – addition and subtraction	Unit 3	Addition and subtraction (2)	11	Inverse operations	Estimate the answer to a calculation and use inverse operations to check answers	
Number – addition and subtraction	Unit 3	Addition and subtraction (2)	12	Problem solving (1)	Solve problems, including missing number problems, using number facts, place value, and more complex addition and subtraction	
Number – addition and subtraction	Unit 3	Addition and subtraction (2)	13	Problem solving (2)	Solve problems, including missing number problems, using number facts, place value, and more complex addition and subtraction	
Number – multiplication and division	Unit 4	Multiplication and division (1)	1	Multiplication – equal groups	Write and calculate mathematical statements for multiplication and division using the multiplication tables that they know, including for two-digit numbers times one-digit numbers, using mental and progressing to formal written methods	Recall and use multiplication and division facts for the 3, 4 and 8 multiplication tables
Number – multiplication and division	Unit 4	Multiplication and division (1)	2	Use arrays	Write and calculate mathematical statements for multiplication and division using the multiplication tables that they know, including for two-digit numbers times one-digit numbers, using mental and progressing to formal written methods	Recall and use multiplication and division facts for the 3, 4 and 8 multiplication tables
Number – multiplication and division	Unit 4	Multiplication and division (1)	3	Multiples of 2	Write and calculate mathematical statements for multiplication and division using the multiplication tables that they know, including for two-digit numbers times one-digit numbers, using mental and progressing to formal written methods	Recall and use multiplication and division facts for the 3, 4 and 8 multiplication tables
Number – multiplication and division	Unit 4	Multiplication and division (1)	4	Multiples of 5 and 10	Write and calculate mathematical statements for multiplication and division using the multiplication tables that they know, including for two-digit numbers times one-digit numbers, using mental and progressing to formal written methods	Recall and use multiplication and division facts for the 3, 4 and 8 multiplication tables
Number – multiplication and division	Unit 4	Multiplication and division (1)	5	Sharing and grouping	Write and calculate mathematical statements for multiplication and division using the multiplication tables that they know, including for two-digit numbers times one-digit numbers, using mental and progressing to formal written methods	Recall and use multiplication and division facts for the 3, 4 and 8 multiplication tables
Number – multiplication and division	Unit 5	Multiplication and division (2)	1	Multiply by 3	Recall and use multiplication and division facts for the 3, 4 and 8 multiplication tables	Write and calculate mathematical statements for multiplication and division using the multiplication tables that they know, including for two-digit numbers times one-digit numbers, using mental and progressing to formal written methods

Strand	Unit		Lesson number	Lesson title	NC Objective 1	NC Objective 2
Number – multiplication and division	Unit 5	Multiplication and division (2)	2	Divide by 3	Recall and use multiplication and division facts for the 3, 4 and 8 multiplication tables	Write and calculate mathematical statements for multiplication and division using the multiplication tables that they know, including for two-digit numbers times one-digit numbers, using mental and progressing to formal written methods
Number – multiplication and division	Unit 5	Multiplication and division (2)	3	The 3 times-table	Recall and use multiplication and division facts for the 3, 4 and 8 multiplication tables	Write and calculate mathematical statements for multiplication and division using the multiplication tables that they know, including for two-digit numbers times one-digit numbers, using mental and progressing to formal written methods
Number – multiplication and division	Unit 5	Multiplication and division (2)	4	Multiply by 4	Recall and use multiplication and division facts for the 3, 4 and 8 multiplication tables	Write and calculate mathematical statements for multiplication and division using the multiplication tables that they know, including for two-digit numbers times one-digit numbers, using mental and progressing to formal written methods
Number – multiplication and division	Unit 5	Multiplication and division (2)	5	Divide by 4	Recall and use multiplication and division facts for the 3, 4 and 8 multiplication tables	Write and calculate mathematical statements for multiplication and division using the multiplication tables that they know, including for two-digit numbers times one-digit numbers, using mental and progressing to formal written methods
Number – multiplication and division	Unit 5	Multiplication and division (2)6	6	The 4 times-table	Recall and use multiplication and division facts for the 3, 4 and 8 multiplication tables	Write and calculate mathematical statements for multiplication and division using the multiplication tables that they know, including for two-digit numbers times one-digit numbers, using mental and progressing to formal written methods
Number – multiplication and division	Unit 5	Multiplication and division (2)	7	Multiply by 8	Recall and use multiplication and division facts for the 3, 4 and 8 multiplication tables	Write and calculate mathematical statements for multiplication and division using the multiplication tables that they know, including for two-digit numbers times one-digit numbers, using mental and progressing to formal written methods
Number – multiplication and division	Unit 5	Multiplication and division (2)	8	Divide by 8	Recall and use multiplication and division facts for the 3, 4 and 8 multiplication tables	Write and calculate mathematical statements for multiplication and division using the multiplication tables that they know, including for two-digit numbers times one-digit numbers, using mental and progressing to formal written methods
Number – multiplication and division	Unit 5	Multiplication and division (2)	9	The 8 times-table	Recall and use multiplication and division facts for the 3, 4 and 8 multiplication tables	write and calculate mathematical statements for multiplication and division using the multiplication tables that they know, including for two-digit numbers times one-digit numbers, using mental and progressing to formal written methods
Number – multiplication and division	Unit 5	Multiplication and division (2)	10	Problem solving – multiplication and division (1)	Solve problems, including missing number problems, involving multiplication and division, including positive integer scaling problems and correspondence problems in which n objects are connected to m objects	Write and calculate mathematical statements for multiplication and division using the multiplication tables that they know, including for two-digit numbers times one-digit numbers, using mental and progressing to formal written methods

Strand	Unit		Lesson number	Lesson title	NC Objective 1	NC Objective 2
Number – multiplication and division	Unit 5	Multiplication and division (2)	11	Problem solving – multiplication and division (2)	Solve problems, including missing number problems, involving multiplication and division, including positive integer scaling problems and correspondence problems in which *n* objects are connected to *m* objects	Write and calculate mathematical statements for multiplication and division using the multiplication tables that they know, including for two-digit numbers times one-digit numbers, using mental and progressing to formal written methods
Number – multiplication and division	Unit 5	Multiplication and division (2)	12	Understand divisibility (1)	Solve problems, including missing number problems, involving multiplication and division, including positive integer scaling problems and correspondence problems in which *n* objects are connected to *m* objects	Write and calculate mathematical statements for multiplication and division using the multiplication tables that they know, including for two-digit numbers times one-digit numbers, using mental and progressing to formal written methods
Number – multiplication and division	Unit 5	Multiplication and division (2)	13	Understand divisibility (2)	Solve problems, including missing number problems, involving multiplication and division, including positive integer scaling problems and correspondence problems in which *n* objects are connected to *m* objects	Write and calculate mathematical statements for multiplication and division using the multiplication tables that they know, including for two-digit numbers times one-digit numbers, using mental and progressing to formal written methods

Mindset: an introduction

Global research and best practice deliver the same message: learning is greatly affected by what learners perceive they can or cannot do. What is more, it is also shaped by what their parents, carers and teachers perceive they can do. Mindset – the thinking that determines our beliefs and behaviours – therefore has a fundamental impact on teaching and learning.

Everyone can!

Power Maths and mastery methods focus on the distinction between 'fixed' and 'growth' mindsets (Dweck, 2007).[1] Those with a fixed mindset believe that their basic qualities (for example, intelligence, talent and ability to learn) are pre-wired or fixed: 'If you have a talent for maths, you will succeed at it. If not, too bad!' By contrast, those with a growth mindset believe that hard work, effort and commitment drive success and that 'smart' is not something you are or are not, but something you become. In short, everyone can do maths!

Key mindset strategies

A growth mindset needs to be actively nurtured and developed. *Power Maths* offers some key strategies for fostering healthy growth mindsets in your classroom.

It is okay to get it wrong

Mistakes are valuable opportunities to re-think and understand more deeply. Learning is richer when children and teachers alike focus on spotting and sharing mistakes as well as solutions.

Praise hard work

Praise is a great motivator, and by focusing on praising effort and learning rather than success, children will be more willing to try harder, take risks and persist for longer.

Mind your language!

The language we use around learners has a profound effect on their mindsets. Make a habit of using growth phrases, such as, 'Everyone can!', 'Mistakes can help you learn' and 'Just try for a little longer'. The king of them all is one little word, 'yet'... I can't solve this...yet!' Encourage parents and carers to use the right language too.

Build in opportunities for success

The step-by-small-step approach enables children to enjoy the experience of success. In addition, avoid ability grouping and encourage every child to answer questions and explain or demonstrate their methods to others.

[1] Dweck, C (2007) *The New Psychology of Success*, Ballantine Books: New York

The *Power Maths* characters

The *Power Maths* characters model the traits of growth mindset learners and encourage resilience by prompting and questioning children as they work. Appearing frequently in the Textbooks and Practice Books, they are your allies in teaching and discussion, helping to model methods, alternatives and misconceptions, and to pose questions. They encourage and support your children, too: they are all hardworking, enthusiastic and unafraid of making and talking about mistakes.

Meet the team!

Creative Flo is open-minded and sometimes indecisive. She likes to think differently and come up with a variety of methods or ideas.

Determined Dexter is resolute, resilient and systematic. He concentrates hard, always tries his best and he'll never give up – even though he doesn't always choose the most efficient methods!

'Let's try again.'

'Mistakes are cool!'

'Have I found all of the solutions?'

'Let's try it this way…'

'Can we do it differently?'

'I've got another way of doing this!'

'I'm going to try this!'

'I know how to do that!'

'Want to share my ideas?'

Curious Ash is eager, interested and inquisitive, and he loves solving puzzles and problems. Ash asks lots of questions but sometimes gets distracted.

'What if we tried this…?'

'I wonder…'

'Is there a pattern here?'

Miaow!

Sparks the Cat

Brave Astrid is confident, willing to take risks and unafraid of failure. She's never scared to jump straight into a problem or question, and although she often makes simple mistakes, she's happy to talk them through with others.

Mathematical language

Traditionally, we in the UK have tended to try simplifying mathematical language to make it easier for young children to understand. By contrast, evidence and experience show that by diluting the correct language, we actually mask concepts and meanings for children. We then wonder why they are confused by new and different terminology later down the line! *Power Maths* is not afraid of 'hard' words and avoids placing any barriers between children and their understanding of mathematical concepts. As a result, we need to be deliberate, precise and thorough in building every child's understanding of the language of maths. Throughout the Teacher Guides you will find support and guidance on how to deliver this, as well as individual explanations throughout the pupil Textbooks.

Use the following key strategies to build children's mathematical vocabulary, understanding and confidence.

Precise and consistent

Everyone in the classroom should use the correct mathematical terms in full, every time. For example, refer to 'equal parts', not 'parts'. Used consistently, precise maths language will be a familiar and non-threatening part of children's everyday experience.

Full sentences

Teachers and children alike need to use full sentences to explain or respond. When children use complete sentences, it both reveals their understanding and embeds their knowledge.

Stem sentences

These important sentences help children express mathematical concepts accurately, and are used throughout the *Power Maths* books. Encourage children to repeat them frequently, whether working independently or with others. Examples of stem sentences are:

'4 is a part, 5 is a part, 9 is the whole.'

'There are .... groups. There are .... in each group.'

Key vocabulary

The unit starters highlight essential vocabulary for every lesson. In the pupil books, characters flag new terminology and the Teacher Guide lists important mathematical language for every unit and lesson. New terms are never introduced without a clear explanation.

Mathematical signs

Mathematical signs are used early on so that children quickly become familiar with them and their meaning. Often, the *Power Maths* characters will highlight the connection between language and particular signs.

The role of talk and discussion

When children learn to talk purposefully together about maths, barriers of fear and anxiety are broken down and they grow in confidence, skills and understanding. Building a healthy culture of 'maths talk' empowers their learning from day one.

Explanation and discussion are integral to the *Power Maths* structure, so by simply following the books your lessons will stimulate structured talk. The following key 'maths talk' strategies will help you strengthen that culture and ensure that every child is included.

Sentences, not words

Encourage children to use full sentences when reasoning, explaining or discussing maths. This helps both speaker and listeners to clarify their own understanding. It also reveals whether or not the speaker truly understands, enabling you to address misconceptions as they arise.

Working together

Working with others in pairs, groups or as a whole class is a great way to support maths talk and discussion. Use different group structures to add variety and challenge. For example, children could take timed turns for talking, work independently alongside a 'discussion buddy', or perhaps play different *Power Maths* character roles within their group.

Think first – then talk

Provide clear opportunities within each lesson for children to think and reflect, so that their talk is purposeful, relevant and focused.

Give every child a voice

Where the 'hands up' model allows only the more confident child to shine, *Power Maths* involves everyone. Make sure that no child dominates and that even the shyest child is encouraged to contribute – and praised when they do.

Assessment strategies

Teaching for mastery demands that you are confident about what each child knows and where their misconceptions lie; therefore, practical and effective assessment is vitally important.

Formative assessment within lessons

The **Think together** section will often reveal any confusions or insecurities; try ironing these out by doing the first **Think together** question as a class. For children who continue to struggle, you or your Teaching Assistant should provide support and enable them to move on.

► Performance in practice can be very revealing: check Practice Books and listen out both during and after practice to identify misconceptions.

► The **Reflect** section is designed to check on the all-important depth of understanding. Be sure to review how the children performed in this final stage before you teach the next lesson.

End of unit check – Textbook

Each unit concludes with a summative check to help you assess quickly and clearly each child's understanding, fluency, reasoning and problem solving skills. Your Teacher Guide will suggest ideal ways of organising a given activity and offer advice and commentary on what children's responses mean. For example, 'What misconception does this reveal?'; 'How can you reinforce this particular concept?'

For younger children, assess in small, teacher-led groups, giving each child time to think and respond while also consolidating correct mathematical language. Assessment with young children should always be an enjoyable activity, so avoid one-to-one individual assessments, which they may find threatening or scary. If you prefer, the End of unit check can be carried out as a whole-class group using whiteboards and Practice Books.

End of unit check – Practice Book

The Practice Book contains further opportunities for assessment, and can be completed by children independently whilst you are carrying out diagnostic assessment with small groups. Your Teacher Guide will advise you on what to do if children struggle to articulate an explanation – or perhaps encourage you to write down something they have explained well. It will also offer insights into children's answers and their implications for next learning steps. It is split into three main sections, outlined below.

My journal is designed to allow children to show their depth of understanding of the unit. It can also serve as a way of checking that children have grasped key mathematical vocabulary. The question children should answer is first presented in the Textbook in the Think! section. This provides an opportunity for you to discuss the question first as a class to ensure children have understood their task. Children should have some time to think about how they want to answer the question, and you could ask them to talk to a partner about their ideas. Then children should write their answer in their Practice Book, using the word bank provided to help them with vocabulary.

The **Power check** allows pupils to self-assess their level of confidence on the topic by colouring in different smiley faces. You may want to introduce the faces as follows:

I am starting to understand. I need more practice

I don't yet know how to do this

I know how to do this, but I will keep practising

Each unit ends with either a Power play or a Power puzzle. This is an activity, puzzle or game that allows children to use their new knowledge in a fun, informal way.

Progress Tests

There are *Power Maths* Progress Tests for each half term and at the end of the year, including an Arithmetic test and Reasoning test in each case. You can enter results in the online markbook to track and analyse results and see the average for all schools' results. The tests use a 6-step scale to show results against age-related expectation.

How to ask diagnostic questions

The diagnostic questions provided in children's Practice Books are carefully structured to identify both understanding and misconceptions (if children answer in a particular way, you will know why). The simple procedure below may be helpful:

Ask the question, offering the selection of answers provided.

Children take time to think about their response.

Each child selects an answer and shares their reasoning with the group.

Give minimal and neutral feedback (for example, 'That's interesting', or 'Okay').

Ask, 'Why did you choose that answer?', then offer an opportunity to change their mind by providing one correct and one incorrect answer.

Note which children responded and reasoned correctly first time and everyone's final choices.

Reflect that together, we can get the right answer.

Keeping the class together

Traditionally, children who learn quickly have been accelerated through the curriculum. As a consequence, their learning may be superficial and will lack the many benefits of enabling children to learn with and from each other.

By contrast, *Power Maths'* mastery approach values real understanding and richer, deeper learning above speed. It sees all children learning the same concept in small, cumulative steps, each finding and mastering challenge at their own level. Remember that when you teach for mastery, EVERYONE can do maths! Those who grasp a concept easily have time to explore and understand that concept at a deeper level. The whole class therefore moves through the curriculum at broadly the same pace via individual learning journeys.

For some teachers, the idea that a whole class can move forward together is revolutionary and challenging. However, the evidence of global good practice clearly shows that this approach drives engagement, confidence, motivation and success for all learners, and not just the high flyers. The strategies below will help you keep your class together on their maths journey.

Mix it up

Do not stick to set groups at each table. Every child should be working on the same concept, and mixing up the groupings widens children's opportunities for exploring, discussing and sharing their understanding with others.

Recycling questions

Reuse the Textbook and Practice Book questions with concrete materials to allow children to explore concepts and relationships and deepen their understanding. This strategy is especially useful for reinforcing learning in same-day interventions.

Strengthen at every opportunity

The next lesson in a *Power Maths* sequence always revises and builds on the previous step to help embed learning. These activities provide golden opportunities for individual children to strengthen their learning with the support of Teaching Assistants.

Prepare to be surprised!

Children may grasp a concept quickly or more slowly. The 'fast graspers' won't always be the same individuals, nor does the speed at which a child understands a concept predict their success in maths. Are they struggling or just working more slowly?

Same-day intervention

Since maths competence depends on mastering concepts one by one in a logical progression, it is important that no gaps in understanding are ever left unfilled. Same-day interventions – either within or after a lesson – are a crucial safety net for any child who has not fully made the small step covered that day. In other words, intervention is always about keeping up, not catching up, so that every child has the skills and understanding they need to tackle the next lesson. That means presenting the same problems used in the lesson, with a variety of concrete materials to help children model their solutions.

We offer two intervention strategies below, but you should feel free to choose others if they work better for your class.

Within-lesson intervention

The **Think together** activity will reveal those who are struggling, so when it is time for practice, bring these children together to work with you on the first practice questions. Observe these children carefully, ask questions, encourage them to use concrete models and check that they reach and can demonstrate their understanding.

After-lesson intervention

You might like to use the **Think together** questions to recap the lesson with children who are working behind expectations during assembly time. Teaching Assistants could also work with these children at other convenient points in the school day. Some children may benefit from revisiting work from the same topic in the previous year group. Note also the suggestion for recycling questions from the Textbook and Practice Book with concrete materials on page 28.

The role of practice

Practice plays a pivotal role in the *Power Maths* approach. It takes place in class groups, smaller groups, pairs, and independently, so that children always have the opportunities for thinking as well as the models and support they need to practise meaningfully and with understanding.

Intelligent practice

In *Power Maths*, practice never equates to the simple repetition of a process. Instead we embrace the concept of intelligent practice, in which all children become fluent in maths through varied, frequent and thoughtful practice that deepens and embeds conceptual understanding in a logical, planned sequence. To see the difference, take a look at the following examples.

Traditional practice

- Repetition can be rote – no need for a child to think hard about what they are doing

- Praise may be misplaced

- Does this prove understanding?

Intelligent practice

- Varied methods – concrete, pictorial and abstract

- Equation expressed in different ways, requiring thought and understanding

- Constructive feedback

All practice questions are designed to move children on and reveal misconceptions.

Simple, logical steps build onto earlier learning.

C-P-A runs throughout – different ways of modelling and understanding the same concept.

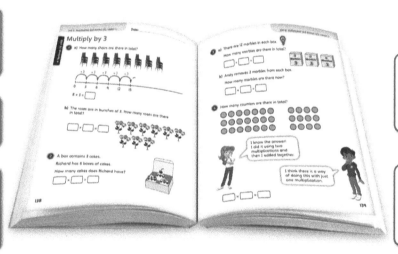

Conceptual variation – children work on different representations of the same maths concept.

Friendly characters offer support and encourage children to try different approaches.

A carefully designed progression

The Practice Books provide just the right amount of intelligent practice for children to complete independently in the final sections of each lesson. It is really important that all children are exposed to the practice questions, and that children are not directed to complete different sections. That is because each question is different and has been designed to challenge children to think about the maths they are doing. The questions become more challenging so children grasping concepts more quickly will start to slow down as they progress. Meanwhile, you have the chance to circulate and spot any misconceptions before they become barriers to further learning.

Homework and the role of parents and carers

While *Power Maths* does not prescribe any particular homework structure, we acknowledge the potential value of practice at home. For example, practising fluency in key facts, such as number bonds and times-tables, is an ideal homework task. You can share the Individual Practice Games for homework (see page 6), or parents and carers could work through uncompleted Practice Book questions with children at either primary stage.

However, it is important to recognise that many parents and carers may themselves lack confidence in maths, and few, if any, will be familiar with mastery methods. A Parents' and Carers' evening that helps them understand the basics of mindsets, mastery and mathematical language is a great way to ensure that children benefit from their homework. It could be a fun opportunity for children to teach their families that everyone can do maths!

Structures and representations

Unlike most other subjects, maths comprises a wide array of abstract concepts – and that is why children and adults so often find it difficult. By taking a concrete-pictorial-abstract (C-P-A) approach, *Power Maths* allows children to tackle concepts in a tangible and more comfortable way.

Concrete

Replacing the traditional approach of a teacher working through a problem in front of the class, the concrete stage introduces real objects that children can use to 'do' the maths – any familiar object that a child can manipulate and move to help bring the maths to life. It is important to appreciate, however, that children must always understand the link between models and the objects they represent. For example, children need to first understand that three cakes could be represented by three pretend cakes, and then by three counters or bricks. Frequent practice helps consolidate this essential insight. Although they can be used at any time, good concrete models are an essential first step in understanding.

Non-linear stages

Pictorial

This stage uses pictorial representations of objects to let children 'see' what particular maths problems look like. It helps them make connections between the concrete and pictorial representations and the abstract maths concept. Children can also create or view a pictorial representation together, enabling discussion and comparisons. The *Power Maths* teaching tools are fantastic for this learning stage, and bar modelling is invaluable for problem solving throughout the primary curriculum.

Abstract

Our ultimate goal is for children to understand abstract mathematical concepts, symbols and notation and of course, some children will reach this stage far more quickly than others. To work with abstract concepts, a child must be comfortable with the meaning of and relationships between concrete, pictorial and abstract models and representations. The C-P-A approach is not linear, and children may need different types of models at different times. However, when a child demonstrates with concrete models and pictorial representations that they have grasped a concept, we can be confident that they are ready to explore or model it with abstract symbols such as numbers and notation.

Use at any time and with any age to support understanding

Variation helps visualisation

Children find it much easier to visualise and grasp concepts if they see them presented in a number of ways, so be prepared to offer and encourage many different representations.

For example, the number six could be represented in various ways:

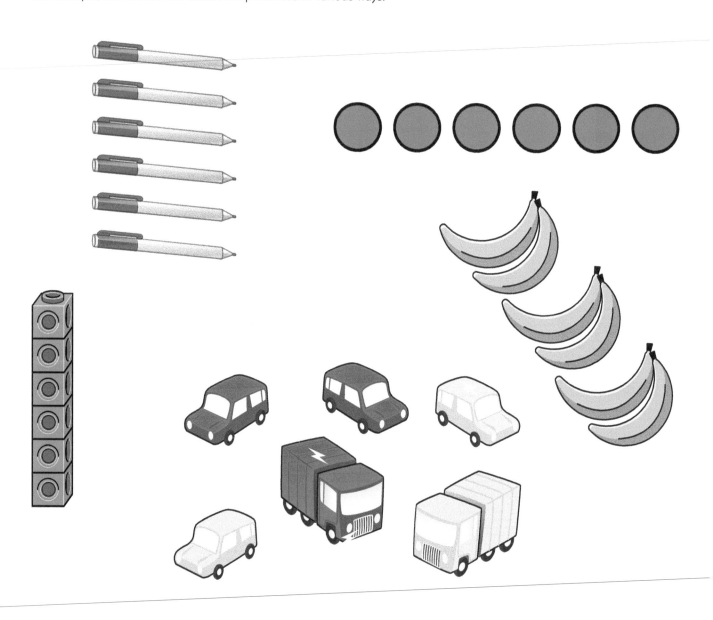

Practical aspects of *Power Maths*

One of the key underlying elements of *Power Maths* is its practical approach, allowing you to make maths real and relevant to your children, no matter their age.

Manipulatives are essential resources for both key stages and *Power Maths* encourages teachers to use these at every opportunity, and to continue the Concrete-Pictorial-Abstract approach right through to Year 6.

The Textbooks and Teacher Guides include lots of opportunities for teaching in a practical way to show children what maths means in real life.

Discover and Share

The **Discover** and **Share** sections of the Textbook give you scope to turn a real-life scenario into a practical and hands-on section of the lesson. Use these sections as inspiration to get active in the classroom. Where appropriate, use the **Discover** contexts as a springboard for your own examples that have particular resonance for your children – and allow them to get their hands dirty trying out the mathematics for themselves.

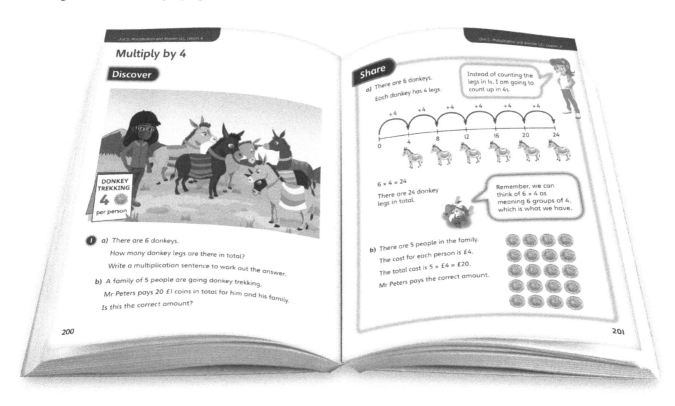

Unit videos

Every term has one unit video which incorporates real-life classroom sequences.

These videos show you how the reasoning behind mathematics can be carried out in a practical manner by showing real children using various concrete and pictorial methods to come to the solution. You can see how using these practical models, such as part-whole and bar models, helps them to find and articulate their answer.

Mastery tips

Mastery Experts give anecdotal advice on where they have used hands-on and real-life elements to inspire their children.

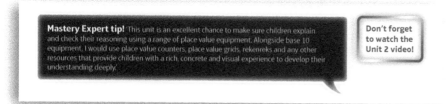

Concrete-Pictorial-Abstract (C-P-A) approach

Each **Share** section uses various methods to explain an answer, helping children to access abstract concepts by using concrete tools, such as counters. Remember, this isn't a linear process, so even children who appear confident using the more abstract method can deepen their knowledge by exploring the concrete representations. Encourage children to use all three methods to really solidify their understanding of a concept.

Pictorial representation – drawing the problem in a logical way that helps children visualise the maths

Concrete representation – using manipulatives to represent the problem. Encourage children to physically use resources to explore the maths.

Abstract representation – using words and calculations to represent the problem.

Practical tips

Every lesson suggests how to draw out the practical side of the **Discover** context.

You'll find these in the **Discover** section of the Teacher Guide for each lesson.

PRACTICAL TIPS Ask children to count aloud in 100s when finding the total number of bricks in each part. For support, children could represent the number of bricks using base 10 equipment.

Resources

Every lesson lists the practical resources you will need or might want to use. There is also a summary of all of the resources used throughout the term on page 40 to help you be prepared.

RESOURCES

Mandatory: base 10 equipment

Optional: place value counters, place value grids

Working with children below age-related expectation

This section offers advice on using *Power Maths* with children who are significantly behind age-related expectation. Teacher judgement will be crucial in terms of where and why children are struggling, and in choosing the right approach. The suggestions can of course be adapted for children with special educational needs, depending on the specific details of those needs.

General approaches to support children who are struggling

Keeping the pace manageable

Remember, you have more teaching days than *Power Maths* lessons so you can cover a lesson over more than one day, and revisit key learning, to ensure all children are ready to move on. You can use the + and – buttons to adjust the time for each unit in the online planning. The NCETM's Ready-to-Progress criteria can be used to help determine what should be highest priority.

Same-day intervention

You could go over the Textbook pages or revisit the previous year's work if necessary (see Addressing gaps). Remember that same-day intervention can be within the lesson, as well as afterwards (see page 29). As children start their independent practice, you can work with those who found the first part of the lesson difficult, checking understanding using manipulatives.

Fluency sessions

Fit in as much practice as you can for number bonds and times-tables, etc., at other times of the day. If you can, plan a short 'maths meeting' for this in the afternoon. You might choose to use a Power Up you haven't used already.

Addressing gaps

Use material from the same topic in the previous year to consolidate or address gaps in learning, e.g. Textbook pages and Strengthen activities. The End of unit check will help gauge children's understanding.

Pre-teaching

Find a 5- to 10-minute slot before the lesson to work with the children you feel would benefit. The afternoon before the lesson can work well, because it gives children time to think in between. Recap previous work on the topic (addressing any gaps you're aware of) and do some fluency practice, targeting number facts etc. that will help children access the learning.

Focusing on the key concepts

If children are a long way behind, it can be helpful to take a step back and think about the key concepts for children to engage with, not just the fine detail of the objective for that year group (e.g. addition with a specific number of columns). Bearing that in mind, how could children advance their understanding of the topic?

Providing extra support within the lesson

Support in the Teacher Guide

First of all, use the Strengthen support in the Teacher Guide for guided and independent work in each lesson, and share this with Teaching Assistants, where relevant. As you read through the lesson content and corresponding Teacher Guide pages before the lesson, ask yourself what key idea or nugget of understanding is at the heart of the lesson. If children are struggling, this should help you decide what's essential for all children before they move on.

Annotating pages

You can annotate questions to provide extra scaffolding or hints if you need to, but aim to build up children's ability to access questions independently wherever you can. Children tend to get used to the style of the *Power Maths* questions over time.

Quick recap as lesson starter

The Quick recap for each lesson in the Teacher Guide is an alternative starter activity to the Power Up. You might choose to use this with some or all children if you feel they will need support accessing the main lesson.

Consolidation questions

If you think some children would benefit from additional questions at the same level before moving on, write one or two similar questions on the board. (This shouldn't be at the expense of reasoning and problem-solving opportunities: take longer over the lesson if you need to.)

Hard copy Textbooks

The Textbooks help children focus in more easily on the mathematical representations, read the text more comfortably, and revisit work from a previous lesson that you are building on, as well as giving children ownership of their learning journey. In main lessons, it can work well to use the e-Textbook for **Discover** and give out the books when discussing the methods in the **Share** section.

Reading support

It's important that all children are exposed to problem solving and reasoning questions, which often involve reading. For whole-class work you can read questions together. For independent practice you could consider annotating pages to help children see what the question is asking, and stem sentences to help structure their answer. A general focus on specific mathematical language and vocabulary will help children access the questions. You could consider pairing weaker readers with stronger readers, or read questions as a group if those who need support are on the same table.

Providing extra depth and challenge with *Power Maths*

Just as prescribed in the National Curriculum, the goal of *Power Maths* is never to accelerate through a topic but rather to gain a clear, deep and broad understanding. Here are some suggestions to help ensure all children are appropriately challenged as you work with the resources.

Overall approaches

First of all, remember that the materials are designed to help you keep the class together, allowing all children to master a concept while those who grasp it quickly have time to explore it in more depth. Use the Deepen support in the Teacher Guide (see below) to challenge children who work through the questions quickly. Here are some questions and ideas to encourage breadth and depth during specific parts of the lesson, or at any time (where no part of the lesson sequence is specified):

- **Discover**: 'Can you demonstrate your solution another way?'

- **Share**: Make sure every child is encouraged to give answers and engage with the discussion, not just the most confident.

- **Think together**: 'Can you model your answers using concrete materials? Can you explain your solution to a partner?'

- Practice: Allow all children to work through the full set of questions, so that they benefit from the logical sequence.

- **Reflect**: 'Is there another way of working out the answer? And another way?'
  'Have you found all the solutions?'
  'Is that always true?'
  'What's different between this question and that question? And what's the same?'

Note that the **Challenge** questions are designed so that all children can access and attempt them, if they have worked through the steps leading up to them. There may be some children in a given lesson who don't manage to do the **Challenge**, but it is not supposed to be a distinct task for a subset of the class. When you look through the lesson materials before teaching, think about what each question is specifically asking, and compare this with the key learning point for the lesson. This will help you decide which questions you feel it's essential for all children to answer, before moving on. You can at least aim for all children to try the **Challenge**!

Deepen activities and support

The Teacher Guide provides valuable support for each stage of the lesson. This includes Deepen tips for the guided and independent practice sections, which will help you provide extra stretch and challenge within your lesson, without having to organise additional tasks. If you have a Teaching Assistant, they can also make use of this advice. There are also suggestions for the lesson as a whole in the 'Going Deeper' section on the first page of the Teacher Guide section for that lesson. Every class is different, so you
can always go a bit further in the direction indicated, if appropriate, and build on the suggestions given.

There is a Deepen activity for each unit. These are designed to follow on from the End of unit check, stretching children who have a firm understanding of the key learning from the unit. Children can work on them independently, which makes it easier for the teacher to facilitate the Strengthen activity for children who need extra support. Deepen activities could also be introduced earlier in the unit if the necessary work has been covered. The Deepen activities are on *ActiveLearn* on the Planning page for each unit, and also on the Resources page).

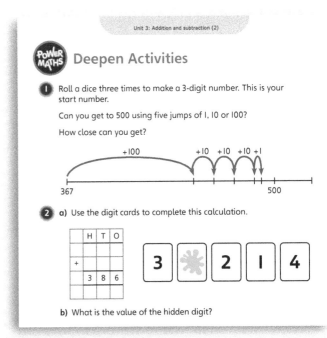

Using the questions flexibly to provide extra challenge

Sometimes you may want to write an extra question on the board or provide this on paper. You can usually do this by tweaking the lesson materials. The questions are designed to form a carefully structured sequence that builds understanding step by step, but, with careful thought about the purpose of each question, you can use the materials flexibly where you need to. Sometimes you might feel that children would benefit from another similar question for consolidation before moving on to the next one, or you might feel that they would benefit from a harder example in the same style. It should be quick and easy to generate 'more of the same' type questions where this is the case.

When you see a question like this one (from Unit 3, Lesson 3), it's easy to make harder examples to do afterwards if you need them. What if the 7 was also blotted out – how many possibilities would there be for the ones digits? For a trickier example, if you set up a similar addition with a tens digit blotted out (and the tens part of the answer given), children would have to factor in the exchanged 10 in order to work out what was missing.

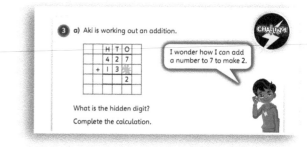

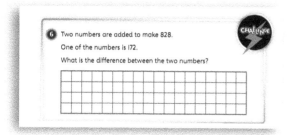

For this example (from Unit 3, Lesson 6), you could ask children to make up their own question(s) for a partner to solve. They could even make up questions to solve themselves! (In fact, for any of these examples you could ask early finishers to create their own question for a partner.)

Here's an example (from Unit 3, Lesson 12) where some of the combinations in the picture feature as questions in the lesson, but others don't. Clearly there are plenty of multi-step problems you could ask using the same information. Children could choose what to buy for their family, or you could tell them about your family and the bikes, helmets and lights you would need. You could also give them a budget to work out the change from buying a list of equipment.

Besides creating additional questions, you should be able to find a question in the lesson that you can adapt into a game or open-ended investigation, if this helps to keep everyone engaged. It could simply be that, instead of answering 5 × 5 etc. on the page, they could build a robot with 5 lots of 5 cubes.

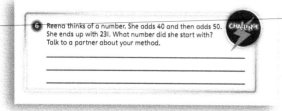

With a question like this (Unit 2, Lesson 9), children could play a game where they say the additions and the end number, and their partner has to work out the start number.

See the bullets above for some general ideas that will help with 'opening out' questions in the books, e.g. 'Can you find all the solutions?' type questions.

Other suggestions

Another way of stretching children is through mixed ability pairs, or via other opportunities for children to explain their understanding in their own way. This is a good way of encouraging children to go deeper into the learning, rather than, for instance, tackling questions that are computationally more challenging but conceptually equivalent in level.

Using *Power Maths* with mixed age classes

Overall approaches

There are many variables between schools that would make it inadvisable to recommend a one-size-fits-all approach to mixed age teaching with *Power Maths*. These include how year groups are merged, availability of Teaching Assistants, experience and preference of teaching staff, range in pupil attainment across years, classroom space and layout, level of flexibility around timetables, and overall organisational structure (whether the school is part of a trust).

Some schools will find it best to timetable separate maths lessons for the different year groups. Others will aim to teach the class together as much as possible using the mixed age planning support on *ActiveLearn* (see the lesson exemplars for ways of organising lessons with strong/medium/weak correlation between year groups). There will also be ways of adapting these general approaches. For example, offset lessons where Year A start their lesson with the teacher, while Year B work independently on the practice from the previous lesson, and then start the next lesson with the teacher while Year A work independently; or teachers may choose to base their provision around the lesson from one year group and tweak the content up/down for the other group.

Key strategies for mixed age teaching

The mixed age teaching webinar on *ActiveLearn* provides advice on all aspects of mixed age teaching, including more detail on the ideas below.

Developing independence over time
Investing time in building up children's independence will pay off in the medium term.

Clear rationale
If someone asked, 'Why did you teach both Unit 3 and 4 in the same lesson/separate lessons?', what would your answer be?

Designing a lesson
1. Identify the core learning for each group
2. Identify any number skills necessary to access the core
3. Consider the flow of concepts and how one core leads to the other

Challenging all children
The questions are designed to build understanding step by step, but with careful thought about the purpose of each question you can tweak them to increase the challenge.

Multiple years combined
With more than two years together, teachers will inevitably need to use the resources flexibly if delivering a single lesson.

Enjoy the positives!

Comparison deepens understanding and there will be lots of opportunities for children, as well as misconceptions to explore. There is also in-built pre-teaching and the chance to build up a concept from its foundations. For teachers there is double the material to draw on! Mixed age teachers require a strong understanding of the progression of ideas across year groups, which is highly valuable for all teachers. Also, it is necessary to engage deeply with the lesson to see how to use the materials flexibly – this is recommended for all teachers and will help you bring your lesson to life!

List of practical resources

Year 3A Mandatory resources

Resource	Lesson
100 square	**Unit 3** Lesson 9 **Unit 4** Lessons 3, 4 **Unit 5** Lesson 6
bar models	**Unit 3** Lesson 13
base 10 equipment	**Unit 1** Lessons 1, 4, 7, 10 **Unit 2** Lessons 1, 2, 3, 4, 6 7, 8, 9, 10 **Unit 3** Lessons 7, 8
base 10 equipment (100s)	**Unit 1** Lessons 3, 5, 6
base 10 equipment (and/or place value counters)	**Unit 3** Lesson 3
base 10 equipment (or other place value equipment)	**Unit 3** Lessons 1, 5
counters	**Unit 1** Lessons 4, 7 **Unit 3** Lesson 9 **Unit 4** Lessons 1, 2, 5 **Unit 5** Lessons 1, 2, 3, 4, 5, 6, 7, 8, 9, 12
counters or cubes	**Unit 5** Lesson 13
cubes	**Unit 4** Lesson 1 **Unit 5** Lessons 1, 2, 3, 4, 5, 6, 7, 8, 9
dice (or dice simulator)	**Unit 1** Lessons 11, 12 **Unit 2** Lesson 10 **Unit 3** Lesson 2
digit cards (0–9)	**Unit 2** Lessons 2, 6 **Unit 3** Lessons 7, 8
digit cards or number cards	**Unit 3** Lesson 10
lollipop sticks or strips of paper to make shapes	**Unit 5** Lesson 12
number lines	**Unit 2** Lesson 10 **Unit 3** Lesson 10
place value counters	**Unit 1** Lessons 4, 7, 10 **Unit 2** Lessons 4, 7 **Unit 3** Lessons 4, 7
place value equipment (e.g. base 10 equipment, place value counters, cards and grids)	**Unit 3** Lessons 2, 5, 6
place value grids	**Unit 2** Lesson 4 **Unit 3** Lessons 3, 4, 7, 8

Year 3A Optional resources

Resource	Lesson
100 square	**Unit 1** Lesson 3 **Unit 2** Lesson 3
100s numbers to 1,000	**Unit 3** Lesson 10
4 times-table flashcards	**Unit 5** Lesson 6
addition scaffolds (blank)	**Unit 3** Lesson 9
bags and boxes of objects in 100s	**Unit 1** Lesson 3
balls	**Unit 5** Lesson 1
base 10 equipment	**Unit 1** Lessons 2, 3, 8, 9, 11, 12, 13 **Unit 3** Lessons 4, 9 **Unit 4** Lesson 4
bead strings	**Unit 1** Lesson 3 **Unit 2** Lesson 8 **Unit 3** Lesson 9
circles (divided into 2, 4 and 8 equal pieces)	**Unit 5** Lesson 7
coins (50p)	**Unit 1** Lesson 13
concrete classroom objects (e.g. counters)	**Unit 3** Lesson 10
counters	**Unit 2** Lesson 6 **Unit 5** Lessons 10, 11
counting sticks	**Unit 1** Lessons 8, 9
cubes	**Unit 3** Lesson 10 **Unit 4** Lesson 5 **Unit 5** Lessons 10, 11, 12
cups	**Unit 4** Lesson 5
dice	**Unit 3** Lesson 2
digit cards	**Unit 3** Lesson 1 **Unit 4** Lessons 3, 4
digit cards (0–20)	**Unit 2** Lesson 7, 8, 9
fruit juice	**Unit 5** Lesson 8
ice lolly moulds	**Unit 5** Lesson 8
number lines	**Unit 1** Lesson 10 **Unit 3** Lessons 2, 5, 6, 11 **Unit 4** Lesson 1
number lines (blank, laminated)	**Unit 2** Lessons 6, 10
paper (circles)	**Unit 5** Lesson 8
paper (scrap)	**Unit 3** Lesson 11
paper (strips)	**Unit 3** Lessons 12, 3
paper clips	**Unit 3** Lessons 2, 10
part-whole models	**Unit 2** Lesson 5 **Unit 3** Lessons 5, 6, 9, 11
part-whole models (laminated)	**Unit 1** Lesson 1
part-whole models (large laminated)	**Unit 1** Lessons 5, 6
pegs	**Unit 1** Lesson 9 **Unit 3** Lesson 10
place value abacus	**Unit 2** Lessons 3, 4
place value counters	**Unit 1** Lessons 12, 13 **Unit 2** Lessons 1, 2 **Unit 3** Lesson 8
place value equipment	**Unit 3** Lesson 13
place value grids	**Unit 2** Lesson 1

Year 3A Optional resources – *continued*

Resource	Lesson
place value grid (laminated)	**Unit 1** Lesson 1 **Unit 2** Lesson 10
place value grid (large laminated)	**Unit 1** Lessons 4, 7
plastic animals	**Unit 5** Lesson 4
plastic cups	**Unit 5** Lesson 1
playing cards (48)	**Unit 5** Lesson 5
rods (coloured)	**Unit 3** Lesson 13
spinners	**Unit 3** Lesson 2
straws	**Unit 4** Lesson 5
string	**Unit 1** Lesson 9 **Unit 3** Lesson 13
string or wool (ball of)	**Unit 2** Lesson 9
ten frame	**Unit 2** Lesson 6
times-tables (written on stairs)	**Unit 5** Lesson 3
toy money	**Unit 1** Lesson 10
washing line	**Unit 3** Lesson 10
whiteboards	**Unit 1** Lesson 9
wooden blocks	**Unit 5** Lesson 11

Getting started with *Power Maths*

As you prepare to put *Power Maths* into action, you might find the tips and advice below helpful.

STEP 1: Train up!

A practical, up-front full day professional development course will give you and your team a brilliant head-start as you begin your *Power Maths* journey. You will learn more about the ethos, how it works and why.

STEP 2: Check out the progression

Take a look at the yearly and termly overviews. Next take a look at the unit overview for the unit you are about to teach in your Teacher Guide, remembering that you can match your lessons and pacing to match your class.

STEP 3: Explore the context

Take a little time to look at the context for this unit: what are the implications for the unit ahead? (Think about key language, common misunderstandings and intervention strategies, for example.) If you have the online subscription, don't forget to watch the corresponding unit video.

STEP 4: Prepare for your first lesson

Familiarise yourself with the objectives, essential questions to ask and the resources you will need. The Teacher Guide offers tips, ideas and guidance on individual lessons to help you anticipate children's misconceptions and challenge those who are ready to think more deeply.

STEP 5: Teach and reflect

Deliver your lesson — and enjoy!

Afterwards, reflect on how it went... Did you cover all five stages? Does the lesson need more time? How could you improve it?

Unit 1
Place value within 1,000

Mastery Expert tip! 'Since understanding of place value underpins the majority of work that they do this year, I made sure to use base 10 equipment until children were confident enough to work with abstract numbers.'

Don't forget to watch the Unit 1 video!

WHY THIS UNIT IS IMPORTANT

This unit is important as it explores 3-digit numbers in depth. For many children, it will be the first time they have met these numbers. This work builds on the place value work that they did in Year 2 and they will extend many of the models and images that they have used previously.

Children begin with learning how to count in 100s. They will learn that a 3-digit number is made up of some 100s, 10s and 1s and they will be able to represent this in many ways (for example, on a place value grid with counters or in a part-whole model). They will extend the number line to 1,000 and know where different numbers lie. They will compare and order 3-digit numbers as well as count in 50s.

This unit underpins a lot of the subsequent work this year and it is essential that children gain a solid understanding of the key concepts within this unit.

WHERE THIS UNIT FITS

→ **Unit 1: Place value within 1,000**
→ Unit 2: Addition and subtraction (1)

This unit builds on children's work in Year 2 on 2-digit numbers. This unit is essential for the work in the rest of this year when they look at the four rules of number, fractions and measure. In the next unit, children move on to adding and subtracting 3-digit numbers.

Before they start this unit, it is expected that children:
• know that a 2-digit number is made up of 10s and 1s
• can represent 2-digit numbers in different ways, such as with base 10 equipment, place value grids and counters, part-whole models and number lines
• can find 1 and 10 more and less than a 2-digit number
• can compare and order 2-digit numbers
• know where a 2-digit number lies on a number line.

ASSESSING MASTERY

Children will know that a number is made up of some 100s, 10s and 1s and will be able to represent numbers in multiple ways. They will be able to find 100, 10 and 1 more or less than a 3-digit number. They will be able to compare and order 3-digit numbers by looking at the place value of each digit. They will understand the number line to 1,000 and start to know where numbers lie on the number line.

COMMON MISCONCEPTIONS	STRENGTHENING UNDERSTANDING	GOING DEEPER
Children may think that 2 tens + 5 hundreds + 7 ones is 257 as opposed to 527.	Use base 10 equipment and place value counters to secure understanding of 3-digit numbers.	Ask children to partition numbers in different ways. For example, 226 is 1×100, 12×10 and 6×1.
Children may compare a 3-digit and a 2-digit number by looking at the first digit rather than the numbers of 100s.	Use base 10 equipment or place value counters in a place value grid to emphasise that they need to compare numbers that have the same place value.	Challenge children to make as many numbers as possible using eight blank counters in an HTO place value grid. Ask them to order their numbers on a number line.

Unit 1: Place value within 1,000

Use these pages to introduce the unit focus to children. You can use the characters to explore different ways of working too.

STRUCTURES AND REPRESENTATIONS

Part-whole model: This model will help children to see how numbers can be partitioned into 100s, 10s and 1s.

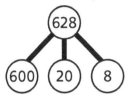

Place value grid, including using base 10 equipment and place value counters: This model will help children organise 3-digit numbers into 100s, 10s and 1s, with both concrete representations and abstract numbers.

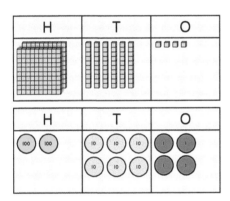

Number line to 1,000: This model will help children to visualise the order of numbers and can help them to compare numbers.

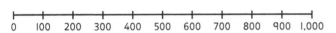

KEY LANGUAGE

There is some key language that children will need to know as part of the learning in this unit.

→ hundreds (100s), tens (10s), ones (1s)
→ place value
→ more, less
→ greater than (>), less than (<), equal to (=)
→ order, compare
→ part-whole model, place value grid, number line
→ estimate, half-way, exchange
→ greatest, smallest, most, least, fewest
→ ascending, descending

PUPIL TEXTBOOK 3A PAGE 6

PUPIL TEXTBOOK 3A PAGE 7

45

Represent and partition numbers to 100

Learning focus

In this lesson, children will recap representing and partitioning numbers to 100 using a variety of representations such as base 10 equipment and part-whole models. This will prepare them for later learning in the unit.

NATIONAL CURRICULUM LINKS

Year 2 Number – number and place value

Recognise the place value of each digit in a two-digit number (tens, ones).

Year 3 Number – number and place value

Identify, represent and estimate numbers using different representations, including the number line.

ASSESSING MASTERY

Children can represent numbers from 0 to 100 in a variety of ways and use these representations to partition numbers into 10s and 1s.

COMMON MISCONCEPTIONS

When completing a part-whole model for a 2-digit number, children may write each digit in a part, for example, writing the parts of 76 as 7 and 6 rather than considering the place value of each digit. Ask:

- *What does the 7 represent?*
- *What does the 6 represent?*

STRENGTHENING UNDERSTANDING

Children who are struggling to partition numbers to 100 first need to place their base 10 equipment representations into the whole of the part-whole model and then physically move the 10s into one part and the 1s into the other part. This will allow them to see the parts of the whole.

GOING DEEPER

Ask children to make different 2-digit numbers using base 10 equipment and partition them. Ask: *Can you make two numbers with the same number of 10s? Can you make two numbers with the same number of 1s?*

KEY LANGUAGE

In lesson: tens (10s), ones (1s), part, whole

Other language to be used by the teacher: represent, partition

STRUCTURES AND REPRESENTATIONS

Part-whole models, place value grid

RESOURCES

Mandatory: base 10 equipment

Optional: laminated part-whole models, laminated place value grid

 In the eTextbook of this lesson, you will find interactive links to a selection of teaching tools.

Discover

Represent and partition numbers to 100

WAYS OF WORKING Pair work

ASK

- Question ① a): *What number is on Emma's card?*
- Question ① b): *How many 10s does Andy have? How many 1s does Andy have? What number has he made?*
- Questions ① a) and b): *What is the same about their numbers? What is different?*

IN FOCUS In question ① a), children use the numerical form to write a number and use Emma's speech to help them recognise the 10s and 1s digits. In question ① b), they then look at base 10 equipment in a place value grid and interpret the number. They connect this with Andy's representation to compare the two numbers.

PRACTICAL TIPS Give children base 10 equipment and a place value grid to make Emma and Andy's numbers and together, physically count the 10s and 1s.

ANSWERS

Question ① a): Children make the number 36 from base 10 equipment. Emma's number is 36.

Question ① b): Andy's number is 46.

Discover

My number has 3 tens and 6 ones.

My number has 1 more ten.

Emma   Andy

① a) Draw or make Emma's number from base 10 equipment. What number has Emma made?

b) What number has Andy made?

8

PUPIL TEXTBOOK 3A PAGE 8

Share

WAYS OF WORKING Whole class teacher led

ASK

- Question ① a): *How many 10s are there? How do you say this? How many 1s are there? What is Emma's number?*
- Question ① b): *Where can you see the 10s in the part-whole model? Where can you see the 1s? What does the whole represent?*

IN FOCUS In question ① a), children are given strategies for counting and recognising numbers made from base 10 equipment. Count slowly with the children so they know to count the 10s first and then the 1s. In question ① b), they are reintroduced to the part-whole model as a way of partitioning 2-digit numbers. Use a place value grid to support children's thinking. Explain that the column headings help determine the place value of each digit in a 2-digit number.

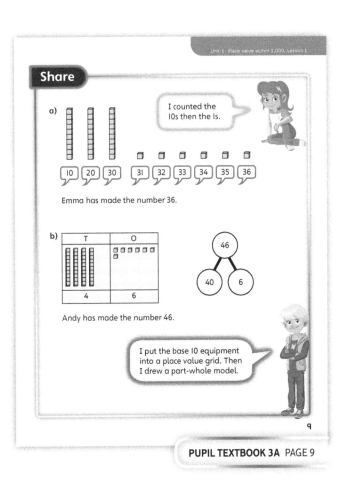

Share

a) I counted the 10s then the 1s.

10 20 30 31 32 33 34 35 36

Emma has made the number 36.

b)

T	O
4	6

46
40 6

Andy has made the number 46.

I put the base 10 equipment into a place value grid. Then I drew a part-whole model.

9

PUPIL TEXTBOOK 3A PAGE 9

Think together

Whole class teacher led (I do, We do, You do)

ASK

- Question **1**: *How many boxes of 10 pencils are there? How many individual pencils are there? How many pencils are there altogether?*
- Question **2**: *How many 10s are there? How many 1s are there? What is the number?*

IN FOCUS In question **1**, children should count the pencils strategically by first counting in 10s and then counting in 1s from 50. In question **2**, children interpret numbers in base 10 equipment in a place value grid. They should identify the number of 10s and the number of 1s in the number. In question **3**, children use the part-whole model with base 10 equipment to complete the part-whole model with digits in. Connections should be made between both part-whole models.

STRENGTHEN Provide children with a laminated place value grid and base 10 equipment to support them in representing and identifying the numbers.

DEEPEN Ask children to make a different number with the same number of 10s or 1s.

ASSESSMENT CHECKPOINT Use questions **1** and **2** to check that children understand representations of number, specifically the number of 10s and the number of 1s. Use question **3** to assess whether children can correctly complete part-whole models. Successfully identifying the misconception in question **3** b) will demonstrate this knowledge.

ANSWERS

Question **1**: 52

Question **2** a): 75

Question **2** b): 28

Question **3** a): 30 and 5

Question **3** b): The 6 should be 60 because the 6 in 63 represents 6 tens not 6 ones.

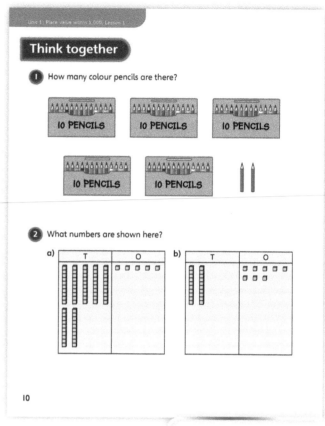

PUPIL TEXTBOOK 3A PAGE 10

PUPIL TEXTBOOK 3A PAGE 11

48

Practice

WAYS OF WORKING Independent thinking

IN FOCUS In questions ① and ② children count objects and should do this strategically by first counting the 10s then continuing in 1s. In question ③ encourage children to identify the number of 10s and 1s to help them find the number represented. In question ④, children complete part-whole models written numerically, filling in both parts and wholes.

STRENGTHEN Provide children with base 10 equipment to represent the numbers to support them with partitioning them into part-whole models. They could organise this into a place value grid to support them further.

DEEPEN In question ⑥ ask children to find other numbers that have the same thing in common. Ask: *Can you create your own set of numbers that have something different in common? Can you partition these numbers?*

ASSESSMENT CHECKPOINT Use questions ① to ③ to assess whether children can interpret representations of number. Use question ④ to assess whether children can identify missing parts and wholes in a part-whole model.

ANSWERS Answers for the **Practice** part of the lesson can be found in the *Power Maths* online subscription.

Reflect

WAYS OF WORKING Independent thinking

IN FOCUS This activity checks that children can make a 2-digit number using base 10 equipment. Children then represent their number in a part-whole model and should be encouraged to link the base 10 equipment to the different parts.

ASSESSMENT CHECKPOINT Check that children can make a number from base 10 equipment and that they draw the correct part-whole model. Ensure they recognise the 10s and 1s in the part-whole model and how they link to the base 10 equipment.

ANSWERS Answers for the **Reflect** part of the lesson can be found in the *Power Maths* online subscription.

After the lesson ⏸

- Can children represent numbers to 100?
- Can children identify the number of 10s and 1s in a 2-digit number?
- Can children partition a 2-digit number using a part-whole model?

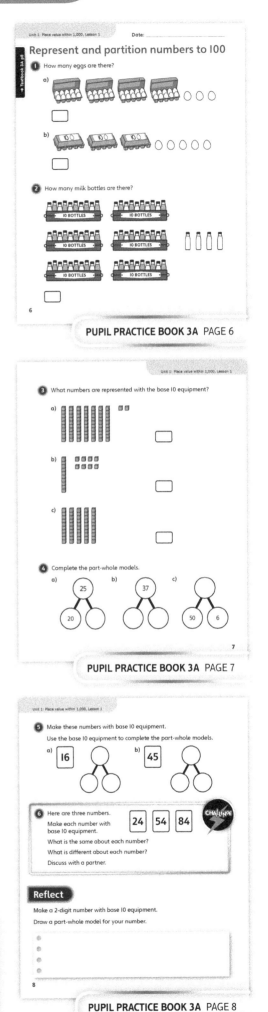

PUPIL PRACTICE BOOK 3A PAGE 6

PUPIL PRACTICE BOOK 3A PAGE 7

PUPIL PRACTICE BOOK 3A PAGE 8

Number line to 100

Learning focus

In this lesson, children will identify and label numbers within 100 on a number line. They should be able to identify and label numbers made up of only 10s as well as numbers with both 10s and 1s.

Before you teach ⏸

- Can children count from 0 to 100 in 10s?
- Can children count on from any number up to 100?
- Do children know what the next 10 is after a number, for example 34?

NATIONAL CURRICULUM LINKS

Year 3 Number – number and place value

Compare and order numbers up to 1,000.

Identify, represent and estimate numbers using different representations, including the number line.

ASSESSING MASTERY

Children can count in 10s from 0 to 100 on a number line. They should understand which numbers lie between two multiples of 10 and, given different start and end points such as 20 and 30, label the numbers in between.

COMMON MISCONCEPTIONS

Children may think that all number lines count in 10s or that all number lines count in 1s. They may only focus on the start point and start counting rather than looking at the difference between the start and end point. Ask:

- *What is the start number?*
- *What is the end number?*
- *What is the difference between these numbers?*
- *What numbers are in between the start and end numbers?*

STRENGTHENING UNDERSTANDING

Children who are struggling should be encouraged to count aloud along the number line to support them in identifying and positioning numbers on the number line. They could use base 10 equipment to make the start and end points to help them find the numbers in between.

GOING DEEPER

Ask children to estimate the position of different numbers on the number line and encourage them to explain their reasoning for their position. They should compare the number to the midpoint of two intervals when explaining.

KEY LANGUAGE

In lesson: number line, one hundred, tens (10s), ones (1s)

Other language to be used by the teacher: estimate, position

STRUCTURES AND REPRESENTATIONS

Number lines, part-whole models

RESOURCES

Optional: base 10 equipment

 In the eTextbook of this lesson, you will find interactive links to a selection of teaching tools.

Quick recap 🔄

Draw a blank part-whole model on the board. Ask a child to write a 2-digit number in the whole. Ask another child to fill in one of the parts and ask another to fill in the other part. Repeat this with more 2-digit numbers.

Discover

Pair work

ASK

- Question ① a): *How long is the race? Where did they start? Who is winning? What does each vertical line represent?*

IN FOCUS In question ① a), children are required to count in 10s on a number line. They could be encouraged to count aloud from 0 to 100 in 10s. In question ① b), children should notice that Runner D is not on a vertical line but instead half-way between two lines. They should be able to reason why this value is 25 m.

PRACTICAL TIPS Consider drawing out a number line in the playground from 0 to 100 and asking children to count aloud as they walk along it.

ANSWERS

Question ① a): Runner B has run 50 m.
Runner C has run 70 m.

Question ① b): Runner D has run 25 m.

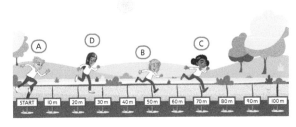

Number line to 100

Discover

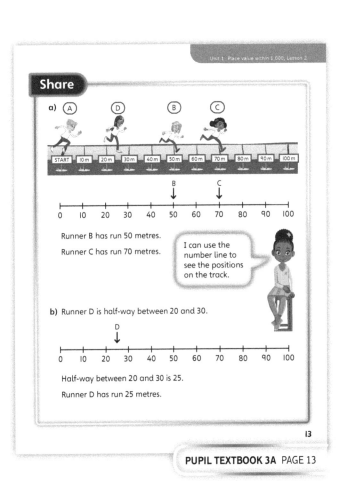

① a) How far has runner B run?
How far has runner C run?

b) How far has runner D run?

12

Share

Whole class teacher led

ASK

- Question ① a): *How can you check that the number line is counting on in 10s? Do you need to start counting from 0 for Runner C's number or can you carry on from Runner B's distance?*
- Question ① b): *Which two markers is Runner D between? What numbers come in between 20 and 30? Runner D is half-way between 20 m and 30 m. How far have they run?*

IN FOCUS The focus here is on exploring the intervals on the number line. Encourage children to check by counting along from the start to the end point in 10s. In question ① b), they should be able to identify all the numbers that lie in this interval and then explain why the answer is 25 m.

Share

a)

Runner B has run 50 metres.

Runner C has run 70 metres.

I can use the number line to see the positions on the track.

b) Runner D is half-way between 20 and 30.

Half-way between 20 and 30 is 25.

Runner D has run 25 metres.

13

51

Think together

Whole class teacher led (I do, We do, You do)

ASK

- Question **1**: *What is the number line counting on in? How do you know? What numbers are the arrows pointing to? Which arrows were easier? Which were harder?*
- Question **2**: *What is the number line going up in? How do you know? How can you check?*

IN FOCUS In question **1**, children focus on a number line from 0 to 100 that is more abstract than the one they saw in the **Discover** section. In question **2**, they look at a number line counting in 1s. They should be encouraged to count aloud along the number line to check they are correct and help them identify the numbers.

STRENGTHEN Children could use base 10 equipment to build the start and end points of the number line and then use this to help them identify the intervals in between. This can help them find the numbers that the arrows are pointing to.

DEEPEN In question **3**, children are required to estimate the position of different numbers on the number line. They should reason about their decisions by comparing the numbers to the midpoint of the intervals.

ASSESSMENT CHECKPOINT Use question **1** to assess whether children can work with a number line from 0 to 100 in 10s. Use question **2** to assess whether children can use a number line counting in 1s starting from a multiple of 10.

ANSWERS

Question **1**: A = 30, B = 75, C = 90.

Question **2**: A = 42, B = 45, C = 47.

Question **3**: Children point to the 10s markings and the 1s markings.

Children point to where 25, 57 and 92 should be on the number line.

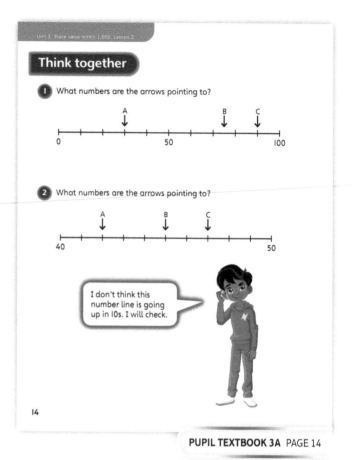

PUPIL TEXTBOOK 3A PAGE 14

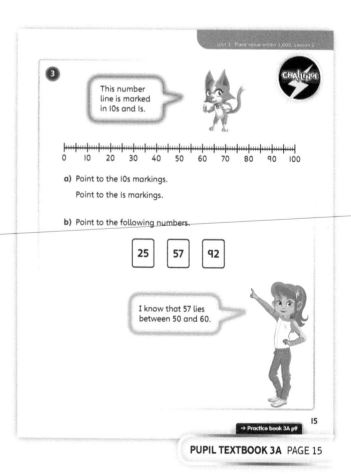

PUPIL TEXTBOOK 3A PAGE 15

Practice

WAYS OF WORKING Independent thinking

IN FOCUS In question **1**, children are required to fill in the intervals on a number line. In question **2**, they need to identify the numbers that arrows on the number lines are pointing to. Question **3** asks children to estimate the number that an arrow is pointing to in the middle of an interval, and question **4** requires them to reason around this idea. In question **5**, children use this acquired knowledge to draw arrows and estimate the position of numbers on a number line.

STRENGTHEN Base 10 equipment could be used to make the start and end points of number lines to support children in identifying intervals. They could also be encouraged to count aloud along the number line.

DEEPEN In questions **6** and **7**, children will be required to reason about the position of the numbers. They should compare with a partner and discuss who is more accurate, regularly comparing the number they are placing with the start, middle and end points of the interval.

ASSESSMENT CHECKPOINT Question **1** checks whether children can label the intervals on a number line. Question **2** assesses whether they can identify numbers on number lines counting in 10s and 1s. Questions **3**, **4** and **5** check whether children can estimate numbers on a number line where arrows are not pointing directly to an interval.

ANSWERS Answers for the **Practice** part of the lesson can be found in the *Power Maths* online subscription.

Reflect

WAYS OF WORKING Independent thinking

IN FOCUS Here children are required to draw their own number line and think about what numbers are easier to label on their number line and which would be more difficult.

ASSESSMENT CHECKPOINT Children should be able to draw a number line, label its start and end points, and identify numbers that are easy to position and others that are not.

ANSWERS Answers for the **Reflect** part of the lesson can be found in the *Power Maths* online subscription.

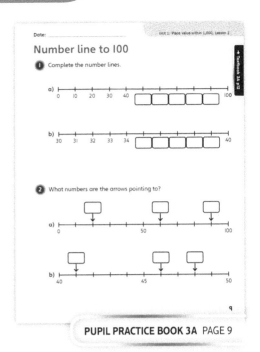

PUPIL PRACTICE BOOK 3A PAGE 9

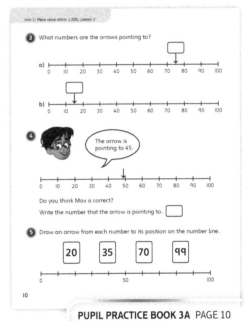

PUPIL PRACTICE BOOK 3A PAGE 10

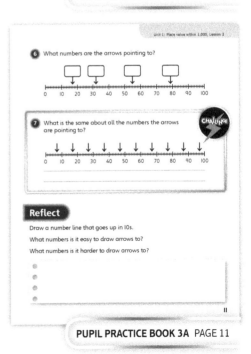

PUPIL PRACTICE BOOK 3A PAGE 11

After the lesson ⏸

- Can children label a number line from 0 to 100 counting in 10s?
- Can children label intervals on a number line counting in 1s between two multiples of 10?
- Can children identify the number that an arrow is pointing to on a number line?

100s

Learning focus

In this lesson, children will learn how to count in 100s from 0 to 1,000. They will write the numbers in both numerals and words.

Before you teach ▮▮

- Can children count from 0 to 100?
- Do they know when to go up to the next 10? Do they know what happens when they get to 99?
- Do children already have an understanding of what 100 is?

NATIONAL CURRICULUM LINKS

Year 3 Number – number and place value

Count from 0 in multiples of 4, 8, 50 and 100; find 10 or 100 more or less than a given number.

Recognise the place value of each digit in a three-digit number (hundreds, tens, ones).

ASSESSING MASTERY

Children can count in 100s from 0 to 1,000 and back again, understand what 100 is and the different ways of representing it. They will write the numbers in both numerals and words.

COMMON MISCONCEPTIONS

Often when children are counting from 0 to 1,000 and they get to 900, they say 'ten hundred' next. Explain that although they have 10 hundreds, we say this as one thousand. Ask:

- *Continue the count 700, 800, 900, … What comes after 900?*

STRENGTHENING UNDERSTANDING

Children who are struggling to count in 100s first need to understand what 100 is. Give them a jar of 100 dice or other objects or get them to count out 100 cubes. Once they have done this give them multiple packs of 100. Use a base 10 100 flat for each pack of 100. The connection between the hundred block and the 100 items should be made clear.

GOING DEEPER

Ask children to count on 500 from 300. They need to keep track of how many 100s they have counted on as well as the number that they reach.

KEY LANGUAGE

In lesson: one thousand, **hundreds (100s)**

Other language to be used by the teacher: count forwards, count backwards, number track

STRUCTURES AND REPRESENTATIONS

Number track

RESOURCES

Mandatory: base 10 equipment (100s)

Optional: bags and boxes of objects in 100s, bead strings, 100 square, base 10 equipment (1s and 10s)

 In the eTextbook of this lesson, you will find interactive links to a selection of teaching tools.

Quick recap

Ask children to count in 10s from 0 to 100. When they get to 100, challenge them to count back in 10s to 0. Make it harder by starting at a different multiple of 10, such as 60.

Discover

100s

Discover

WAYS OF WORKING Pair work

ASK

- Question ① a): *How many dice are in one jar? How many are in three jars?*

IN FOCUS Question ① a) is used to ensure children know what 100 is. It also helps check that children can count in 100s when faced with a small number of 100s. Then children can move to question ① b), with a greater number of 100s.

PRACTICAL TIPS Make sure each pair or group has 100 objects (for example, cubes) that they can use to represent one jar of dice or counters in the picture.

ANSWERS

Question ① a): There are 300 dice.

Question ① b): There are 600 counters.

① a) How many dice?

b) How many counters?

16

PUPIL TEXTBOOK 3A PAGE 16

Share

WAYS OF WORKING Whole class teacher led

ASK

- Question ① a): *How can you represent 100? What different ways do you know?*
- Question ① b): *Is there a fast way to count the counters?*

IN FOCUS Questions ① a) and b) get children to think about counting in 100s and check whether, beyond 100, they can count in 100s instead of in 1s or 10s.

STRENGTHEN Count with children from 0 to 100 in 10s and pretend to lose count. Draw out how it helps to put the objects into groups of 10, in case they lose count. Draw the connection between counting in 1s and counting in 100s. Use a base 10 100 flat to represent the 100 dice in a jar and then count on: 'One hundred, two hundred, …'.

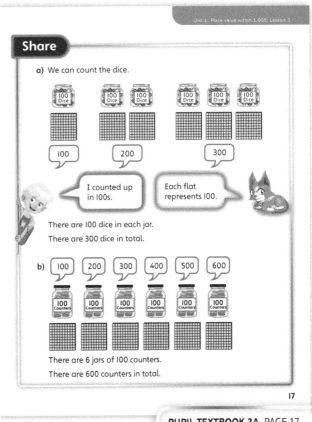

PUPIL TEXTBOOK 3A PAGE 17

Think together

WAYS OF WORKING Whole class teacher led (I do, We do, You do)

ASK

- Question **2**: *What are these sequences going up in? Which sequences are going forwards? Which sequences are going backwards? How can counting in 100s help you find the answers?*
- Question **3**: *700, 800, 900 … What comes next?*

IN FOCUS Question **1** shows the count from 0 to 800 in 100s and shows all the numbers represented as numerals and words. It is important that children know these numbers. Question **2** applies their knowledge to number tracks and sequences. There is a mix of sequences to check that they can count both forwards and backwards. Question **3** encourages children to think what word comes after 900. Some children may think that the next 100 after 900 is ten hundred. Introduce the term 'one thousand'.

STRENGTHEN Use bags of 100 objects and show them alongside base 10 100 flats. Ask children to count the 100 flats and show them this alongside the numbers. Count both forwards (by adding flats) and backwards (by removing flats).

DEEPEN Ask children to count on 200 from 500, or 600 less than 800. This requires them to keep track of where they are and how many 100s they have counted forwards or backwards.

ASSESSMENT CHECKPOINT In questions **2** and **3**, assess whether children can count forwards and backwards in 100s from any number.

ANSWERS

Question **1**: 400, four hundred
500, five hundred
600, six hundred
700, seven hundred
800, eight hundred

Question **2** a): 400, 500

Question **2** b): 300, 100

Question **2** c): 500, 700, 900

Question **3**: There are 1,000 marbles.
There are one thousand marbles.

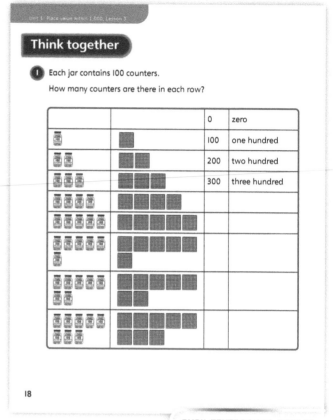

PUPIL TEXTBOOK 3A PAGE 18

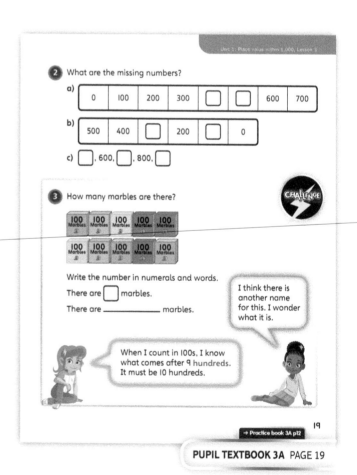

PUPIL TEXTBOOK 3A PAGE 19

Practice

WAYS OF WORKING Independent thinking

IN FOCUS Question ❶ checks whether children can count in 100s from 0 to 500. Question ❹ is more abstract and checks whether children know the forwards and backwards count from 0 to 1,000. Children have to work out the missing 100s. Question ❺ starts with the total number and asks how many 100s. Children need to count on in 100s until they reach 700. They also need to keep track of how many 100s they have counted.

STRENGTHEN Use bags of 100 objects to give children practice in counting in 100s. Then introduce 100 base 10 equipment, explaining that they represent the bags of objects. Ask children to count the 100 flats and show them this alongside the numbers in both numerals and words. Count both forwards (by adding flats) and backwards (by removing flats).

DEEPEN Ask children to work out 300 more than 700 or 400 less than 900. They need to understand that 300 more than 700 means counting on 300 from 700, so they need to count on 3 hundreds. They need to keep track of where they are and how many 100s they have counted.

ASSESSMENT CHECKPOINT Use question ❶ to assess whether children know the numbers in words and numerals for 100 to 1,000.

Use question ❹ to assess whether children can count forwards and backwards in 100s from any number.

ANSWERS Answers for the **Practice** part of the lesson can be found in the *Power Maths* online subscription.

Reflect

WAYS OF WORKING Pair work

IN FOCUS The questions check whether children can count in 100s both forwards and backwards. They also check their understanding of how and when to say 1,000.

ASSESSMENT CHECKPOINT Check whether children are counting in 100s. Do this by getting children to do the counts individually first, then in pairs and then as a whole class. Look for children who are counting backwards instead of forwards, and vice versa.

ANSWERS Answers for the **Reflect** part of the lesson can be found in the *Power Maths* online subscription.

After the lesson ⏸

- Do children know the count from 0 to 1,000 in 100s?
- Do children know that 1,000 is one thousand and not said as ten hundred?
- Can children represent a number of 100s in base 10 equipment and in other ways?

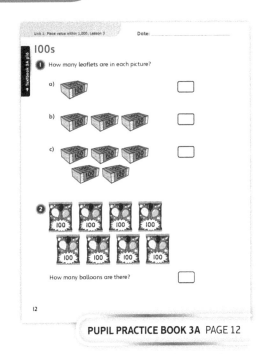

PUPIL PRACTICE BOOK 3A PAGE 12

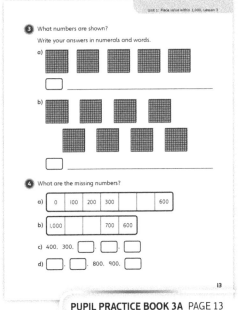

PUPIL PRACTICE BOOK 3A PAGE 13

PUPIL PRACTICE BOOK 3A PAGE 14

Represent numbers to 1,000

Learning focus

In this lesson, children will represent numbers in place value grids using counters. They will write numbers represented with counters in a place value grid.

Before you teach

- Can children make a 3-digit number using base 10 equipment?
- Do they know what each digit in a 3-digit number represents?

NATIONAL CURRICULUM LINKS

Year 3 Number – number and place value

Identify, represent and estimate numbers using different representations, including the number line.

Recognise the place value of each digit in a three-digit number (hundreds, tens, ones).

ASSESSING MASTERY

Children can represent numbers on a place value grid using labelled and blank counters. They can identify a number from its representation with counters.

COMMON MISCONCEPTIONS

Children may not know what to do when there is 0 in one of the columns. Ask:

- *How many 10s in 605? How do you show this on a place value grid?*
- *What happens if a place value grid has 0 in the ones column? What does this tell you about the number?*

When representing a 2-digit number on an HTO place value grid, a common mistake is to fill in the columns from left to right. Ask:

- *How many 10s and 1s are there in 76? Which column do these go in on the place value grid?*
- *What goes in the hundreds column?*

STRENGTHENING UNDERSTANDING

Encourage children to organise their work in a clearly labelled place value grid. To strengthen understanding, ask children to count aloud as they place counters on the grid. Provide further scaffolding by representing a number with base 10 equipment before using place value counters, so that they understand that a 100 counter represents a 100 flat of base 10 equipment and a 10 counter represents a 10 rod of base 10 equipment and so on. This helps children understand that the physical size of the counter is not significant.

GOING DEEPER

Ask children to make as many numbers as they can using five counters. Ask how many numbers have at least one counter in every column of the HTO place value grid. Children should start to develop systematic approaches to answering this type of question.

KEY LANGUAGE

In lesson: hundreds (100s), tens (10s), ones (1s), place value grid

STRUCTURES AND REPRESENTATIONS

Place value grid, part-whole model, number track

RESOURCES

Mandatory: place value counters, blank counters, base 10 equipment

Optional: large laminated place value grid

 In the eTextbook of this lesson, you will find interactive links to a selection of teaching tools.

Quick recap 🔁

Draw a 10 square number track on the board. Put 100 in the first box and 200 in the next box. Ask children to come up, one at a time, and fill in one box each. Challenge children not to just fill in the next box.

Discover

Represent numbers to 1,000

WAYS OF WORKING Pair work

ASK

- Question ❶ a): *Which boxes should Toshi use first? Why? What can you use to record your counting?*

IN FOCUS Question ❶ a) checks whether children can work out how many 100s, 10s and 1s make up a number, as in the previous lesson. Some children may try to start with single bulbs instead of boxes of 100. Discuss why it is better to start with the 100s (for example, it is quicker). Question ❶ b) further develops the concept of working out the number of 100s, 10s and 1s by asking children to make the number using base 10 equipment.

PRACTICAL TIPS Ensure that children have sufficient base 10 equipment to be able to make the numbers.

ANSWERS

Question ❶ a): 215 bulbs need to be planted.

Question ❶ b): Children make 215 with base 10 equipment using 2 hundreds, 1 ten and 5 ones.

Discover

I need to plant all these bulbs.

❶ a) How many bulbs need to be planted?

b) Make this number with base 10 equipment. How many 100s, 10s and 1s did you use?

20

PUPIL TEXTBOOK 3A PAGE 20

Share

WAYS OF WORKING Whole class teacher led

ASK

- Question ❶ a): *How many boxes of 100 bulbs are there? How many boxes of 10 bulbs are there? How many individual bulbs are there? How many bulbs are there altogether?*
- Question ❶ b): *How does the number of each piece of base 10 equipment relate to the number of each size box? How many 100s, 10s and 1s are there?*

IN FOCUS The two parts of the question develop from what children have done previously. They should recognise that a 100 flat relates to a box of 100 bulbs, the 10 rod relates to the box of 10 bulbs, and the 1 cube relates to the individual bulbs. Ensure the connection between the number of boxes and number of base 10 equipment is made explicit.

Share

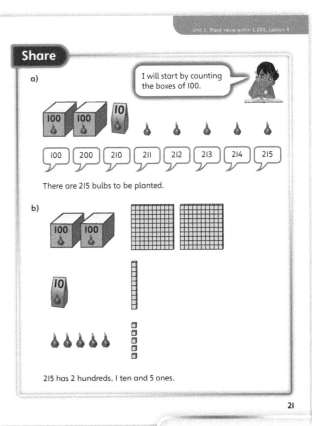

a)

I will start by counting the boxes of 100.

100, 200, 210, 211, 212, 213, 214, 215

There are 215 bulbs to be planted.

b)

215 has 2 hundreds, 1 ten and 5 ones.

21

PUPIL TEXTBOOK 3A PAGE 21

Think together

WAYS OF WORKING Whole class teacher led (I do, We do, You do)

ASK

- Question **1**: *How many packets of 100 sunflower seeds are there? How many packets of 10 sunflower seeds are there? How many individual seeds are there? How many are there altogether?*
- Question **2** b): *What do you notice about the order of base 10 equipment? What can you do first?*

IN FOCUS Question **1** builds on the **Discover** and **Share** sections. Children should be encouraged to build the number of sunflower seeds with base 10 equipment to continue to make connections between each piece and the different quantities of seeds. In question **2**, children work abstractly looking at numbers made from base 10 equipment and should be encouraged to identify the number of 100s, 10s and 1s. In part b) they should be careful to ensure they start with the highest place value.

STRENGTHEN A place value grid could be used to support children, by arranging their base 10 equipment into the correct columns to help them interpret the numbers. They could be encouraged to count aloud in 100s, then 10s, then 1s to find the number.

DEEPEN In question **3**, encourage children to find all the possible 3-digit numbers and interpret the different numbers of 100s, 10s and 1s each number has. Ask: *Which numbers have the most hundreds? Which numbers have the fewest hundreds?*

ASSESSMENT CHECKPOINT Use question **2** to check whether children can correctly interpret numbers represented by base 10 equipment, and question **3** to ensure they can interpret numbers written in abstract form, even if they need to build them for support.

ANSWERS

Question **1**: 743

Question **2** a): 526

Question **2** b): 246

Question **3** a): Children should use base 10 equipment to make the numbers 347, 374, 437, 473, 734, 743.

Question **3** b)–d): Answers depend on the way the digits are arranged:

347: 3 hundreds, 4 tens and 7 ones
      300   40   7
374: 3 hundreds 7 tens and 4 ones
      300   70   4
437: 4 hundreds, 3 tens and 7 ones
      400   30   7
473: 4 hundreds, 7 tens and 3 ones
      400   70   3
743: 7 hundreds, 4 tens and 3 ones
      700   40   3
734: 7 hundreds, 3 tens and 4 ones
      700   30   4

Think together

1 How many sunflower seeds are there?

100 100 100 100 100 100 100

2 What numbers are represented by this base 10 equipment?

a)

b)

22

PUPIL TEXTBOOK 3A PAGE 22

3 Here are three digit cards:

CHALLENGE

3 4 7

Use the cards to make some 3-digit numbers.

a) Use base 10 equipment to make your numbers.
b) How many 100s does each number have?
c) How many 10s does each number have?
d) How many 1s does each number have?

I wonder how many numbers I can make with these cards.

→ Practice book 3A p15

23

PUPIL TEXTBOOK 3A PAGE 23

Practice

WAYS OF WORKING Independent thinking

IN FOCUS Questions ❶ and ❷ aim to consolidate children's understanding of 100s, 10s and 1s in a real-life context. Question ❸ then gives children opportunity to consolidate working abstractly with base 10 equipment. In question ❹, they make a number using base 10 equipment, relating it to the number of 100s, 10s and 1s in the number.

STRENGTHEN In questions ❶ and ❷, children could be given base 10 equipment and a place value grid to support them in identifying how many items there are.

DEEPEN In question ❻, children should be encouraged to work systematically to find all possible numbers. This can be explored further by giving children four digit cards and asking them how many different 3-digit numbers they can make. Ask: *How do children know they have made all the possible numbers?*

THINK DIFFERENTLY Question ❺ asks children to identify a number from an abstract description.

ASSESSMENT CHECKPOINT Use question ❸ to assess whether children can write 3-digit numbers represented using base 10 equipment. Use question ❹ to assess whether children can identify the number of 100s, 10s and 1s in a number written in digits.

ANSWERS Answers for the **Practice** part of the lesson can be found in the *Power Maths* online subscription.

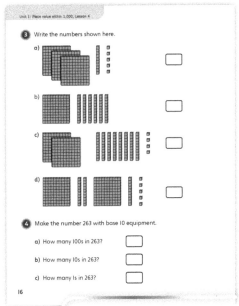

PUPIL PRACTICE BOOK 3A PAGE 15

PUPIL PRACTICE BOOK 3A PAGE 16

Reflect

WAYS OF WORKING Independent thinking

IN FOCUS This question checks whether children can interpret the number of 100s, 10s and 1s in a number.

ASSESSMENT CHECKPOINT Children should be able to explain their answers and could use base 10 equipment to support their explanations.

ANSWERS Answers for the **Reflect** part of the lesson can be found in the *Power Maths* online subscription.

After the lesson ⏸

- Can children represent a 3-digit number using base 10 equipment?
- Can children work out what numbers are represented by base 10 equipment?
- Do children understand that in 3-digit numbers the 100s come first
- and then the 10s and then the 1s?

PUPIL PRACTICE BOOK 3A PAGE 17

Partition numbers to 1,000

Learning focus

In this lesson, children will understand that a number up to 1,000 is made up of some 100s, some 10s and some 1s. They will use base 10 equipment and part-whole models to represent numbers.

NATIONAL CURRICULUM LINKS

Year 3 Number – number and place value

Recognise the place value of each digit in a three-digit number (100s, 10s, 1s).

Identify, represent and estimate numbers using different representations, including the number line.

ASSESSING MASTERY

Children can represent 3-digit numbers using base 10 equipment and write them onto a part-whole model. Children can recognise that a 3-digit number is made up of some 100s, some 10s and some 1s.

COMMON MISCONCEPTIONS

A common mistake that children make is that they do not understand the place value and size. They may, for example, see 4 hundred flats and 6 ones cubes and think this is 460 as they are just going left to right. Show children representations not in order. Ask:

- *Can you represent 248 using base 10 equipment? What happens if I move the base 10 equipment around? What is my number now?*

STRENGTHENING UNDERSTANDING

First recap representing 2-digit numbers. Ask children to show a 2-digit number, for example, 58. Explain that they can now add some 100s to this and this gives a 3-digit number. Encourage children to write out the numbers (for example, 348 is 3 hundreds, 4 tens and 8 ones) and to make each part with base 10 equipment.

GOING DEEPER

Ask if all 3-digit numbers are made up of three parts. Ask children to make numbers that can be represented on a part-whole model that just has two parts. Ask children if they can represent the number 228 in different ways using base 10 equipment (for example, 1 hundred, 12 tens and 8 ones) and to explain why these are equal.

KEY LANGUAGE

In lesson: hundreds (100s), tens (10s), ones (1s), part-whole model, digit

Other language to be used by the teacher: partition

STRUCTURES AND REPRESENTATIONS

Part-whole model

RESOURCES

Mandatory: base 10 equipment (including 100s)

Optional: large laminated part-whole model

 In the eTextbook of this lesson, you will find interactive links to a selection of teaching tools.

Discover

Unit 1: Place value within 1,000, Lesson 5

Partition numbers to 1,000

WAYS OF WORKING Pair work

ASK

- Question ① a): *What do the different base 10 equipment represent? What is a number between 100 and 999 made up of? How can you tell how many 100s, how many 10s and how many 1s there are?*
- Question ① b): *How can you use a part-whole model to show this? Why do you think that there are three parts?*

IN FOCUS Question ① a) asks children to make 235 with concrete base 10 equipment. They should start to see that numbers between 100 and 999 are made up of some 100s, some 10s and some 1s. Question ① b) asks children to represent the number on a part-whole model.

PRACTICAL TIPS Make sure children have all the necessary base 10 equipment. Say that there are 235 children in the school and ask children to represent this using the equipment.

ANSWERS

Question ① a): 235 using base 10 equipment

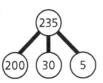

Question ① b):

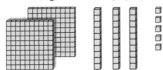

Discover

There are 235 children in our school.

235

① a) Use base 10 equipment to represent 235.

b) Complete the part-whole model.

24

PUPIL TEXTBOOK 3A PAGE 24

Share

WAYS OF WORKING Whole class teacher led

ASK

- Question ① a): *How many children are in the school? How can you represent 235 using base 10 equipment? Which equipment do you use for each representation? Why do you use 2 hundred flats?*
- Question ① b): *How can you show this using a part-whole model? Why do you need to use three parts? Why do you not use just two parts?*

IN FOCUS Question ① a) encourages children to represent objects using base 10 equipment. It is important they recognise which base 10 equipment represents each part of the number. By the end of **Share** they should have an understanding of how to represent 3-digit numbers in base 10 equipment and on part-whole models, and should recognise numbers written in this form.

Share

a) There are 235 children in the school.

100 200 210 220 230 231 232 233 234 235

I counted the 100s first, then the 10s and then the 1s.

I put the base 10 equipment in a part-whole model to work out the parts.

b)

235

200 30 5

25

PUPIL TEXTBOOK 3A PAGE 25

Think together

Whole class teacher led (I do, We do, You do)

ASK

- Question **1**: *How can you use base 10 equipment to represent this number? Is there any other equipment you can use?*
- Question **3**: *Does it matter what order you put the parts in? Will they still make the same number? How can you write your partition as an addition?*

IN FOCUS Questions **1** and **2** provide further practice of what children have been doing in the **Discover** section. They look at different ways of representing 3-digit numbers, using base 10 equipment and part-whole models. Question **3** looks into more abstract partitioning of 3-digit numbers. Children will see that numbers can be partitioned into 100s, 10s and 1s. Some numbers do not have certain parts. They should write their partitions as an addition too.

STRENGTHEN Represent concrete objects, such as 245 toys, using base 10 equipment. Separate the 100s from the 10s and the 10s from the 1s. Clearly associate each group of objects with the particular base 10 equipment. Put the base 10 equipment into a part-whole model before writing in the digits. This will help children understand the numbers.

DEEPEN Provide numbers in base 10 equipment that are not given in the order of 100s, 10s and 1s. This will help children understand that the order does not matter; 100s will always be 100s. Ask: *Are there any 3-digit numbers that are made up of just two parts?* Ask for examples and establish what is the same and what is different about these numbers. Ask: *Are there any 3-digit numbers that only have one part?*

ASSESSMENT CHECKPOINT Use question **1** to assess whether children can represent numbers up to 1,000 using base 10 equipment and part-whole models. Use question **2** to assess whether children can identify the number from its representation. Children need to understand that a 3-digit number is made up of some 100s, some 10s and some 1s. Question **3** assesses whether children can write a number in abstract form and as an addition.

ANSWERS

Question **1** a): Children build 251 using base 10 equipment with 2 flats, 5 rods and 1 cube.

Question **1** b):

```
       (251)
      /  |  \
  (200)(50)(1)
```

Question **2** a):
```
       (364)
      /  |  \
  (300)(60)(4)
```
b)
```
       (137)
      /  |  \
  (100)(30)(7)
```

Question **3**:
```
    (615)          (293)          (304)        (340)
   /  |  \        /  |  \        /   \        /   \
(600)(10)(5)  (200)(90)(3)   (300)(4)    (300)(40)
```
615 = 600 + 10 + 5   293 = 200 + 90 + 3   304 = 300 + 4   340 = 300 + 40

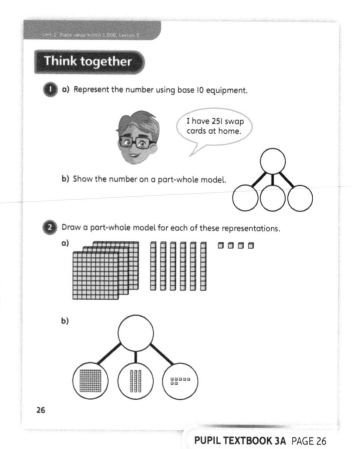

PUPIL TEXTBOOK 3A PAGE 26

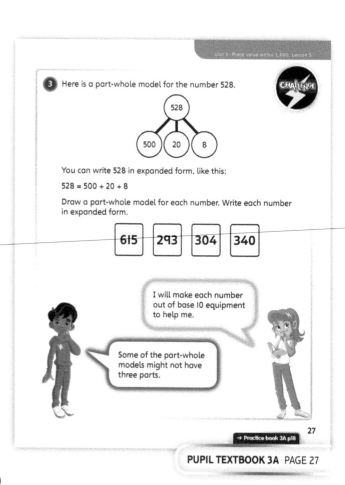

PUPIL TEXTBOOK 3A PAGE 27

Practice

WAYS OF WORKING Independent thinking

IN FOCUS Question ① consolidates children's understanding of 3-digit numbers represented in base 10 equipment, reinforcing understanding that a number is made up of 100s, 10s and 1s.

Questions ③, ④ and ⑤ then ask children to partition 3-digit numbers. They look at a variety of examples, including some that only have two parts. Questions ⑥ and ⑦ take children through the process of showing a partition as an addition. Children start to realise that the parts can be in any order.

STRENGTHEN In questions ② and ③, children can use base 10 equipment to replace the abstract parts. This will help them understand the place value of each number. When blocks are not given in the order of 100s, 10s and 1s, encourage children to put them in this order.

DEEPEN Ask children to reflect on the partitions from question ④ onwards and if there are other ways they can partition each number. For example, could they partition 350 into 200 + 150? Also ask them to reflect on 200 + 60 + 5 is equal to 60 + 200 + 5, so the order of the partition does not matter.

ASSESSMENT CHECKPOINT Questions ① to ③ assess whether children can partition numbers using a part-whole model. Question ⑤ onwards assesses children's understanding of the related addition partition.

ANSWERS Answers for the **Practice** part of the lesson can be found in the *Power Maths* online subscription.

Reflect

WAYS OF WORKING Independent thinking

IN FOCUS This question checks children's understanding of place value and if they can partition the number into 100s, 10s and 1s.

ASSESSMENT CHECKPOINT Check that children realise that although they can see the digits 4, 6 and 8, the value of the digits in Phil's partition are incorrect. They should articulate that the 6 means there are 6 tens or 60.

ANSWERS Answers for the **Reflect** part of the lesson can be found in the *Power Maths* online subscription.

After the lesson ⏸

- Can children represent a 3-digit number using base 10 equipment and on a part-whole model?
- Can children work out what numbers are represented by these representations?
- Do children understand that in 3-digit numbers the 100s come first and then the 10s and then the 1s?

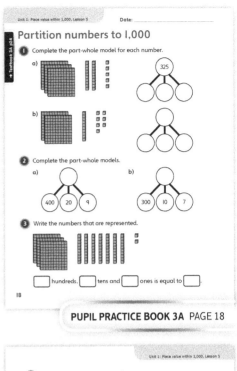

PUPIL PRACTICE BOOK 3A PAGE 18

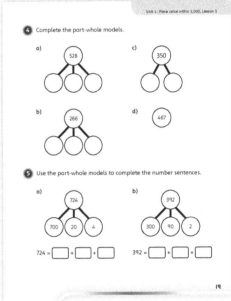

PUPIL PRACTICE BOOK 3A PAGE 19

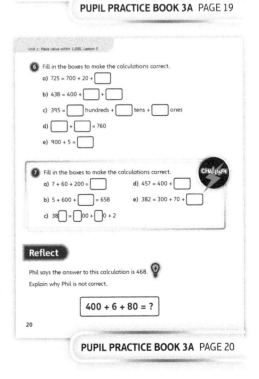

PUPIL PRACTICE BOOK 3A PAGE 20

Partition numbers to 1,000 flexibly

Learning focus

In this lesson, children build on their understanding of 3-digit numbers and learn that they can be partitioned in different ways. They will use base 10 equipment and part-whole models to represent 3-digit numbers.

NATIONAL CURRICULUM LINKS

Year 3 Number – number and place value

Recognise the place value of each digit in a 3-digit number (100s, 10s, 1s).

ASSESSING MASTERY

Children can represent 3-digit numbers in different ways using base 10 equipment and part-whole models. Children recognise that a 3-digit number can be partitioned in different ways.

COMMON MISCONCEPTIONS

A common misconception is that children think numbers can only be partitioned in one way as they do not understand the relationship between the different place values or that they can have some 100s in more than one part. Ask:

- *Is there another way to partition the number?*
- *Can you use base 10 equipment to help you?*

STRENGTHENING UNDERSTANDING

First recap standard partitioning of numbers to 1,000. Ask children to partition a 3-digit number, such as 345, into 100s, 10s and 1s using base 10 equipment. Ask children to move one piece of base 10 equipment to a different pile. Is the number still the same? Can they explain why they are equal?

GOING DEEPER

Ask children if all 3-digit numbers have to be made up of exactly three parts. Challenge children to make numbers that can be represented on a part-whole model that has just two parts. Give children a part-whole model with a 3-digit number as the whole, such as 547, and ask them to partition it into two parts. Ask: *Can you partition it into more than two parts?*

KEY LANGUAGE

In lesson: hundreds (100s), tens (10s), ones (1s), part-whole model, digit

Other language to be used by the teacher: partition, flexibly

STRUCTURES AND REPRESENTATIONS

Part-whole model

RESOURCES

Mandatory: base 10 equipment (including 100s flats)

Optional: large laminated part-whole model

 In the eTextbook of this lesson, you will find interactive links to a selection of teaching tools.

Quick recap 🔎

Say you are thinking of a number, such as 'My number has 2 hundreds, 5 tens and 3 ones'. Ask children to make your number and tell you what number you were thinking of. Repeat for more 3-digit numbers. How many can they get right? Challenge children by having numbers with 0 tens or 0 ones.

Discover

Unit 1: Place value within 1,000, Lesson 6

Partition numbers to 1,000 flexibly

WAYS OF WORKING Pair work

ASK

- Question ❶ a): *What pieces of base 10 equipment does Lexi have? What is the value of each piece of base 10 equipment? What number has Lexi made?*

IN FOCUS Question ❶ a) builds on children's previous learning of recognising a 3-digit number made from base 10 equipment and representing it in a part-whole model using standard partitioning. In question ❶ b), they then make the number in a different way to introduce them to flexible partitioning. Practical movement of base 10 equipment can support this.

PRACTICAL TIPS Ensure children have sufficient base 10 equipment to make Lexi's number and represent it in three clear piles, first showing 100s, 10s, and 1s. Then ask them to move one piece to a different pile. Now what are the parts?

ANSWERS

Question ❶ a): Lexi has made 262.

Question ❶ b): Answers will vary, for example:

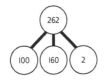 

Share

WAYS OF WORKING Whole class teacher led

ASK

- Question ❶ a): *How many 100s, 10s and 1s are there? How do you show this in a part-whole model?*
- Question ❶ b): *If you move 1 hundred into the 10s pile, does the number change? How do you know? Now what are the parts? How can you show this in a part-whole model?*

IN FOCUS Question ❶ b) introduces children to flexible partitioning of 3-digit numbers. It is important they recognise that as long as they do not remove any of the base 10 equipment, then the number does not change; they are simply changing the value of the parts.

❶ a) What number has Lexi made?
   Represent this number in a part-whole model.

b) Find a different way to partition Lexi's number into three parts.
   Show your answer in a part-whole model.

28

PUPIL TEXTBOOK 3A PAGE 28

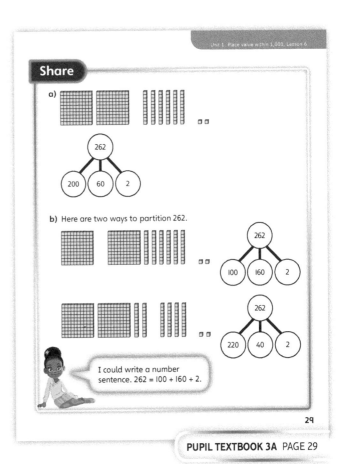

Think together

Whole class teacher led (I do, We do, You do)

ASK

- Question **1**: *What is the value of each part? How do you know? How can you show this in a part-whole model? What is the same about all the part-whole models? What is different?*
- Question **2**: *What is the whole? What parts do you know? What is missing? How do you know?*
- Question **2** d), e) and f): *How can you show this in a part-whole model? How does this help to find the missing part?*

IN FOCUS Question **1** focuses on children being able to represent numbers that have been partitioned flexibly in a part-whole model. They should recognise that the whole is the same each time. In question **2**, they look at different ways of partitioning 526 and fill in number sentences to support their partitioning.

STRENGTHEN Ensure children have access to base 10 equipment to make the numbers to support them in flexibly partitioning numbers. If they have enough pieces to make the whole, then they can use them to make the parts and use the remaining pieces to identify missing parts.

DEEPEN Prompt children to see if they can partition the numbers in any other ways and explain how they know. Encourage them to work systematically to check they have found them all.

ASSESSMENT CHECKPOINT Use question **1** to assess whether children can complete a part-whole model to represent a partition, and question **2** to ensure they can flexibly partition numbers.

ANSWERS

Question **1** a):

354 → 200, 150, 4

Question **1** b):

354 → 100, 250, 4

Question **1** c):

354 → 200, 140, 14

Question **2** a): 420

Question **2** b): 320

Question **2** c): 220

Question **2** d): 120

Question **2** e): 16

Question **2** f): 310

Question **3**: 357

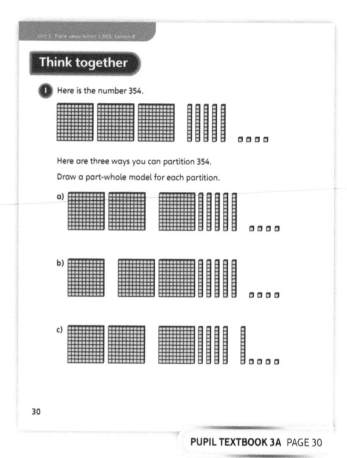

PUPIL TEXTBOOK 3A PAGE 30

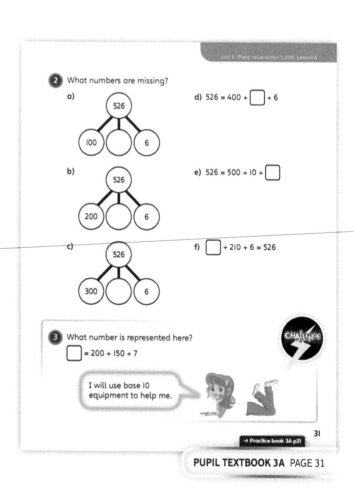

PUPIL TEXTBOOK 3A PAGE 31

Practice

WAYS OF WORKING Independent thinking

IN FOCUS In question ❶, base 10 equipment is clearly separated into parts and children are required to complete part-whole models to represent this. They should be asked how they know it is still 352 in each case. In question ❷, they repeat this approach but partition the 10s instead. In question ❸, children are required to identify missing parts in part-whole models.

STRENGTHEN Ensure children have access to base 10 equipment to support their thinking and help them to partition numbers flexibly. If they have enough pieces to make the whole, they can use them to make the parts and use the remaining pieces to identify missing parts.

DEEPEN In question ❺, ask children if there are any other ways to partition 748, working systematically first to find the ways to partition the number into three parts and then challenge them to partition it into more than three parts.

THINK DIFFERENTLY In question ❹, children are given a part-whole model with five parts and asked to find the whole. Provide them with base 10 equipment to make the parts before combining them to find the whole.

ASSESSMENT CHECKPOINT Use questions ❶ and ❷ to check that children can complete part-whole models to represent partitions and question ❸ to ensure they can partition a number in different ways.

ANSWERS Answers for the **Practice** part of the lesson can be found in the *Power Maths* online subscription.

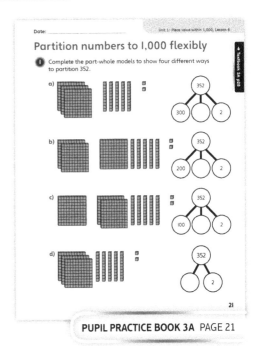

PUPIL PRACTICE BOOK 3A PAGE 21

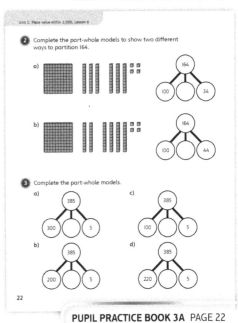

PUPIL PRACTICE BOOK 3A PAGE 22

Reflect

WAYS OF WORKING Independent thinking

IN FOCUS This question checks that children can partition a number in more than one way.

ASSESSMENT CHECKPOINT Check that children have identified three correct ways of partitioning the number 524.

ANSWERS Answers for the **Reflect** part of the lesson can be found in the *Power Maths* online subscription.

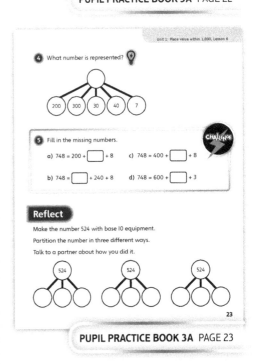

PUPIL PRACTICE BOOK 3A PAGE 23

After the lesson ⏸

- Can children represent 3-digit numbers using base 10 equipment and part-whole models?
- Can children represent the same number in different ways using base 10 equipment and part-whole models?
- Can children identify missing parts in part-whole models where the number has been partitioned flexibly?

100s, 10s and 1s

Learning focus

In this lesson, children will represent numbers in place value grids using counters. They will write numbers represented with counters in a place value grid.

Before you teach

- Can children make a 3-digit number using base 10 equipment?
- Do they know what each digit in a 3-digit number represents?

NATIONAL CURRICULUM LINKS

Year 3 Number – number and place value

Recognise the place value of each digit in a three-digit number (100s, 10s, 1s).

Identify, represent and estimate numbers using different representations, including the number line.

ASSESSING MASTERY

Children can represent numbers on a place value grid using labelled and blank counters. They can identify a number from its representation with counters.

COMMON MISCONCEPTIONS

Children may not know what to do when there is 0 in one of the columns. Ask:

- *How many 10s are there in 605? How do you show this on a place value grid?*
- *What happens if a place value grid has 0 in the ones column? What does this tell you about the number?*

When representing a 2-digit number on an HTO place value grid, a common mistake is to fill in the columns from left to right. Ask:

- *How many 10s and 1s are there in 76? Which column do these go in on the place value grid?*
- *What goes in the hundreds column?*

STRENGTHENING UNDERSTANDING

Encourage children to organise their work in a clearly labelled place value grid. To strengthen understanding, ask children to count aloud as they place counters on the grid. Provide further scaffolding by representing a number with base 10 equipment before using place value counters, so that they understand that a 100 counter represents a 100 flat of base 10 equipment and a 10 counter represents a 10 rod of base 10 equipment, and so on. This helps children understand that the physical size of the counter is not significant.

GOING DEEPER

Ask children to make as many numbers as they can using five counters. Ask: *How many numbers have at least one counter in every column of the HTO place value grid?* Children should start to develop systematic approaches to answering this type of question.

KEY LANGUAGE

In lesson: hundreds (100s), tens (10s), ones (1s), place value grid

STRUCTURES AND REPRESENTATIONS

Place value grid, part-whole model

RESOURCES

Mandatory: place value counters, blank counters, base 10 equipment

Optional: large laminated place value grid

 In the eTextbook of this lesson, you will find interactive links to a selection of teaching tools.

Quick recap 🔎

Ask children to write down a 3-digit number with 5 tens. Get them to show each other. What do they notice? Ask children to write down a 3-digit number with 7 ones. What do they notice when they look at the answers as a class?

Discover

Unit 1: Place value within 1,000, Lesson 7

100s, 10s and 1s

WAYS OF WORKING Pair work

ASK

- Question ① a): *What number is Miss Hall holding? How can Richard make this number?*
- Question ① b): *How can Richard write this number into a place value grid?*

IN FOCUS Question ① a) checks that children can represent numbers using base 10 equipment whilst question ① b) focuses on the use of place value counters. When moving onto part b), children should recognise that the physical size of the equipment does not reflect their value, as it has done previously. Encourage them to read aloud the number of 100s, 10s and 1s in the number to support them in representing them.

PRACTICAL TIPS Ensure that children have sufficient practical equipment available to make the numbers. Ask them to make the numbers using base 10 equipment and place value counters side by side, and consider what is the same and what is different about each representation.

ANSWERS

Question ① a): Children build 425 with base 10 equipment using 4 flats, 2 rods and 5 cubes.

Question ① b): Children build 425 with place value counters using 4 hundreds, 2 tens and 5 ones.

Discover

① a) Make 425 using base 10 equipment.

  b) Make 425 using place value counters.
     How many of each counter did you use?

32

PUPIL TEXTBOOK 3A PAGE 32

Share

WAYS OF WORKING Whole class teacher led

ASK

- Question ① a): *How many 100s, 10s and 1s are there? How can you see this in the part-whole model?*
- Question ① b): *How many 100s, 10s and 1s are there? What is the same and what is different about the place value counters and the base 10 equipment?*

IN FOCUS The two parts of this question develop children's understanding of different representations of numbers. Use the examples on this page to explore more ways of writing 425. Ensure children understand how all of the different representations link to each other and how the number of 100s, 10s and 1s in the number can be seen in each representation.

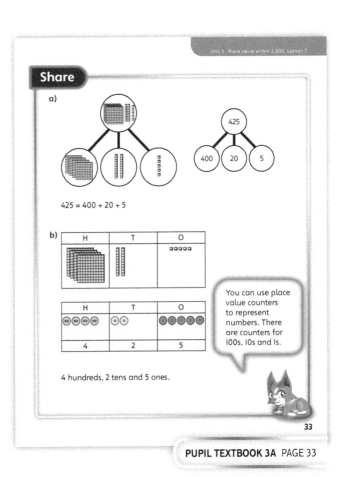

PUPIL TEXTBOOK 3A PAGE 33

71

Think together

WAYS OF WORKING Whole class teacher led (I do, We do, You do)

ASK

- Question **1**: *How many 100s counters do you need? How many 10s counters? How many 1s counters? How do you know? Which column do you put each counter in? How do you know?*
- Question **3** a): *Which of these numbers has counters in all the columns? How can you tell?*

IN FOCUS Question **2** asks children to work out what numbers are represented by place value counters, starting with a number where all three digits are not 0 and then moving on to numbers where place holders are needed. Question **3** includes all the representations of 3-digit numbers that children have encountered in this unit and shows how they can all be represented on a place value grid. Question **3** b) uses plain counters and it is important for children to see that what matters is how many counters there are in each column and that this assigns their value.

STRENGTHEN Ask children to count aloud in 100s first until they get to the correct number of 100s, and then count in 10s and then in 1s. As they count, ask them to select the correct place value counters. Show explicitly the connections between the number of each counter and the digit in the written number.

DEEPEN Building on from question **3** b), give children eight blank counters and ask them to make as many numbers as they can by placing them on a place value grid. Ask: *Can you make a number that has the same digit in each column? Can you make a number with 4 tens?*

ASSESSMENT CHECKPOINT Use questions **1** and **2** to assess whether children can represent numbers using place value grids and place value counters. Use question **3** to assess whether children can understand different representations of numbers. Check they understand what to do when there are no counters in a column.

ANSWERS

Question **1** a):

H	T	O
⑩⑩ ⑩⑩	⑩ ⑩ ⑩	① ① ① ① ① ①

Question **1** b): 2 hundreds counters, 3 tens counters and 6 ones counters.

Question **2** a): 135

Question **2** b): 308

Question **3** a):

H	T	O
●●●●● ●	●●	●●●●● ●●

H	T	O
●	●●●●	●●●

H	T	O
●	●●●●●	

H	T	O
	●●●●● ●●●	●●●●●

Question **3** b): 627

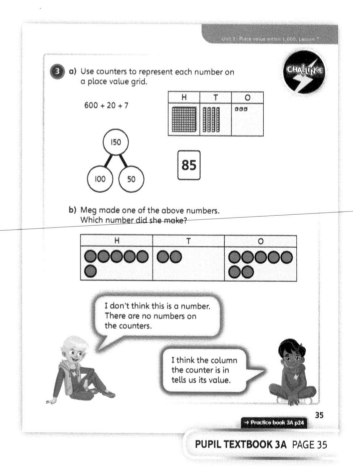

Think together

1 a) Make the number 236 with place value counters.

H	T	O

b) How many of each counter did you use?

2 What numbers are represented in the place value grids?

a)

H	T	O
⑩⑩	⑩ ⑩ ⑩	① ① ① ① ①

b)

H	T	O
⑩ ⑩ ⑩		① ① ① ① ① ① ① ①

34

PUPIL TEXTBOOK 3A PAGE 34

3 a) Use counters to represent each number on a place value grid.

600 + 20 + 7

H	T	O
▦	▥▥	▫▫▫

150 → 100, 50

85

b) Meg made one of the above numbers. Which number did she make?

H	T	O
●●●●● ●	●●	●●●●● ●●

I don't think this is a number. There are no numbers on the counters.

I think the column the counter is in tells us its value.

35

→ Practice book 3A p24

PUPIL TEXTBOOK 3A PAGE 35

Practice

WAYS OF WORKING Independent thinking

IN FOCUS In question ①, children are given place value counters in place value grids, including examples of where there are 0 counters in certain columns. Question ② then turns the question round, giving children the number and asking them to make it using place value counters. This also includes a number where children need to leave one column empty. Question ③ develops this further by using different representations of numbers. In question ⑤, children have to start using the knowledge that 10 tens is equal to 100 and use this to explain why the two numbers are identical.

STRENGTHEN To strengthen understanding, use base 10 equipment alongside the counters. In question ①, ask children to create the place value grids using base 10 equipment instead of counters. Then arrange the base 10 equipment from the place value grid in a row, 100s then 10s then 1s. Ask children to count aloud in 100s, then 10s, then 1s. Now ask children to count the counters in the same way.

DEEPEN Ask children to make numbers using seven blank counters on a place value grid. Ask: *How can you be confident that you have found all the possible numbers?* They should be able to develop a strategy for finding all the numbers.

THINK DIFFERENTLY In question ④, children start to use blank counters instead of place value counters. For children who struggle with this concept, create Tim's place value grid using place value counters. Then replace each place value counter with a blank counter, explaining that its value stays the same because the counter is still in the same column.

ASSESSMENT CHECKPOINT Use questions ① and ② to assess whether children can work out what numbers are represented by place value counters on a place value grid, and whether they can represent numbers on a place value grid with place value counters. Assess whether children understand what to do when there is 0 in a particular column. Use question ④ to assess whether children can use blank counters in a place value grid.

ANSWERS Answers for the **Practice** part of the lesson can be found in the *Power Maths* online subscription.

Reflect

WAYS OF WORKING Pair work

IN FOCUS Children record the numbers they can make using six blank counters and reflect on any strategy that they use to ensure they make all the numbers. For example, do they just do trial and improvement, or do they develop a systematic approach?

ASSESSMENT CHECKPOINT Check whether children understand what happens when there are no counters in a column, particularly the first column.

ANSWERS Answers for the **Reflect** part of the lesson can be found in the *Power Maths* online subscription.

After the lesson ⏸

- Can children represent numbers (written in different forms) with counters on a place value grid?
- Do children know how many 100s, 10s and 1s there are in numbers represented on a place value grid? Can they write the numbers?
- Do children understand what it means to have 0 in one of the columns, and when they need to record the 0, and when they do not?

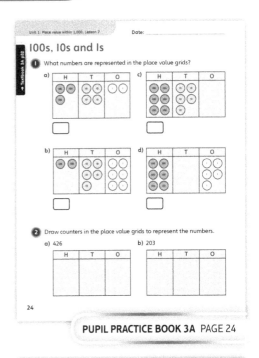

PUPIL PRACTICE BOOK 3A PAGE 24

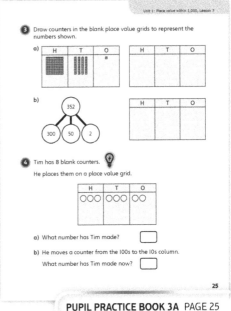

PUPIL PRACTICE BOOK 3A PAGE 25

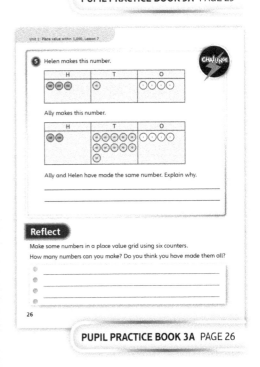

PUPIL PRACTICE BOOK 3A PAGE 26

Use a number line to 1,000

Learning focus

In this lesson, children will identify values and mark points on number lines that go up in 100s, 10s and 1s.

Before you teach

- Can children work out missing values on number lines that go up in 10s and 1s within 100?
- Can children count on in 100s, 10s and 1s within 1,000?

NATIONAL CURRICULUM LINKS

Year 3 Number – number and place value

Identify, represent and estimate numbers using different representations, including the number line.

Recognise the place value of each digit in a three-digit number (100s, 10s, 1s).

ASSESSING MASTERY

Children can work out whether a number line goes up in 100s, 10s or 1s. They can identify values on a number line and mark on given values.

COMMON MISCONCEPTIONS

Children may expect all number lines to go up in 100s or fail to identify correctly what the number line does go up in. Encourage children to count aloud whilst pointing to each mark. Ask:

- *What will you count on in? Have you said the correct end point?*
- *What else could you count on in?*

Children may also mark numbers such as 980 half-way between 900 and 1,000. Ask:

- *Is 980 closer to 900 or 1,000? How could you check?*

STRENGTHENING UNDERSTANDING

Encourage children to label numbers at every mark, to reinforce counting in 100s, 10s and 1s. Counting aloud to check whether a number goes up in 100s, 10s and 1s will also help children. Consider also using base 10 equipment next to the numbers on a large number line, to help children understand the meaning of the numbers.

GOING DEEPER

Using a number line going up in 100s from 0 to 1,000, ask children to mark points such as 50, 110, 295, 500 and 770. Ask: *What happens if the number line does not have any markings on it? Can you still mark the positions of these numbers?*

KEY LANGUAGE

In lesson: estimate, hundreds (100s), tens (10s), ones (1s), number line

STRUCTURES AND REPRESENTATIONS

Number line

RESOURCES

Optional: base 10 equipment, counting stick

 In the eTextbook of this lesson, you will find interactive links to a selection of teaching tools.

Quick recap

Draw on the board or give children a number line with 0 at the start and 10 at the end. The number line will need to have 10 intervals. Can children fill in the missing numbers? Challenge children by changing the end number to, for example, 20 or 30.

Discover

WAYS OF WORKING Pair work

ASK

- Question ① a): *How long is the course? What is the distance between the markings? How can you work out how far each boat has travelled?*

IN FOCUS In this question, children extend the number line to 1,000, going up in 100s in the context of a boat race. Question ① a) asks children to work out where each boat is and how far it has travelled. This involves their knowledge of counting on in 100s. Boat C also involves them estimating a position half-way between two known marks.

PRACTICAL TIPS The number line is extended to 1,000. Consider laying out the classroom, hall or playground so that there are 11 posts or chairs, labelled with every 100 m from 0 m to 1,000 m. Ask children to stand in pairs where they think the boats are. Do not give them any additional information that is not in the picture. A counting stick could also be used, where the markings go up in 100s.

ANSWERS

Question ① a): Boat A has travelled 300 metres. Boat B has travelled 600 metres. Boat C has travelled approximately 750 metres.

Question ① b): Point to 900 metres for Boat D. Boat D will be ahead of Boat C, on the 900 m line.

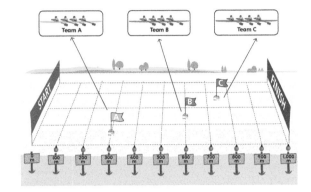

Use a number line to 1,000

Discover

① a) How far has boat A travelled?
How far has boat B travelled?
Estimate how far boat C has travelled.

b) Boat D has travelled 900 metres.
Where will it be?

36

PUPIL TEXTBOOK 3A PAGE 36

Share

WAYS OF WORKING Whole class teacher led

ASK

- Question ① a): *Can you see how the number line matches the boat course? What does the number line go up in? What is half-way between each pair of marks on the number line?*
- Question ① b): *How can you work out where boat D is? Did you count from the left? Can you work it out by counting from the right?*

IN FOCUS Question ① b) encourages children to identify steps of 100 and to use them to solve a problem. Ensure that children understand that the marks are every 100 metres. Get them to check by counting on in 100s from 0 to 1,000, recapping the work that they did in the first lesson. Discuss Sparks's comment about 750 lying half-way between 700 and 800 and why 750 is an estimate. Ask children to find half-way points between any other two marks.

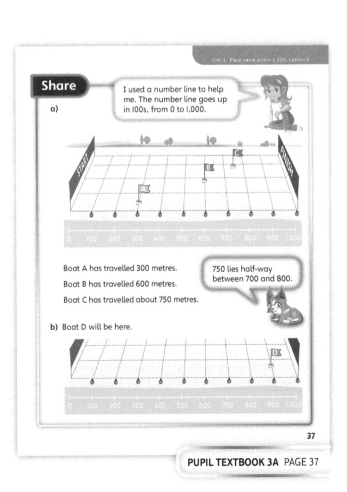

Share

I used a number line to help me. The number line goes up in 100s, from 0 to 1,000.

a)

Boat A has travelled 300 metres.
Boat B has travelled 600 metres.
Boat C has travelled about 750 metres.

750 lies half-way between 700 and 800.

b) Boat D will be here.

37

PUPIL TEXTBOOK 3A PAGE 37

Think together

Whole class teacher led (I do, We do, You do)

ASK

- Question **2** a): *Does each arrow point to a mark on the number line? How can you work out the number?*
- Question **2** b): *What does this number line go up in? How can you work out the missing numbers?*

IN FOCUS All the questions focus on finding and marking 3-digit values on number lines. Questions **1** and **2** both use number lines that go up in 100s. Question **1** uses the familiar context of the boat race from **Discover**, whilst question **2** includes values that do not lie on marks on the number line. Encourage children to check first that the number line does increase in 100s. Question **3** extends work from Year 2, where children met number lines that go up in 10s and 1s, to include 3-digit numbers. Children should try and work out what the number lines go up in by counting and checking that the start and end points match.

STRENGTHEN Count aloud with children from 0 to 1,000 in 100s, pointing to each mark. You may additionally support children by placing base 10 equipment underneath each number to show that each time it is 100 more. Similarly, when using number lines that go up in 10s or 1s, count aloud whilst pointing to each mark.

DEEPEN Ask children to draw their own number line from 0 to 1,000 and place numbers, such as 10, 375, 402, 798 and 925, on it. Ask them to describe where the numbers will be before marking them on the number line. For example, do they know that 375 is three-quarters of the way between 300 and 400?

ASSESSMENT CHECKPOINT Use question **2** a) to assess whether children can identify the position of arrows on 100s and at half-way points between two numbers. Use question **3** to assess whether children can work out what the lines go up in (for example, 10s). Children should also be able to draw an arrow to particular points on number lines that go up in 100s, 10s and 1s.

ANSWERS

Question **1** a): 100, 200, 500, 700, 900

Question **1** b): Boat A has travelled 100 metres.

Question **1** c): Boat C has travelled 800 metres.

Question **1** d): Boat B has travelled about 450 metres.

Question **2** a): 100, 550, 800

Question **2** b): Children point to 300, 500 and 990.

Question **3** a): 270, 271, 272, 274, 275, 276, 277, 278, 279, 210, 220, 230, 240, 260, 270, 280, 290, 300

Question **3** b): Children point to 275 on each line.

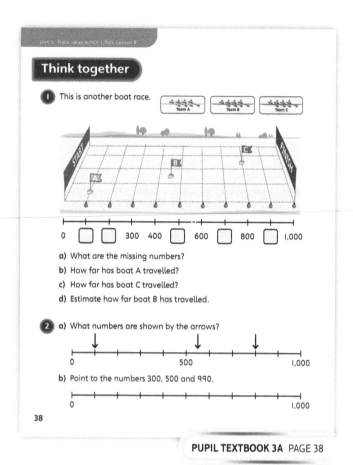

Think together

1 This is another boat race.

a) What are the missing numbers?
b) How far has boat A travelled?
c) How far has boat C travelled?
d) Estimate how far boat B has travelled.

2 a) What numbers are shown by the arrows?

b) Point to the numbers 300, 500 and 990.

38

PUPIL TEXTBOOK 3A PAGE 38

3 a) Work out all the missing numbers.

b) Point to 275 on each number line.

I don't think these number lines go up in 100s this time. I will try a number in the first blank box and see if it works as I count up to 250.

Last year we saw number lines that went up in 10s and 1s. I wonder if any of these do that.

→ Practice book 3A p27

39

PUPIL TEXTBOOK 3A PAGE 39

Practice

WAYS OF WORKING Independent thinking

IN FOCUS In questions **2** and **3**, children need to reason with number lines, finding missing values and writing on values. Children first need to work out what the number lines go up in, so that they can work out the correct position of each number.

STRENGTHEN Ask children what the number lines could go up in. Ask: *How can you check you are correct?* Count aloud with children to help them check. Encourage children to write on the values at each mark to help their understanding. A counting stick may be useful for questions **2** and **3**.

DEEPEN Ask children to add their own arrows to the number line in question **5** and challenge a partner. This encourages them to think about what makes a number easy or difficult to mark.

THINK DIFFERENTLY Question **4** provides a blank number line, where only the start and end values are marked. Children need to reason about what the missing number could be. Children should start to realise that the missing number will depend on where the number line ends or what it goes up in.

ASSESSMENT CHECKPOINT Use question **2** to assess whether children can identify missing numbers on number lines that go up in 100s and 10s. Questions **3** and **5** assess whether they can identify numbers on number lines that go up in 100s, 10s and 1s, including half-way numbers on lines going up in 100s or 10s, and draw arrows to numbers.

ANSWERS Answers for the **Practice** part of the lesson can be found in the *Power Maths* online subscription.

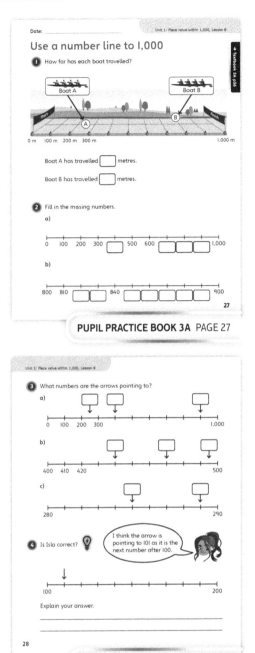

PUPIL PRACTICE BOOK 3A PAGE 27

PUPIL PRACTICE BOOK 3A PAGE 28

Reflect

WAYS OF WORKING Pair work

IN FOCUS Marking 650 on each number line practises the main part of the lesson. However, children then have to reason why the number will not be in the same position each time, showing their understanding of number lines that go up in different steps and with different start and end points.

ASSESSMENT CHECKPOINT Check that children can identify what each number line goes up in and can then place 650 accurately on each line.

ANSWERS Answers for the **Reflect** part of the lesson can be found in the *Power Maths* online subscription.

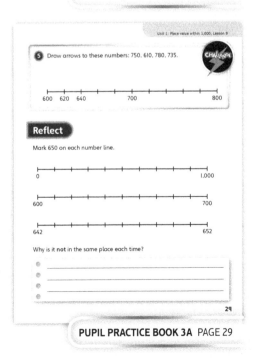

PUPIL PRACTICE BOOK 3A PAGE 29

After the lesson ⏸

- Can children identify whether a number line goes up in 100s, 10s or 1s?
- Can they mark points on the number line or identify the values that arrows are pointing to?
- Do children know how to find numbers like 550 on a number line that goes up in 100s?

Estimate on a number line to 1,000

Learning focus

In this lesson, children will begin to understand where numbers lie on a number line. They will identify numbers that lie between two points.

Before you teach

- Can children work out missing values on number lines that go up in 100s, 10s or 1s within 1,000?
- Do children know the place value of each digit in a number?

NATIONAL CURRICULUM LINKS

Year 3 Number – number and place value

Identify, represent and estimate numbers using different representations including the number line.

ASSESSING MASTERY

Children can work out points that lie between two points on a number line. They can also identify start and/or end points on a number line given a set of points in between.

COMMON MISCONCEPTIONS

Children may often just look at the first digit to work out if a number lies between two points on a line. For example, they identify 72 as lying between 700 and 800 because it starts with the digit 7. Ask:

- *What does the 7 represent in 72? What does the 7 represent in 700?*

Making the numbers with base 10 equipment may also help them see that 72 does not lie between 700 and 800.

STRENGTHENING UNDERSTANDING

Ask children to make the start and end numbers using base 10 equipment. They can then make each of the given numbers and compare them to work out whether the new number lies between the two points.

GOING DEEPER

Ask children to draw their own number line between two points and identify numbers that lie between the two points. Alternatively, give children a set of numbers for them to work out two possible numbers that the numbers could lie between. Ask: *How can you work out where the points will go?*

KEY LANGUAGE

In lesson: number line, half-way

Other language to be used by the teacher: hundreds (100s), tens (10s), ones (1s)

STRUCTURES AND REPRESENTATIONS

Number line

RESOURCES

Optional: base 10 equipment, counting sticks, whiteboards, string, pegs

 In the eTextbook of this lesson, you will find interactive links to a selection of teaching tools.

Quick recap 🔍

Ask children to draw four straight lines on their whiteboard. Ask them to divide their line into two, then into three, then four and five. Which did they find easier? Were the intervals the same distance? They can repeat the activity for extra practice.

Discover

Unit 1: Place value within 1,000, Lesson 9

Estimate on a number line to 1,000

WAYS OF WORKING Pair work

ASK

- Question ❶ a): *Is 500 greater than 0? Is 500 less than 1,000?*
- Question ❶ b): *Where does the number line start and end? Is 395 closer to the start number or the end number?*

IN FOCUS Each number line in the **Discover** section is different. The aim of both questions ❶ a) and ❶ b) is for children to start to get a feel for where numbers lie on a number line.

PRACTICAL TIPS This lesson builds on the previous lesson by looking more at open number lines. An open number line gives children more freedom, as it can be blank or just have the start or end point. There are no markings indicating points in between. You might find it useful to have the four lines from the picture as string around the classroom with numbers pegged on the ends. You could encourage children to come and peg numbers to these lines during the lesson.

ANSWERS

Question ❶ a): You can peg 500 to lines A and C.

Question ❶ b): 395 should go just before the 400 mark.

Discover

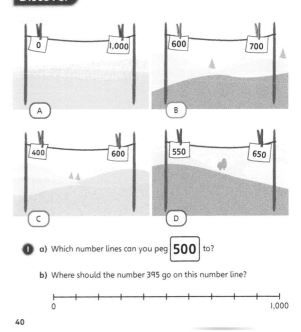

❶ a) Which number lines can you peg **500** to?

b) Where should the number 395 go on this number line?

0 ——————————— 1,000

40

PUPIL TEXTBOOK 3A PAGE 40

Share

WAYS OF WORKING Whole class teacher led

ASK

- Question ❶ a): *How do you know which washing lines you can peg 500 to?*
- Question ❶ b): *What is the number line counting on in? How do you know? Can you check? Between which two intervals will 395 go? How do you know? Which number is it closer to?*

IN FOCUS In question ❶ a), children need to realise that they must look at the start and end points of each washing line to see whether 500 will go on the line. In question ❶ b), children should count in 100s to support them in labelling the number line before identifying where 395 will go.

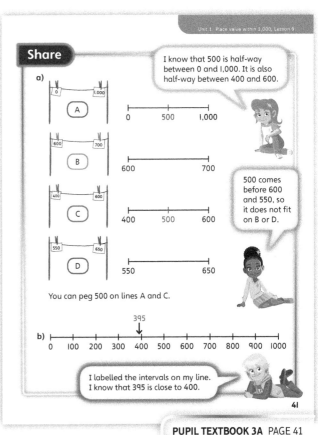

PUPIL TEXTBOOK 3A PAGE 41

Think together

Whole class teacher led (I do, We do, You do)

ASK

- Question **1**: *What is the start point of the number line? What is the end point? How can you use this to decide which numbers can go on the line? How do you know that 450 does not go on the number line?*
- Question **2**: *What is the number between? How do you know? What number is in the middle?*
- Question **3**: *What number is in the middle? How do you know? How does this help?*

IN FOCUS In question **2**, children should be encouraged to label the intervals on the number line to support them in estimating the values. Discuss how there could be multiple answers of a similar value because they are only estimates. In question **3**, children should be encouraged to work out a strategy to answer the question. They could start by marking the intervals on the number line.

STRENGTHEN To strengthen understanding, ask children to read and count aloud along the number line to find the number they are looking for. Using base 10 equipment to make the start and end numbers could support them in seeing the middle values.

DEEPEN Ask children to draw their own number line between two points, such as 300 and 400, then estimate where different numbers belong on the number line.

ASSESSMENT CHECKPOINT Use questions **1** and **2** to assess whether children know which numbers lie between any two given numbers. Also check that they can work out approximately where the numbers lie.

ANSWERS

Question **1**: 220, 222, 250 and 275

Question **2** a): 650

Question **2** b): 278

Question **2** c): 519

Question **3** a):

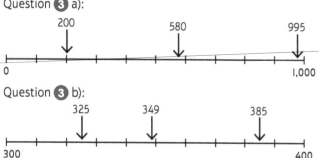

Question **3** b):

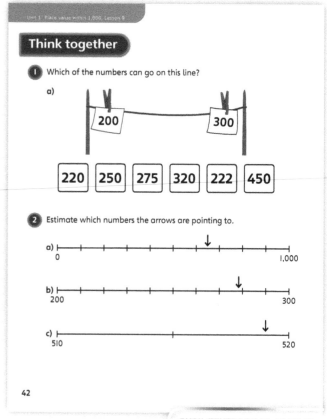

Practice

WAYS OF WORKING Independent thinking

IN FOCUS Questions ❶ to ❸ help children to develop a sense of where numbers lie on a number line. Questions ❶ and ❷ give them the division of the number line, whilst questions ❸ and ❹ only show the start and end points.

STRENGTHEN In question ❶, support children by asking them to count aloud and label the intervals on the number line to see if they can work out what each number line goes up in and where the numbers will be. In the other questions, ask them to use base 10 equipment to make the start and end numbers to help them identify the numbers in between.

DEEPEN In question ❸, ask children to make the start and end numbers using base 10 equipment to support them in identifying numbers in this interval. In question ❹, encourage them to draw on the intervals to help them identify the position of numbers.

ASSESSMENT CHECKPOINT Use questions ❶ and ❷ to assess whether children can recognise what a number line is counting on in and where numbers lie. Use question ❸ to see whether children can list some numbers that lie between two points on a number line.

ANSWERS Answers for the **Practice** part of the lesson can be found in the *Power Maths* online subscription.

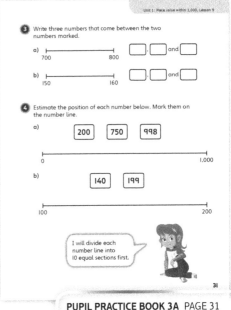

PUPIL PRACTICE BOOK 3A PAGE 30

PUPIL PRACTICE BOOK 3A PAGE 31

Reflect

WAYS OF WORKING Independent thinking

IN FOCUS Children need to draw and interpret their own number line. Encourage discussion about whether the length of the number line affects the answer if the start and end points do not change.

ASSESSMENT CHECKPOINT Check that children can interpret their number line and use it to make sensible estimates for the labelled values.

ANSWERS Answers for the **Reflect** part of the lesson can be found in the *Power Maths* online subscription.

PUPIL PRACTICE BOOK 3A PAGE 32

After the lesson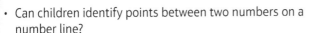

- Can children identify points between two numbers on a number line?
- Can they identify where a number line starts and/or ends based on numbers that are marked on the line?

Find 1, 10 and 100 more or less

Learning focus

In this lesson, children will find 1, 10 and 100 more or less than a given number (including cases that require an exchange). They will also find the original number given the increase or decrease.

Before you teach

- Can children represent a 3-digit number using base 10 equipment and on a place value grid with counters?

NATIONAL CURRICULUM LINKS

Year 3 Number – number and place value

Count from 0 in multiples of 4, 8, 50 and 100; find 10 or 100 more or less than a given number.

Recognise the place value of each digit in a three-digit number (100s, 10s, 1s).

ASSESSING MASTERY

Children can find 1, 10 and 100 more or less than a given number, including cases that involve an exchange. Children recognise which digit(s) will change. Children can also find the original number following an increase or decrease of 1, 10 and 100 by considering the inverse.

COMMON MISCONCEPTIONS

When answering a question such as 527 is 100 more than ☐, children often think they have to find 100 more. Children need to see that 527 is the value after the increase, therefore they need to do the inverse. Ask:

- *Have you been given the original number or the end number? Should your answer be higher or lower than the given number?*

STRENGTHENING UNDERSTANDING

Some children may find it helpful to use base 10 equipment or place value counters, particularly with questions that involve exchange. Model the question using base 10 equipment or place value counters, then use a place value grid. When counting objects, encourage children to count the 100s, then the 10s and then the 1s. Some children may use a number line to help track the counting.

GOING DEEPER

Ask children questions about which digits can change if they find 1, 10, or 100 more. Ask: *Is it possible for all the digits to change when you find 1 less or 1 more?* Repeat for 10 and 100 more or less. Encourage children to experiment with different numbers.

KEY LANGUAGE

In lesson: more, less, **exchange**

Other language to be used by the teacher: inverse

STRUCTURES AND REPRESENTATIONS

Place value grid

RESOURCES

Mandatory: base 10 equipment, place value counters

Optional: number lines, toy money

 In the eTextbook of this lesson, you will find interactive links to a selection of teaching tools.

Quick recap 🔎

Ask children to draw three boxes on their whiteboards with the number 47 in the middle box. Ask them to write down 1 more than 47 in the right-hand box and 1 less than 47 in the other.

Repeat the game for different numbers. Change it by asking them to find 10 more and 10 less. Briefly discuss what digit they notice changes.

Discover

Unit 1: Place value within 1,000, Lesson 10

Find 1, 10 and 100 more or less

WAYS OF WORKING Pair work

ASK

- Question ① a): *What can you use to help you work out how many points Amal has? Which digit will change?*
- Question ① b): *How can you work out 1 less than 204?*

IN FOCUS In question ① a), children can use their knowledge of place value to work out how many points the player has. They then need to work out which digit changes when Amal receives 100 more points. Discuss how the points can be used like place value counters, as they are in 100s, 10s and 1s.

PRACTICAL TIPS The context for this lesson is a board game where players have notes with points on. You could give children toy money from a game.

ANSWERS

Question ① a): Amal has 253 points.
          Amal adds 100 points, so he now has 353.

Question ① b): Holly now has 203 points.

Discover

① **a)** How many points does Amal have?

Amal receives 100 more.

How many points does he have now?

b) Holly has 204 points.

She loses 1 point.

How many points does she have now?

44

PUPIL TEXTBOOK 3A PAGE 44

Share

WAYS OF WORKING Whole class teacher led

ASK

- Question ① a): *What did Dexter use to represent the points?*
- Question ① b): *How is 1 less shown on a place value grid?*

IN FOCUS Question ① b) involves finding 1 less than a given number. Ensure that children understand that losing 1 point means that the player has 1 less.

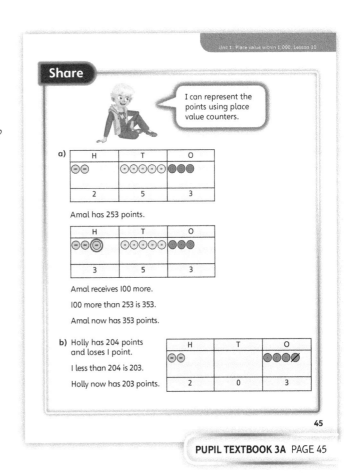

45

PUPIL TEXTBOOK 3A PAGE 45

Think together

WAYS OF WORKING Whole class teacher led (I do, We do, You do)

ASK

- Question ❶: *What are the starting numbers? Do you have to find 1, 10, 100 more or less?*
- Question ❸: *Do you need to exchange anything? How do you know?*

IN FOCUS Questions ❶ and ❷ give children practice in finding 1, 10 and 100 more or less than given numbers, moving from numbers presented in pictorial form to abstract numbers. In question ❸, children need to realise that they do not have enough 10s and so they need to exchange. This question also addresses some common misconceptions that children make.

STRENGTHEN Provide children with base 10 equipment or place value counters to find 1, 10 and 100 more or less. Encourage them to be systematic in their counting, counting the 100s, then the 10s and then the 1s. In questions that involve exchange, tell them to change 1 of their 100 flats for 10 ten rods, or 1 ten rod for 10 ones cubes.

DEEPEN After each part in question ❸, ask children to work out which digits changed. Give children some similar problems and ask them to predict which digits will change.

ASSESSMENT CHECKPOINT Use questions ❶, ❷ and ❸ to assess whether children can find 1, 10, and 100 more or less than any number, from both pictorial representations and abstract numbers. Check that children can identify when they need to carry out an exchange first.

ANSWERS

Question ❶ a): 10 more than 365 is 375.

Question ❶ b): 100 less than 648 is 548.

Question ❶ c): 1 more than 248 is 249.

Question ❷ a): 663, 463.

Question ❷ b): 238, 218.

Question ❷ c): 719, 717.

Question ❸ a): 10 more than 195 is 205. Kate needs to exchange 10 tens for 100.

Question ❸ b): Ebo has found 100 more than 457 rather than 100 less.

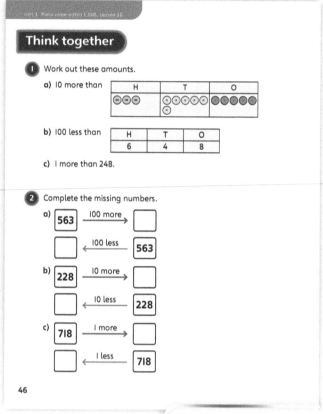

PUPIL TEXTBOOK 3A PAGE 46

PUPIL TEXTBOOK 3A PAGE 47

Practice

WAYS OF WORKING Independent thinking

IN FOCUS Question ❶ provides opportunity for children to consolidate finding 1, 10 and 100 more than a given number. Question ❷ is similar with a focus on 1, 10 and 100 less. In question ❸, children find a mixture of 1, 10 and 100 more or less than given numbers that are written abstractly. In question ❹, children are required to find the missing number given a value that is 1, 10 or 100 more or less.

STRENGTHEN Provide children with base 10 equipment to make the numbers and use this to find 1, 10 or 100 more or less. When finding more, children should recognise that they need to add a piece, whilst when finding less they need to remove one. In question ❹, encourage children to read their answer aloud to check it makes sense.

DEEPEN Question ❻ gives further practice in moving between numbers using the result of an increase or decrease. Ask: *How do you know if your answers are correct? Can you make up a similar question of your own?*

THINK DIFFERENTLY In question ❺, children are given an increase of Mo's number and need to use this to find Mo's number. They then use Mo's number to complete a table. Ask: *Which section did you find easiest to fill in? Which section did you find the hardest?*

ASSESSMENT CHECKPOINT Use question ❸ to assess whether children can find 1, 10 and 100 more or less than any number including cases that require an exchange. Use question ❹ to assess whether children can use the increased/decreased value to find the original number.

ANSWERS Answers for the **Practice** part of the lesson can be found in the *Power Maths* online subscription.

Reflect

WAYS OF WORKING Pair work

IN FOCUS The question involves children finding 1, 10 and 100 more or less and working out the original number given the increase or decrease by using the inverse.

ASSESSMENT CHECKPOINT Check that children can find 1, 10, and 100 more and less than the number they have generated. Also, check they can work out the original number given an increase or decrease.

ANSWERS Answers for the **Reflect** part of the lesson can be found in the *Power Maths* online subscription.

After the lesson ⏸

- Can children find 1, 10 and 100 more or less than a given number?
- Can they find the original number given an increase or decrease?
- Do children know which digit to look at and which ones will change?
- Can children find 1, 10 and 100 more or less than numbers that involve exchange?

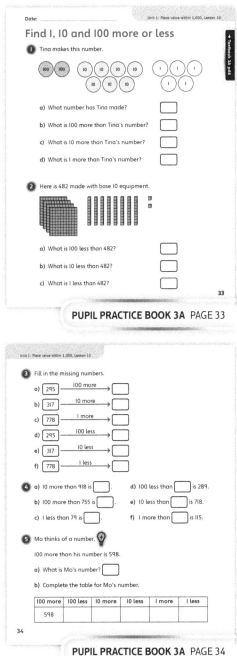

PUPIL PRACTICE BOOK 3A PAGE 33

PUPIL PRACTICE BOOK 3A PAGE 34

PUPIL PRACTICE BOOK 3A PAGE 35

Compare numbers to 1,000

Learning focus

In this lesson, children will compare two 3-digit numbers. Children will also be able to work out missing digits to make an inequality statement correct.

Before you teach

- Do children know what the signs <, > and = mean?
- Can children use base 10 equipment to compare two 3-digit numbers?
- Can children explain how to compare two 2-digit numbers?

NATIONAL CURRICULUM LINKS

Year 3 Number – number and place value

Compare and order numbers up to 1,000.

Identify, represent and estimate numbers using different representations, including the number line.

ASSESSING MASTERY

Children can compare two 3-digit numbers using <, > and = signs. Children should compare the numbers by first considering the 100s and then the 10s (if needed) and then the 1s. Children can also work out the missing digits in numbers.

COMMON MISCONCEPTIONS

Children sometimes do not think about the place value of a digit. For example, when comparing 586 and 84 they think that 84 is greater than 586 as the first digit of 84 is 8 and the first digit of 586 is 5. Encourage children to think about the place value of each digit. Ask:

- *What does the 8 represent in 84? What does the 5 represent in 586? How many 100s are there in 84? Which number has more 100s?*

STRENGTHENING UNDERSTANDING

Ask children to make the numbers using base 10 equipment and put them in a place value grid. Then ask them to compare the 100s by lining them up. If they have the same amount then ask them to compare the 10s, and so on. Each time, link the base 10 equipment to the numbers in the place value grid.

GOING DEEPER

Ask children to explain the greatest and least number of comparisons that need to be made to compare 3-digit numbers. Challenge them to write down two numbers where they need to make one, two and three comparisons.

KEY LANGUAGE

In lesson: more than, less than, greater than, smaller than, is equal to, <, >, =

STRUCTURES AND REPRESENTATIONS

Place value grid, number line

RESOURCES

Mandatory: dice

Optional: base 10 equipment

 In the eTextbook of this lesson, you will find interactive links to a selection of teaching tools.

Quick recap

Play a game in pairs. Ask player 1 to roll two dice and make a 2-digit number. Ask player 2 to do the same. The winner is the person who makes the larger number.

Play several times. Change half-way through to the winner being the one with the smaller number. Ask children to write down a statement that includes <, > or = for their numbers.

Discover

Unit 1: Place value within 1,000, Lesson 11

Compare numbers to 1,000

WAYS OF WORKING Pair work

ASK

- Question ① a): *What are the numbers? Do you need to make the numbers out of base 10 equipment or counters to compare? Is there a way of working out which is larger by just comparing the digits? Which digits should you compare first? What happens if they are the same?*
- Question ① b): *Is there a way of working out which is smaller by just comparing the digits?*

IN FOCUS In questions ① a) and ① b), children compare pairs of 3-digit numbers. This time the numbers are abstract (i.e. there are not 240 objects). Although some children may make the numbers using base 10 equipment, they should be looking to form a method that does not always require them to make the number. Eventually children should explore the numbers to realise that they can compare the 100s first, and then the 10s and 1s if necessary. They should compare the digits from left to right.

PRACTICAL TIPS Make the number cards for children to hold up at the front of the classroom. Have place value grids for children to put the numbers in. Some children may need to use base 10 equipment for support.

ANSWERS

Question ① a): 395 is the greater number.

Question ① b): 542 is the smaller number.

Discover

① a) Which is the greater number?

　　240　395

　　b) Which is the smaller number?

　　542　589

48

PUPIL TEXTBOOK 3A PAGE 48

Share

WAYS OF WORKING Whole class teacher led

ASK

- Question ① a): *What did you compare first? How many comparisons did you need to make? Why do you compare from left to right? How can you use a number line to compare the numbers?*

IN FOCUS For both questions, listen to children discuss their method. Highlight the method of comparing numbers without using base 10 equipment. Ensure children understand that they can compare the digits from left to right. In question ① a), one comparison is needed to compare 240 and 395. A number line is used to compare 240 and 395, which is an alternative method that children may not have thought of. In question ① b), two comparisons are needed to compare 542 and 589.

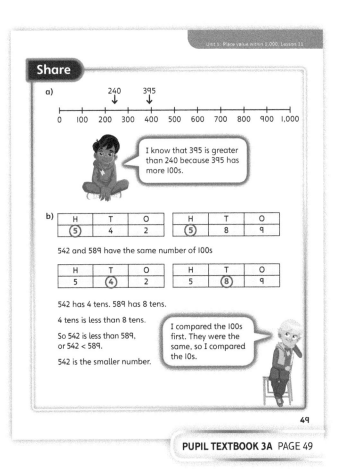

PUPIL TEXTBOOK 3A PAGE 49

Think together

WAYS OF WORKING Whole class teacher led (I do, We do, You do)

ASK

- Question ❶: *Which number has more 100s?*
- Question ❷: *How can you compare the numbers? What do you look at first? Then what do you look at?*
- Question ❸: *How many answers can you find?*

IN FOCUS Question ❶ asks children to compare two 3-digit numbers and a 3-digit number with a 2-digit number. 183 and 92 could cause confusion as children may not recognise the difference in place value and think 183 is smaller than 92, because the first digit in 183 is smaller than the first digit in 92. Encourage children to work from left to right, looking at the 100s first. In question ❷, children start to make use of the <, > and = symbols that they have met before. Children may need reminding what the symbols represent.

STRENGTHEN Throughout, if children need more support to compare numbers ask them to make the numbers using base 10 equipment and put them in a place value grid. Then use the base 10 equipment to compare the 100s, 10s and 1s, as necessary, by lining them up.

DEEPEN In question ❸, ask children to work out how many different solutions there are for each question. Ask: *What is the smallest digit you can use in each number? What is the largest? Can you explain why?*

ASSESSMENT CHECKPOINT Use question ❷ to assess whether children can compare two numbers by comparing the 100s first, then the 10s if necessary and then the 1s.

ANSWERS

Question ❶ a): 542

Question ❶ b): 92

Question ❷ a): 948 > 820

Question ❷ b): 385 > 368

Question ❷ c): 600 < 950

Question ❷ d): 392 = 300 + 90 + 2

Question ❸ a): 0, 1, 2 or 3

Question ❸ b): 1, 2, 3 or 4

Question ❸ c): 1, 2, or 3

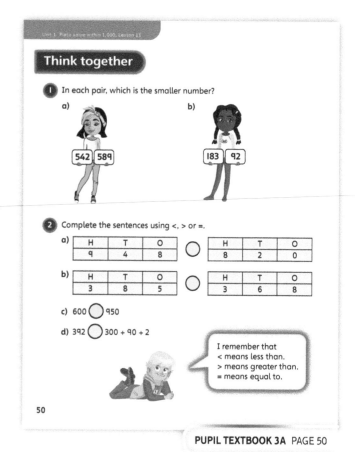

PUPIL TEXTBOOK 3A PAGE 50

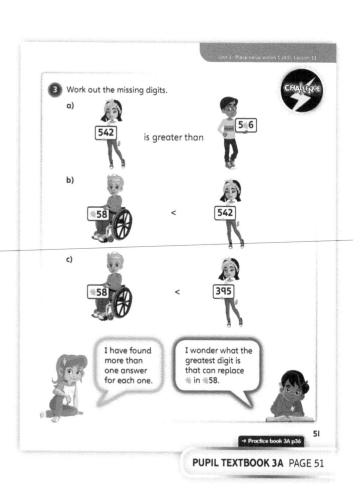

PUPIL TEXTBOOK 3A PAGE 51

Practice

Independent thinking

IN FOCUS Questions **1**, **2** and **3** ask children to work out greater and smaller numbers. In question **4**, children need to use the <, > and = signs correctly. There is variation in how the numbers are presented to reinforce earlier work. In question **5**, where children have to work out missing digits to make the inequality statements correct, they should realise that there is more than one answer. Question **6** provides descriptions of three numbers. Children need to work out if they have enough information to be able to make the comparisons and give reasons for their answers, based on the information provided.

STRENGTHEN In question **5**, ask children to make the parts of the numbers that they know using base 10 equipment and put them in a place value grid. Ask: *What could you put in the missing columns to make the statement true?* Ask them to write the place value grid comparison that they have made using base 10 equipment.

DEEPEN In question **6**, ask: *How much more information do you need to be able to compare Reena and Zac's numbers?*

ASSESSMENT CHECKPOINT Use question **3** to assess whether children can compare any two 3-digit numbers to determine which is greater and which is smaller. Children should accurately use the > and < signs to compare numbers.

ANSWERS Answers for the **Practice** part of the lesson can be found in the *Power Maths* online subscription.

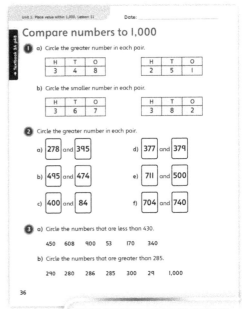

PUPIL PRACTICE BOOK 3A PAGE 36

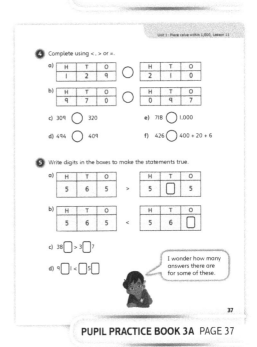

PUPIL PRACTICE BOOK 3A PAGE 37

Reflect

WAYS OF WORKING Pair work

IN FOCUS Children are given two numbers to compare. They need to clearly explain the steps involved in comparing two 3-digit numbers.

ASSESSMENT CHECKPOINT Check that children know that they need to compare the 100s first and then, if these are the same, the 10s and finally, if necessary, the 1s. Children should use this method to compare any two 3-digit numbers.

ANSWERS Answers for the **Reflect** part of the lesson can be found in the *Power Maths* online subscription.

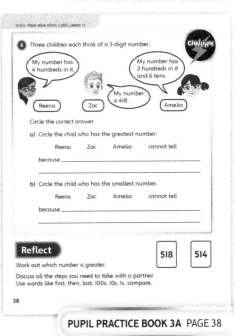

PUPIL PRACTICE BOOK 3A PAGE 38

After the lesson ⏸

- Can children compare two 3-digit numbers by comparing the greatest place value first?
- Can children find missing numbers to ensure a mathematical statement is correct?
- Do children use the < and > signs accurately?

89

Order numbers to 1,000

Learning focus

In this lesson, children will order three or more numbers up to 3-digits. They will also work out missing digits in lists of ordered numbers.

NATIONAL CURRICULUM LINKS

Year 3 Number – number and place value

Compare and order numbers up to 1,000.

Recognise the place value of each digit in a three-digit number (100s, 10s, 1s).

ASSESSING MASTERY

Children can order three or more 3-digit numbers. Children can also work out the missing digits in numbers listed in order.

COMMON MISCONCEPTIONS

Children sometimes do not think about the place value of a digit. For example, when ordering 850, 98 and 700 they may think that 98 is the greatest number as this is the one with the greatest first digit. Ask:

- *What does the 8 represent in 850? What does the 9 represent in 95? What about the 7 in 700? Which number has more 100s?*

Children may think that when finding missing digits they only have to look at adjacent numbers. This is not always the case and numbers later in the list could have an impact on the missing digits. Ask:

- *Compare the first two numbers. What are the possible digits for the second number? Now compare the second number with the third number. What are the possible digits for the second number? Which digits are in both lists?*

STRENGTHENING UNDERSTANDING

Children can make the numbers using base 10 equipment to help them understand the size or use counters on a place value grid. This will help them to see which number has the greatest number of 100s, then 10s and then 1s.

GOING DEEPER

Ask children to explain the greatest and least number of comparisons that need to be made to order three 3-digit numbers.

KEY LANGUAGE

In lesson: ascending, descending, greater than (>), taller, tallest, longest, shortest, greatest, smallest, most, least, fewest

Other language to be used by the teacher: more, less than (<), equal to (=)

STRUCTURES AND REPRESENTATIONS

Place value grid, number line

RESOURCES

Mandatory: dice

Optional: base 10 equipment, place value counters

 In the eTextbook of this lesson, you will find interactive links to a selection of teaching tools.

Quick recap 🔁

Play a game in pairs. Ask player 1 to roll three dice and make the largest number they can. Ask player 2 to do the same. The winner is the player who has made the larger number. Play several times. Change half-way through to the winner being the one with the smaller number.

Discover

Order numbers to 1,000

Discover

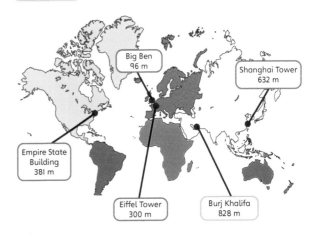

WAYS OF WORKING Pair work

ASK

- Question ① a): *How can you compare the two numbers? How many 100s are there in each number? Do you need to compare more than the 100s?*
- Question ① b): *How can you put these buildings in height order? Which building is the shortest? How do you know?*

IN FOCUS Question ① a) reinforces the work from the previous lesson. The question highlights a key misconception, with some children thinking that Big Ben is taller because 9 is greater than 3. Remind children that the place value of the digit is important. Question ① b) prompts children to think of different strategies for ordering the buildings. Some children may want to compare two buildings at a time, some may use a number line, and others may consider all the buildings at once.

PRACTICAL TIPS Make cards of the buildings and their heights and peg these on to a line. Encourage children to move their positions on the line until they are in the correct order.

ANSWERS

Question ① a): The Empire State Building is taller than Big Ben.

Question ① b): Big Ben, Eiffel Tower, Empire State Building, Shanghai Tower, Burj Khalifa

① a) Which is taller, the Empire State Building or Big Ben?

b) Put the buildings in order of height.
   Start with the shortest.

52

PUPIL TEXTBOOK 3A PAGE 52

Share

WAYS OF WORKING Whole class teacher led

ASK

- Question ① b): *How does the table help you to compare the numbers? Which column should you look at first? Why? What happens if the 100s are the same? What column do you look at then? Why?*

IN FOCUS Question ① a) highlights the fact that there are 0 hundreds in 96 metres. This will help children understand that 381 is greater than 96 and so the Empire State Building is taller than Big Ben. Discuss if the number line is a helpful way to compare numbers. In question ① b), the table shows the building heights in a place value grid. Discuss Astrid's method for ordering numbers. Talk children through the example. You may find it easier to write the numbers on the board and cross out numbers as they are used.

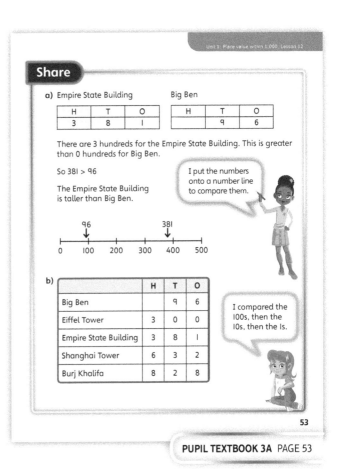

PUPIL TEXTBOOK 3A PAGE 53

Think together

WAYS OF WORKING Whole class teacher led (I do, We do, You do)

ASK

- Question **1**: *How do you know which number is the smallest? What digit did you look at first?*
- Question **2**: *What numbers are shown? Which number is greater? Which digit did you look at first? Which inequality sign could you use? Can you say it in a sentence?*
- Question **3**: *How many answers can you find?*

IN FOCUS Question **1** requires children to identify the smallest number from a group of three numbers. This only requires them to compare the 100s. In question **2**, they are required to compare 100s, 10s and 1s. Children may find it useful to represent the numbers in a place value grid.

STRENGTHEN Provide children with base 10 equipment and/or place value counters to support them in comparing numbers. Ensure they always start with the digit that has the highest place value.

DEEPEN In question **3** b), ask children to work out how many different solutions there are for each question. Ask: *What is the smallest digit that can be used in each number? What is the largest? Can you explain why?*

ASSESSMENT CHECKPOINT Use question **2** to assess whether children can compare two numbers by comparing the 100s first, then the 10s if necessary, then the 1s.

ANSWERS

Question **1**: Ferry, Cruise ship, Container ship

Question **2**: B, D, A, C

Question **3** a): 188, 276, 300, 712

Question **3** b): Any number between 0 and 9; 9; 2; any number between 0 and 9.

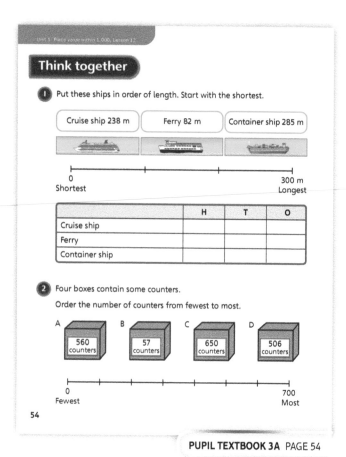

PUPIL TEXTBOOK 3A PAGE 54

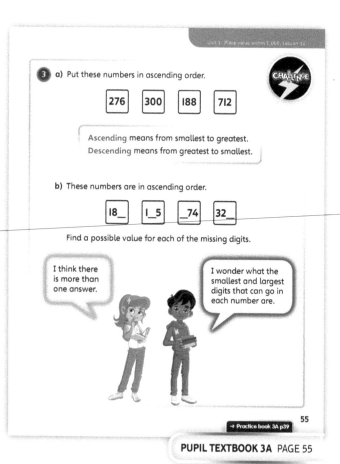

PUPIL TEXTBOOK 3A PAGE 55

Practice

WAYS OF WORKING Independent thinking

IN FOCUS Question ① mirrors the **Discover** task and asks children to compare heights of towers. The table should be used as demonstrated in **Share**, with children looking for the tower with the smallest number of 100s. In question ④, children are not given the prompt of a place value grid. Encourage children to use a place value grid if it helps them, although some may be able to work out the order by just looking at the digits. Parts c) and d) include numbers that have a different number of digits, so children need to understand the importance of place value. Question ⑤ asks children to reason more deeply by working out missing digits. Children need to understand that there are lots of different possible answers and their answers may depend on their choices for other digits.

STRENGTHEN In questions ①, ② and ③, children could make the numbers using base 10 equipment to help them understand their size. They may then find it easier to see when they need to compare just the 100s, and when they need to also compare 10s and 1s. In question ④, encourage children to draw a place value grid like in the earlier questions to help them order the numbers, or provide them with blank grids to fill in.

DEEPEN In question ⑤, ask children how many different answers they can find. Encourage them to work systematically, to be sure that they have found all the possible solutions.

ASSESSMENT CHECKPOINT Use question ④ to assess whether children can order three or more numbers by first comparing the 100s, then the 10s and then the 1s, if needed. Use question ⑤ to assess whether children can find missing digits to make a set of numbers in order.

ANSWERS Answers for the **Practice** part of the lesson can be found in the *Power Maths* online subscription.

Reflect

WAYS OF WORKING Pair work

IN FOCUS Children use their understanding from this lesson to order a set of numbers. They can use whatever method they feel most comfortable with, but they need to explain in words or show a clear method for what they did. For example, they may draw a number line or create a place value grid.

ASSESSMENT CHECKPOINT Check that children have a method for ordering numbers. Check that children know that there are two different orders that they can put the numbers in.

ANSWERS Answers for the **Reflect** part of the lesson can be found in the *Power Maths* online subscription.

After the lesson ⏸

- Can children order a set of three or more numbers from smallest to largest, or largest to smallest?
- Do children know how to accurately use the <, > and = signs in mathematical statements?

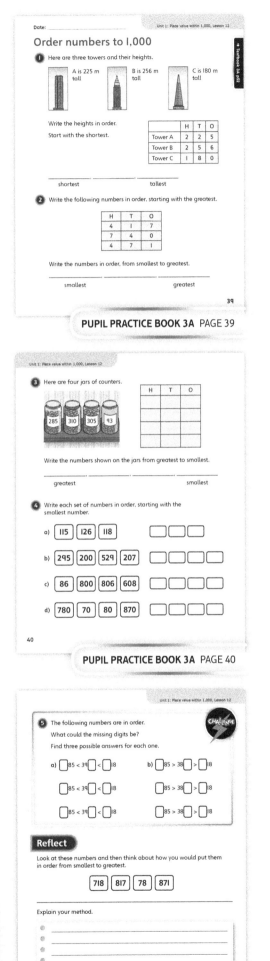

PUPIL PRACTICE BOOK 3A PAGE 39

PUPIL PRACTICE BOOK 3A PAGE 40

PUPIL PRACTICE BOOK 3A PAGE 41

93

Count in 50s

Learning focus

In this lesson, children will count on and back in 50s from 0 to 1,000 and count from any multiple of 50. They will work out how many 50s there are in a number.

Before you teach

- Can children count on and back in 10s?
- Do children know that 10 tens make 100?
- Do children know the different coins and amounts and know how many pence make £1?

NATIONAL CURRICULUM LINKS

Year 3 Number – number and place value

Count from 0 in multiples of 4, 8, 50 and 100; find 10 or 100 more or less than a given number.

ASSESSING MASTERY

Children can count on and back in 50s from 0 to 1,000. Children should be able to start at any multiple of 50. They can work out how many 50s there are in a number by counting on to that number. Children can start to spot patterns and be able to identify numbers that are in the pattern.

COMMON MISCONCEPTIONS

Children may miscount because they are trying to keep track of too many things. For example, when counting on 7 fifties from 300 they may struggle to keep track of how many 50s they have counted and where they are. Encourage them to use a number line to keep track. Ask:

- *What number are you starting from? Count on 1 fifty. What number are you at now? How many 50s have you counted? How can you keep track of how many you have counted?*

STRENGTHENING UNDERSTANDING

To strengthen understanding, give children base 10 equipment to help them. Start with 5 tens rods and count in 10s with them. Explain that this is 50. Ask them to take another 50 and count the total, which is 100. Ask them to count aloud from 0 and then show them how they could have counted in 50s: 0, 50, 100. Ask: *What can you swap 10 tens for?* Continue doing this, counting on, and children will start to notice the pattern.

GOING DEEPER

Give children some problems involving money. Ask: *How many 50p coins are there in £6? How much money do you have in pounds and pence if you have 15 50p coins?* Tell them that banks keep 50p pieces in bags of 20. Ask: *How much is this in pounds?*

KEY LANGUAGE

In lesson: fifty, 50s

Other language to be used by the teacher: count on, count back

STRUCTURES AND REPRESENTATIONS

Number line

RESOURCES

Optional: base 10 equipment, place value counters, 50p coins

 In the eTextbook of this lesson, you will find interactive links to a selection of teaching tools.

Quick recap

How far can children count in 5s to 100? How far can children count back in 5s to 0? Count as a class by going from one child to the next. Go back to the start if a child gets it wrong.

Discover

Unit 1: Place value within 1,000, Lesson 13

Count in 50s

WAYS OF WORKING Pair work

ASK

- Question ❶ a): *Do you recognise this flag? How many stars are on the flag? How did you count the number of stars? How can you find the number of stars on 4 flags?*
- Question ❶ b): *How could you work out the number of flags? Is there a quicker way?*

IN FOCUS In question ❶ a), children first identify that there are 50 stars on each flag. Some may already know this, but encourage them to check. Encourage children to look for counting strategies other than just counting in 1s. Once children have worked out there are 50 stars on a flag, they need to look at methods for working out how many are on 4 flags. Children may count in 1s, 5s, 10s or 50s, but encourage them to find the most efficient method (i.e. counting in 50s). They could use a number line and base 10 equipment to help them. Encourage them to look for patterns. In question ❶ b), children need to count in 50s until they get to 350. They need to develop a system for keeping count of how many 50s they have counted.

PRACTICAL TIPS Count aloud in 5s up to 50. Ask children to stand at the front of the classroom holding number cards in 50s in order from 0 to 500. Count aloud in 50s.

ANSWERS

Question ❶ a): There are 50 stars on each flag.
There are 200 stars on 4 flags.

Question ❶ b): Sylvie counted 350 stars on 7 flags.

Discover

❶ a) How many stars are on each flag?
How many stars are on 4 flags?

b) Sylvie counts 350 stars.
How many flags has she counted?

56

PUPIL TEXTBOOK 3A PAGE 56

Share

WAYS OF WORKING Whole class teacher led

ASK

- Question ❶ a): *How did Astrid count the stars? Did you count in the same way? What other ways could you have counted? What did you count in to work out how many stars are on 4 flags? Did you use a number line? Did you notice any patterns in the numbers?*
- Question ❶ b): *Why did you count in 50s? Does the number line help? How can you keep track of how many 50s you have counted? Will you get the same answer if you count back from 350? Why?*

IN FOCUS Question ❶ a) shows a different method for counting, rather than in 1s. Children counted in 5s in Year 1, so this is a useful opportunity to revisit that before counting in 50s. Question ❶ b) looks at the opposite to what they did in part a), counting on in 50s to get to a specific number.

Share

57

PUPIL TEXTBOOK 3A PAGE 57

Think together

Whole class teacher led (I do, We do, You do)

ASK

- Question **1**: *What does the table tell you? What do the numbers go up in on the bottom row? Why?*
- Question **2**: *How did you work out part a)? Can you use your table in question* **1** *to help you? Do you have to start again from 0 for 16 flags?*
- Question **3**: *How do you know which numbers will appear in the count? Is there a way you can tell without counting on in 50s from 0 to 1,000? How?*

IN FOCUS Question **1** looks at the sequence of multiples of 50. By this time, children should be learning the sequence, but they can link it to a number line to help them if they need to. Children should start noticing patterns in the last two digits. Question **2** looks to see if they can extend the table themselves and find the number of stars on 15 flags. Some children may need to write all the numbers down, whereas some may be able to keep track in their heads. Discuss whether they need to start at 0 each time for the 16 and 17 flags. Children should start to realise they just need to count on 50 more from the previous total. Question **3** looks at the pattern of the last two digits and gets children to think about how to identify numbers in the count without actually counting on in 50s from 0 to 1,000.

STRENGTHEN To strengthen understanding, give children place value counters. Start with 5 ten counters and count in 10s with them. Take another 5 ten counters and count the total, which is 100. Ask them to count aloud from 0 to 100 in 10s, counting the separate counters. Then put the counters in piles of 5 and count aloud in 50s. Swap the 10 ten counters for 1 hundred counter. Continue doing this, counting on, and children will start to notice the pattern.

DEEPEN Ask: *How many 50s are there between 350 and 800? Can you explain how you worked this out?*

ASSESSMENT CHECKPOINT Use question **3** to assess whether children can count on and back in 50s within 1,000. Check that children can recognise numbers that are multiples of 50.

ANSWERS

Question **1**: 150, 200, 250, 300, 350, 400, 450, 500

Question **2** a): There are 750 stars on 15 flags.

Question **2** b): There are 800 stars on the 16 flags.

Question **2** c): There are 850 stars on 17 flags.

Question **3**: 750, 650, 400, 50, 250
The last two digits are 00 or 50.

PUPIL TEXTBOOK 3A PAGE 58

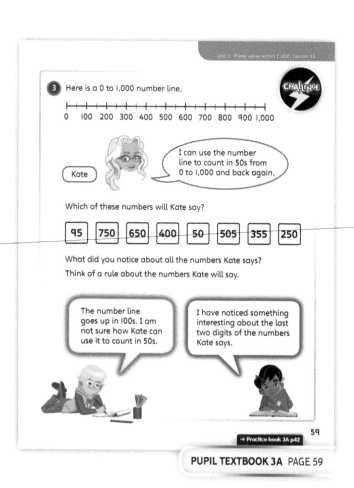

PUPIL TEXTBOOK 3A PAGE 59

Practice

WAYS OF WORKING Independent thinking

IN FOCUS Question ❶ reinforces counting in 50s. The table does not contain all the numbers from 1 to 10, so children should be careful when filling in the later values. Question ❺ links what children did in Year 2 on money to counting on in 50p coins. They first need to convert £7 to pence and then count on in 50s to make 700. Children may need to be reminded that there are 100 pence in a pound.

STRENGTHEN In question ❺, children could use 50p coins to count on to £7. Provide a number line for them to put their 50p coins on. Ensure that they understand that two 50p coins make 100p, which is the same as £1.

DEEPEN Give children some more money problems. For example, Bella has twelve 50p coins and Jack has seven £1 coins. Ask: *Who has more money? Can you write your answer as an inequality?*

THINK DIFFERENTLY Question ❹ requires children to count in mixed amounts, mixing counting in 100s and counting in 50s. Children could count in 50s by considering the 100s as two 50s or they could count the 100s first and then start counting on in 50s.

ASSESSMENT CHECKPOINT Use questions ❶, ❷ and ❸ to assess whether children can count on and back in 50s from any multiple of 50 between 0 and 1,000. Children should be able to apply their knowledge in a variety of contexts, such as money, number tracks and number lines.

ANSWERS Answers for the **Practice** part of the lesson can be found in the *Power Maths* online subscription.

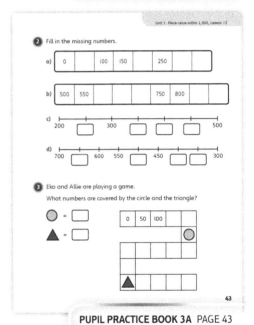

PUPIL PRACTICE BOOK 3A PAGE 42

PUPIL PRACTICE BOOK 3A PAGE 43

Reflect

WAYS OF WORKING Pair work

IN FOCUS Children try to spot the pattern of the numbers that will be in the count from 0 to 1,000 in 50s. They should count on from 0 to 1,000 and then look for any similarities between the numbers.

ASSESSMENT CHECKPOINT Check that children can count on in 50s and ensure that they can identify that every number ends in either 00 or 50.

ANSWERS Answers for the **Reflect** part of the lesson can be found in the *Power Maths* online subscription.

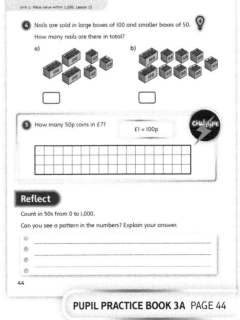

PUPIL PRACTICE BOOK 3A PAGE 44

After the lesson ⏸

- Can children count in 50s on and back between 0 and 1,000?
- Can children count in 50s on and back from any multiple of 50?
- Do children know that all the multiples of 50 end in 00 or 50?

End of unit check

Don't forget the unit assessment grid in your *Power Maths* online subscription.

WAYS OF WORKING Group work adult led

IN FOCUS In question ❷, children must accurately read from a number line, knowing when a number is half-way between two markers. Question ❸ is designed to check that they know when to add 10 to the given number and when to do the inverse operation. Question ❼ is a SATS-style problem and interweaves several topics within this unit. Children must know how to find 100 more and how to order numbers. Look for different approaches that children take to a two-step problem.

ANSWERS AND COMMENTARY Children who have mastered the concepts in this unit will be able to write down a 3-digit number given a representation in base 10 equipment. They will also realise that it does not matter which order the base 10 equipment appears in. Children will show that they can work out 1, 10 and 100 more or less than a given number and can work out the original number given an increase or decrease. Children will compare and order numbers, using the correct inequality signs where appropriate.

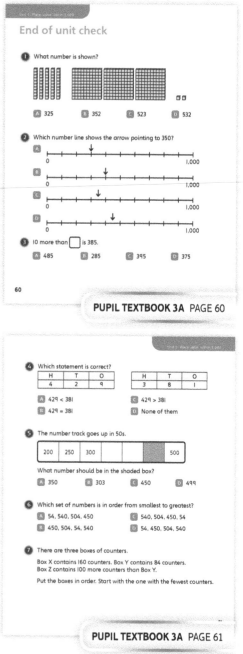

PUPIL TEXTBOOK 3A PAGE 60

PUPIL TEXTBOOK 3A PAGE 61

Q	A	WRONG ANSWERS AND MISCONCEPTIONS	STRENGTHENING UNDERSTANDING
1	B	A, C or D indicates that children do not know that the order of the digits is 100s, 10s or 1s or do not know the size of the base 10 equipment.	For question 1, emphasise and show using base 10 equipment that 100s come first, then 10s and then 1s. Show this on a place value grid and next to this a part-whole model to secure understanding.
2	C	A or B could indicate a lack of understanding of numbers half-way between two marks on a number line.	
3	D	A suggests that children have found 100 more. Choosing C indicates a lack of understanding of how to find the original number.	When finding 1, 10 and 100 more, ask children to focus on the order of the language. For example, '285 is 10 more' is different to 'what is 10 more than 285?' Ask them to make the number with base 10 equipment or place value counters if necessary and ask: *Is this 10 more or do we need to work out 10 more?*
4	C	A, B or D indicates confusion with inequality signs.	
5	C	B suggests that they counted on from 300 in 1s. D suggests that they have found 1 less than 500.	
6	D	A or C indicates a lack of understanding of place value.	
7	Y, X, Z	Some children may think box Y contains the most counters as it starts with 8 and they do not understand that there are 0 hundreds.	

My journal

WAYS OF WORKING Independent thinking

ANSWERS AND COMMENTARY

In question **1**, children may:
- say the number (for example, 415 or 4 hundred and 15)
- describe how the number is made (4 hundreds, 1 ten and 5 ones or 400 + 10 + 5)
- make a comparison (for example, 415 is 100 more then 315)
- show the number on a number line
- show the number using place value counters
- show the number using a part-whole model.

Encourage children to do as many different variations as they can.

In question **2**, children should be able to make:
- 502, 511, 520, 601, 610 using seven counters
- 503, 512, 521, 530, 602, 620 using eight counters.

Some children may include 700. Explain that this is not less than 700. If children are struggling, ask them how many 100s the number must have if it lies between 500 and 700.

To extend this activity, ask children to put the numbers in order or represent them on a number line. Ask: *What would happen if you had 9 counters? What strategy are you using to solve this?*

Power check

WAYS OF WORKING Independent thinking

ASK
- *How many ways do you think you can represent 859?*
- *How confident are you at partitioning a number into 100s, 10s and 1s?*
- *What is 1, 10 and 100 more than a number?*
- *How confident do you feel comparing and ordering two or more numbers?*

Power play

WAYS OF WORKING Pair work

IN FOCUS Use this Power play to assess whether children understand the key concepts in this unit. The criteria in the table make children think about the place value of the numbers. For example, a number greater than 200 must have 2 or more counters in the hundreds column, an odd number must have an odd number of counters in the ones column, and so on.

ANSWERS AND COMMENTARY Largest number: 600; smallest number: 6; odd number greater than 200: 213, 231, 303, 321, 411 or 501; even number less than 200: 6, 24, 42, 60, 114, 132 or 150; same number of 100s and 1s: 60, 141, 222 or 303; 10 more is 241: 231.

For questions where there is more than one answer, encourage children to find as many answers as they can. Suggest that they make up their own questions for their partner. Ask: *What do the digits add up to in each number? Why is this the case?*

After the unit ⏸

- Can children represent 3-digit numbers in multiple ways?
- Do children display flexibility with 3-digit numbers and are they able to make numbers to fit certain criteria, and to compare and order numbers?

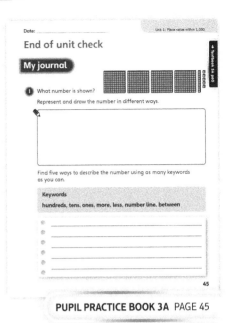

PUPIL PRACTICE BOOK 3A PAGE 45

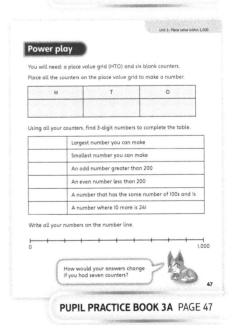

PUPIL PRACTICE BOOK 3A PAGE 46

PUPIL PRACTICE BOOK 3A PAGE 47

Strengthen and **Deepen** activities for this unit can be found in the *Power Maths* online subscription.

Unit 2
Addition and subtraction ①

Mastery Expert tip! 'This unit is an excellent chance to make sure children explain and check their reasoning using a range of place value equipment. Alongside base 10 equipment, I would use place value counters, place value grids, rekenreks and any other resources that provide children with a rich, concrete and visual experience to develop their understanding deeply.'

Don't forget to watch the Unit 2 video!

WHY THIS UNIT IS IMPORTANT

This unit is important because it builds on the strong foundation of place value from Unit 1 to develop key concepts in addition and subtraction. Children explore additions and subtractions gradually, by considering in detail the adding of 1s, 10s and 100s separately. They then explore the need to exchange where addition or subtraction may cross the next place value column. This unit and the next unit are the real foundations for addition and subtraction for children's whole mathematics career. This is why it is essential that time is taken to discuss the ideas and problems. This unit focuses heavily on the use of efficient mental strategies to answer problems and prepares children to understand mental methods ahead of using more formal strategies in the next unit.

WHERE THIS UNIT FITS

→ Unit 1: Place value within 1,000
→ **Unit 2: Addition and subtraction (1)**
→ Unit 3: Addition and subtraction (2)

This unit builds upon the previous work children have done on place value within 1,000. It is essential they have an understanding of this concept as it will make this unit much simpler to grasp. The main focus of this unit is on mental strategies for addition and subtraction, before moving on to more formal methods in the next unit. Developing these strategies at an early stage will help children in the future to take a step back from a problem and decide the best method and strategy to solve it. This unit also develops children's reasoning and justifying skills, which they are developing throughout the year.

Before they start this unit, it is expected that children:
- know the value of each digit in a 3-digit number
- know how to represent a number using place value equipment
- can partition a number as an addition or on a part-whole model, including partitioning flexibly
- know the location of a number on a number line
- know how parts and wholes are related in additions and subtractions.

ASSESSING MASTERY

By the end of this unit, children should become increasingly confident at adding and subtracting 1s, 10s and 100s to a 3-digit number in their head, including where they may need to exchange and the answer crosses a 10 or 100. They should be able to see patterns and explain which place value columns are likely to change before they tackle a question.

COMMON MISCONCEPTIONS	STRENGTHENING UNDERSTANDING	GOING DEEPER
Children may resort to counting strategies rather than using known number bonds to add digits within an addition.	Use part-whole models to display partially completed number bonds, to support children to think of these and use them effectively in their calculations.	Explore problems with multiple solutions and challenge children to explain how they have found all possible solutions in terms of the number bonds they used.
Children may struggle to cross the 10 or 100 in their head when adding and subtracting.	Look at connections between numbers. For example, 5 + 7 = 12, 50 + 70 = 120. Therefore, when adding 127 and 5 in their head, they know the answer must end in a 2.	Ask children to write down questions that require no exchanges and then questions that require one exchange. Ask them to consider if it is possible to write down an addition that has two exchanges.

UNIT STARTER PAGES

Use these pages to introduce the unit focus to children as a whole class. You can use the different characters to explore different ways of working, and to begin to discuss and develop their reasoning skills relating to addition and subtraction with a focus on mental methods.

STRUCTURES AND REPRESENTATIONS

Place value equipment: Can be manipulated by children to model the differences between addition and subtraction and also to help them understand why sometimes you need to exchange.

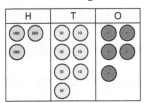

Number lines: This is a very useful model which in this unit enables children to understand how exchange is related to crossing 10s and 100s. Number lines allow children to form a mental picture of what they have to do.

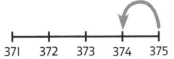

371 372 373 374 375

Part-whole models: This model is vital for children to be able to visualise how number bonds are related to the calculations involving 100s, 10s and 1s, and also for representing the flexible partitioning of numbers as it relates to exchange.

KEY LANGUAGE

There is some key language that children will need to know as part of the learning in this unit.

→ add, addition, total, altogether, sum

→ subtract, subtraction, take away, difference

→ exchange, across

→ 1s, 10s, 100s

→ mental method

→ part-whole model

→ partition

→ place value

→ number bonds to 100

PUPIL TEXTBOOK 2A PAGE 62

PUPIL TEXTBOOK 2A PAGE 63

Use known number bonds

Learning focus

In this lesson, children will use their knowledge of number bonds within 10 to add and subtract multiples of 100, up to 1,000.

Before you teach

- Can children find related additions and subtractions to a number fact such as $3 + 4 = 7$?
- Can children use place value equipment to represent 3 hundreds?
- Can children count up and back in 100s?

NATIONAL CURRICULUM LINKS

Year 2 Number – number and place value

Recognise the place value of each digit in a 2-digit number.

Year 3 Number – addition and subtraction

Add and subtract numbers mentally, including: a three-digit number and ones, a three-digit number and tens, a three-digit number and hundreds.

ASSESSING MASTERY

Children can use their knowledge of number bonds within 10 to add and subtract multiples of 100 up to 1,000.

COMMON MISCONCEPTIONS

Children may try to use a counting strategy rather than their knowledge of number bonds to derive additions and subtractions. Ask:

- *How can you use number bonds within 10 to add and subtract 100s?*

Children may struggle to unitise 100. Ask:
- *What number fact could help you answer $200 + \boxed{} = 500$?*

STRENGTHENING UNDERSTANDING

Use place value equipment such as base 10 equipment or place value counters to model the link between $2 + 3 = 5$ and $200 + 300 = 500$. Explore with children how this works with other part-whole relationships within 10. Children may find it helps to use a part-whole model alongside the place value counters.

GOING DEEPER

Challenge children to derive a family of eight addition and subtraction facts for multiples of 100 related to $2 + 3 = 5$, or other number bonds within 10. Then, challenge children to find eight related facts for multiples of 10 as well.

KEY LANGUAGE

In lesson: number bond, addition, subtraction, left, hundreds (100s), total, altogether, fact family, part-whole model

Other language to be used by the teacher: related facts, take away

STRUCTURES AND REPRESENTATIONS

Part-whole model, number line

RESOURCES

Mandatory: base 10 equipment

Optional: place value counters, place value grids

 In the eTextbook of this lesson, you will find interactive links to a selection of teaching tools.

Quick recap

Say a number from 0 to 10. Ask children to write down as many bonds to that number as possible in 30 seconds. For example, number bonds to 5. They get 1 point for each bond they write. Extend by including subtraction facts such as $9 - 4 = 5$.

Discover

WAYS OF WORKING Pair work

ASK

- Question ① a): *Can you represent the number of bricks using base 10 equipment? How does counting in 100s help you find the number of bricks? Are there more or fewer bricks on the lorry than on the ground? How do you know?*

IN FOCUS Discuss the fact that there are 100 bricks in each set and that this can be represented using base 10 equipment with a 10×10 grid. When finding the total number of bricks, encourage children to unitise the 100s by asking: *There are 3 hundreds and then 4 hundreds – how many 100s are there in total?*

PRACTICAL TIPS Ask children to count aloud in 100s when finding the total number of bricks in each part. For support, children could represent the number of bricks using base 10 equipment.

ANSWERS

Question ① a): There are 300 bricks on the ground. There are 400 bricks on the lorry.

Question ① b): 300 + 400 = 700
There are 700 bricks in total.

Unit 2: Addition and subtraction (1), Lesson 1

Use known number bonds

Discover

① a) How many bricks are on the ground?
How many bricks are on the lorry?

b) How many bricks are there in total?

64

PUPIL TEXTBOOK 3A PAGE 64

Share

WAYS OF WORKING Whole class teacher led

ASK

- Question ① a): *How does counting in 100s help you find the number of bricks on the ground and on the lorry?*
- Question ① b): *How do the part-whole models represent the number of bricks? How does the number line help to find the total number of bricks?*

IN FOCUS In question ① b), ensure children make the connection between what each number represents when counting in 1s in the number bond to 7 and counting in 100s. What similarities can children see in each representation?

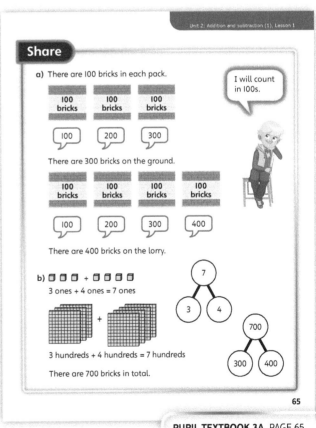

PUPIL TEXTBOOK 3A PAGE 65

Think together

Unit 2: Addition and subtraction (1), Lesson 1

WAYS OF WORKING Whole class teacher led (I do, We do, You do)

ASK

- Questions ❶ and ❷: *Which bond within 10 will help solve each calculation?*
- Questions ❶ and ❷: *Will you need to find a whole by addition, or a part using subtraction? How do you know?*
- Question ❷: *How can you represent these calculations using place value equipment?*

IN FOCUS In Questions ❶ and ❷, children use known number bonds to 10 to help them add hundreds. Use the stem sentences to help children see the connection 3 ones + 4 ones = 7 ones, and how that shows that 3 hundreds + 4 hundreds = 7 hundreds. Question ❸ really prompts children to recognise that knowing bonds is a good way to derive and find a number of related facts. For each of the questions in the **Think together** section, encourage children to recognise the relationship between addition and subtraction within 10, and the related facts for multiples of 100s.

STRENGTHEN Use part-whole models and place value equipment to model the relationships between number bonds within 10 and related facts for multiples of 100. Model the language to unitise 100s: 3 hundreds add 4 hundreds is 7 hundreds, so 300 + 400 = 700.

DEEPEN Challenge children to convince you or a partner that they have found all eight facts in a fact family. Ask: *How can you be sure that you have found all eight?*

ASSESSMENT CHECKPOINT Can children explain how the part-whole relationships for bonds within 10 are related to adding and subtracting 100s? Can children explain the place value of their additions and subtractions? For example, 3 hundreds add 5 hundreds is 8 hundreds, because 3 + 5 = 8.

ANSWERS

Question ❶ a): 5 ones

Question ❶ b): 5 hundreds

Question ❶ c): 6 ones

Question ❶ d): 6 hundreds

Question ❷ a): 2 + 6 = **8**, 200 + 600 = **800**

Question ❷ b): 5 + 1 = **6**, 500 + 100 = **600**

Question ❷ c): 7 – 5 = **2**, 700 – 500 = **200**

Question ❷ d): 8 – 3 = **5**, 800 – 300 = **500**

Question ❸: 900 = 700 + 200
900 = 200 + 700
900 – 700 = 200
200 = 900 – 700
700 = 900 – 200

PUPIL TEXTBOOK 3A PAGE 66

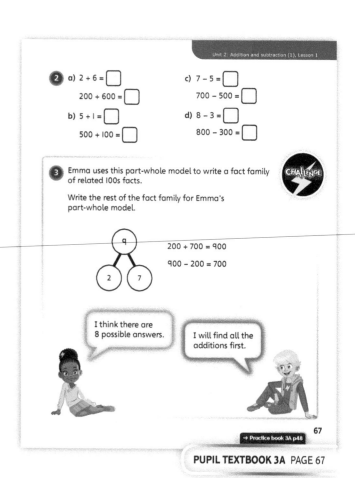

PUPIL TEXTBOOK 3A PAGE 67

Practice

WAYS OF WORKING Independent thinking

IN FOCUS In questions ❶ and ❷, children focus on the idea of unitising to support them with additions and subtractions involving 100s. The base 10 equipment supports children's understanding of this. They use unitising to support them in answering abstract additions and subtractions in question ❹. Discuss the connections between each pair of questions. In question ❺, ask children what they notice about the numbers in the questions and answers. In question ❻, children have to identify the correct numbers to work with to help them answer the questions.

STRENGTHEN Provide children with base 10 equipment to support them in representing questions and using this to find the answer. For further support, encourage them to count aloud to find the answers if needed.

DEEPEN In question ❼, encourage children to think carefully about the order in which they attempt the question then explain their reasoning for this. Can they create their own question similar to this for a partner?

ASSESSMENT CHECKPOINT Use questions ❶ to ❸ to assess whether children can calculate using 100s where pictorial representations are used to support them. Use questions ❹ and ❺ to assess whether children can calculate abstractly using 100s. Use question ❻ to assess whether children can identify the correct information from a question in order to answer it.

ANSWERS Answers for the **Practice** part of the lesson can be found in the *Power Maths* online subscription.

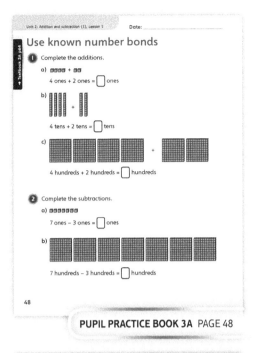

PUPIL PRACTICE BOOK 3A PAGE 48

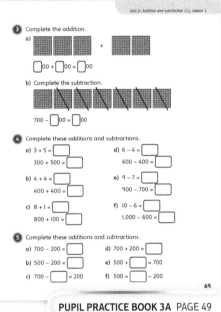

PUPIL PRACTICE BOOK 3A PAGE 49

Reflect

WAYS OF WORKING Pair work

IN FOCUS Children should work independently to create their fact families and then compare them with their partner to see if their fact families agree. Can they convince one another they have found all eight facts?

ASSESSMENT CHECKPOINT Can children explain the relationship between the bond within 10 and the addition and subtraction of 100s?

ANSWERS Answers for the **Reflect** part of the lesson can be found in the *Power Maths* online subscription.

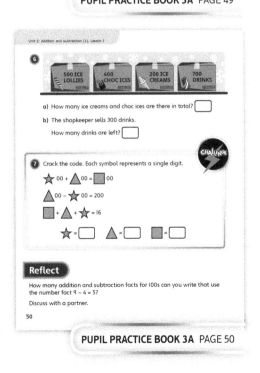

PUPIL PRACTICE BOOK 3A PAGE 50

After the lesson ⏸

- Can children use number bonds efficiently to add and subtract 100s?
- Are children able to represent the calculations using place value equipment?

Add/subtract 1s

Learning focus

In this lesson, children will add and subtract a 1-digit number to and from a 3-digit number, using their understanding of place value. In this lesson, children will not be required to cross 10s.

Before you teach

- Do children understand what a 3-digit number means?
- Do children know the difference between a digit and a number?
- Can children represent a 3-digit number using place value equipment?

NATIONAL CURRICULUM LINKS

Number – addition and subtraction

Add and subtract numbers mentally, including: a 3-digit number and 1s, a 3-digit number and 10s, a 3-digit number and 100s.

Solve problems, including missing number problems, using number facts, place value, and more complex addition and subtraction.

ASSESSING MASTERY

Children can add 1-digit numbers to a 3-digit number by adding the 1s of both numbers. Children can subtract 1-digit numbers from a 3-digit number by taking away 1-digit numbers from the 1s of 3-digit numbers in calculations that do not require children to cross a 10.

COMMON MISCONCEPTIONS

Children may use a counting on strategy, where adding the 1s digits is a more efficient strategy. Ask:
- *How many 1s in the total? Can you work it out without counting on or back? What bond will help you here?*

Children may struggle if their understanding of place value above 100 is not secure. Ask:
- *What three numbers come next: 207, 208, 209 ...?*

STRENGTHENING UNDERSTANDING

Children should represent the calculations using place value equipment to reinforce the concept of using place value to make their methods efficient.

GOING DEEPER

Open questions, such as: *I add a 3-digit number and a 1-digit number. The total is 489. What could my numbers have been?* should prompt children to explore multiple solutions and develop systematic problem solving.

KEY LANGUAGE

In lesson: addition, subtraction, number bonds, hundreds (100s), tens (10s), ones (1s), solutions, total, altogether, number statements

STRUCTURES AND REPRESENTATIONS

Number line, place value grid

RESOURCES

Mandatory: base 10 equipment, 0–9 digit cards

Optional: place value counters

 In the eTextbook of this lesson, you will find interactive links to a selection of teaching tools.

Quick recap

Write the number 62 on the board. Ask children to add 3 to the number. Discuss how they did it. Encourage children to see that they just have to add the 2 and 3 together in their head. Repeat by adding on 4, 5, etc., and with different starting numbers.

Discover

Unit 2: Addition and subtraction (1), Lesson 2

Add/subtract 1s

Discover

1 a) These people are outside the museum. Once they go in, how many visitors will be in the museum in total?

b) One person then leaves.

How many people are left in the museum?

68

PUPIL TEXTBOOK 3A PAGE 68

WAYS OF WORKING Pair work

ASK

- Question 1 a): *How could you arrange place value equipment to represent the place value clearly?*
- Question 1 a): *How many 1s, 10s and 100s will you have in total?*

IN FOCUS In question 1 a), children should discuss how the number being added is only a 1-digit number. They may see that you can combine the 1s rather than having to add any 10s or 100s. Encourage children to use their knowledge of bonds rather than counting on.

PRACTICAL TIPS Children should represent the number of people in the museum and also the number of people about to enter the museum using place value equipment.

ANSWERS

Question 1 a): 245 + 4 = 249

Question 1 b): 249 − 1 = 248
248 people are left.

Share

WAYS OF WORKING Whole class teacher led

ASK

- Question 1 a): *Can you see how the base 10 equipment has been organised to show the addition clearly?*
- Question 1 a): *Why do the 10s digit and the 100s digit not change?*
- Question 1 a): *What mistakes might you make if you did not organise the base 10 equipment clearly?*
- Question 1 a) and b): *Could you still do the calculations in the same way if you did not have base 10 equipment?*

IN FOCUS The important concept in this part of the lesson is to notice how the calculation can be solved by using knowledge of bonds within 10. The place value equipment represents the concept of adding the 1s digits, but the calculation can be solved without resorting to counting.

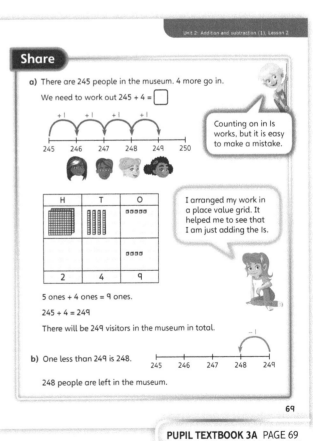

Share

a) There are 245 people in the museum. 4 more go in.

We need to work out 245 + 4 = ☐

245   246   247   248   249   250

Counting on in 1s works, but it is easy to make a mistake.

H	T	O
		ooooo
		oooo
2	4	9

I arranged my work in a place value grid. It helped me to see that I am just adding the 1s.

5 ones + 4 ones = 9 ones.

245 + 4 = 249

There will be 249 visitors in the museum in total.

b) One less than 249 is 248.

245   246   247   248   249

248 people are left in the museum.

69

PUPIL TEXTBOOK 3A PAGE 69

Think together

Whole class teacher led (I do, We do, You do)

ASK

- Question **1** a): *First, how many people were in the museum? Then, what happened? Which column in the place value grid will change? How do you know?*
- Question **1** b): *Which size pieces of base 10 equipment do you need to add? How do you know? Which digits do not change? Which do? Why?*
- Question **2**: *What do you notice? Why does this happen?*

IN FOCUS In question **1**, children focus on adding and subtracting 1s to and from a 3-digit number where there is no exchange. They should recognise that the only digit that changes in these examples is the 1s digit and should be encouraged to explain this. In question **2** a), children spot patterns when adding 1s, noticing that when you increase the number you are adding by 1, your answer increases by 1. In question **2** b), children spot patterns when subtracting 1s, noticing that when the number you are starting with has the same 1s digit each time and you are subtracting the same amount, the 1s digit of the answer is the same.

STRENGTHEN Provide children with base 10 equipment and a place value grid to represent calculations and support them in finding answers.

DEEPEN In question **3**, encourage children to explain their reasoning and method as to how they approached the question. Ask them if there are multiple possible answers and, if so, how do they know they have found them all?

ASSESSMENT CHECKPOINT Use questions **1** and **2** to assess whether children can add 1-digit numbers to 3-digit numbers and subtract 1-digit numbers from 3-digit numbers where the calculations do not require an exchange. Use question **3** to assess whether children understand the relationship between the 1s and can use this to find missing digits.

ANSWERS

Question **1** a): 319 − 7 = 312

There were 312 people left in the museum.

Question **1** b): 291 + 6 = 297

Question **2** a): 352 + 3 = 355, 352 + 4 = 356, 352 + 5 = 357, 352 + 6 = 358, 352 + 7 = 359

Question **2** b): 718 − 3 = 715, 248 − 3 = 245, 178 − 3 = 175, 438 − 3 = 435, 908 − 3 = 905

Question **3** a): There are six possible solutions:
430 + 5 = 435, 431 + 4 = 435, 432 + 3 = 435, 433 + 2 = 435, 434 + 1 = 435, 435 + 0 = 435

Question **3** b): There are five possible solutions:
439 − 4 = 435, 438 − 3 = 435, 437 − 2 = 435, 436 − 1 = 435, 435 − 0 = 435

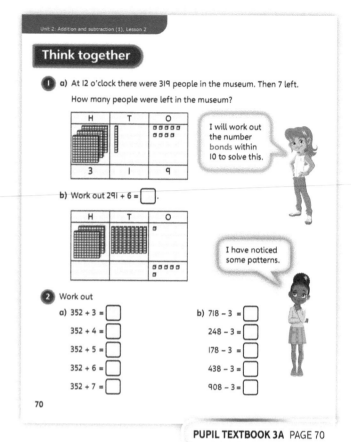

PUPIL TEXTBOOK 3A PAGE 70

PUPIL TEXTBOOK 3A PAGE 71

Practice

WAYS OF WORKING Independent thinking

IN FOCUS In question ❶, children use base 10 representations to support them with adding and subtracting 1s. In questions ❷ and ❸, children are encouraged to make connections between adding and subtracting consecutive 1-digit numbers.

STRENGTHEN Provide children with base 10 equipment and a place value grid to support them in answering the calculations.

DEEPEN Ask children to explain the connections they notice in question ❸ and why this happens. In question ❻, encourage them to think aloud with their approach to the questions.

THINK DIFFERENTLY In question ❺, children use the patterns they have noticed so far to find missing numbers.

ASSESSMENT CHECKPOINT Use questions ❶ to ❸ to assess whether children can add and subtract 1-digit numbers to and from 3-digit numbers. Use questions ❺ and ❻ to assess whether children can identify missing numbers and digits in calculations.

ANSWERS Answers for the **Practice** part of the lesson can be found in the *Power Maths* online subscription.

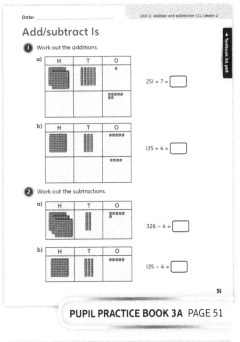

PUPIL PRACTICE BOOK 3A PAGE 51

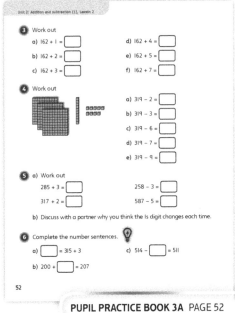

PUPIL PRACTICE BOOK 3A PAGE 52

Reflect

WAYS OF WORKING Independent thinking

IN FOCUS Notice whether children can organise their drawings to represent place value accurately. Do they use base 10 equipment? Do they resort to counting on a number line or number track?

ASSESSMENT CHECKPOINT Do children's drawings represent the addition and subtraction of 1s digits?

ANSWERS Answers for the **Reflect** part of the lesson can be found in the *Power Maths* online subscription.

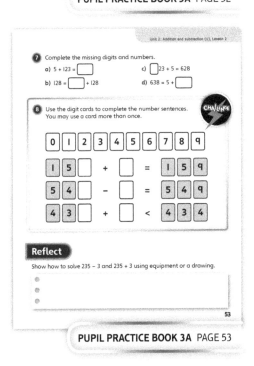

PUPIL PRACTICE BOOK 3A PAGE 53

After the lesson ⏸

- Do children recognise how place value helps them solve these calculations efficiently?
- Can children justify their answers to themselves and others?
- Are children confident applying known number bonds to these calculations?

Add/subtract 10s

Learning focus

In this lesson, children will add and subtract multiples of 10 to and from a 3-digit number by using their knowledge of number bonds to add and subtract the 10s digits.

Before you teach

- Can children count in 10s from 0 to 100 and back again?
- Can children count on in 10s from 34? Do they know what stays the same and what changes?

NATIONAL CURRICULUM LINKS

Number – addition and subtraction

Add and subtract numbers mentally, including: a 3-digit number and 1s, a 3-digit number and 10s, a 3-digit number and 100s.

ASSESSING MASTERY

Children can explain how to add and subtract a multiple of 10 to or from a 3-digit number in terms of the place value of the digits. Children can explain why, for example, when you add 30 to 654, the 10s digit increases by 3.

COMMON MISCONCEPTIONS

Children may resort to counting in 10s, rather than using knowledge of bonds and place value. Ask:
- *How do you solve 35 + 60 = ☐?*

Children may struggle where there is a zero in the 1s or 10s place of the 3-digit number they are adding to. Ask:
- *What is the same and what is different about 302, 312 and 320?*

STRENGTHENING UNDERSTANDING

Encourage children to represent the calculation as jumps on a number line in order to illustrate whether the unknown number is greater or less than the number given. The number line need not be used as a counting strategy, but it can model the structure of the calculations. Display or create part-whole models to support the use of number bonds.

GOING DEEPER

Challenge children to solve missing number problems, especially when worded as: *20 more than my number is 245. What is my number?* or *345 is 30 less than what number?* Encourage children to create and represent these missing number problems on a number line.

KEY LANGUAGE

In lesson: subtraction, tens (10s), addition, part-whole model, zero (0), column, fewer

Other language to be used by the teacher: place value, multiple of 10, ones (1s), hundreds (100s), more, place value grids

STRUCTURES AND REPRESENTATIONS

Number line, place value grid

RESOURCES

Mandatory: base 10 equipment

Optional: place value abacus, 100 square

 In the eTextbook of this lesson, you will find interactive links to a selection of teaching tools.

Quick recap

Start with the number 15. Go round the class and ask how far children can count in 10s. Can they count to 95 in 10s? Can they go any higher? Can they start at 15 and count up in 20s, 30s? How far can they count?

Discover

Add/subtract 10s

Discover

WAYS OF WORKING Pair work

ASK

- Question ① a): *What number has Aki made? How do you know?*
- Question ① b): *What is he adding? How do you know? Which column will he add the beads to? What number will he have then? How do you know?*

IN FOCUS In question ① b), the key point is that children connect the idea of adding 30 with adding 3 tens. This will lead to a development in their understanding of more formal column methods over the next series of lessons.

PRACTICAL TIPS Children could represent the place value abacus context by using base 10 equipment or place value counters placed on three lines, each line labelled 'H', 'T' and 'O' respectively. They could also represent this in a place value grid.

ANSWERS

Question ① a): 351: 3 hundreds, 5 tens, 1 one

Question ① b): 351 + 30 = 381

Aki

① **a)** What number has Aki made?

b) Aki adds 3 beads to the 10s.

Write this as an addition and find the new total.

72

PUPIL TEXTBOOK 3A PAGE 72

Share

WAYS OF WORKING Whole class teacher led

ASK

- Question ① a): *Where can you see Aki's 3 hundreds? Where can you see the 5 tens? Where can you see the 1 one?*
- Question ① b): *Where will Aki add the 3 tens to? Which digit will change? Which digits will stay the same? How do you know? What number has Aki made? What calculation is this?*

IN FOCUS The important point is for children to connect the addition of 10s with the place value of the numbers being added. Children should discuss why only the 10s digit changes for this calculation, and how they can use their knowledge of bonds within 10 to solve the calculation accurately.

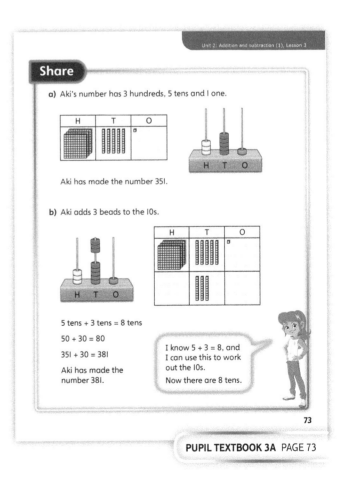

Share

a) Aki's number has 3 hundreds, 5 tens and 1 one.

Aki has made the number 351.

b) Aki adds 3 beads to the 10s.

5 tens + 3 tens = 8 tens

50 + 30 = 80

351 + 30 = 381

Aki has made the number 381.

I know 5 + 3 = 8, and I can use this to work out the 10s.

Now there are 8 tens.

73

PUPIL TEXTBOOK 3A PAGE 73

Think together

WAYS OF WORKING Whole class teacher led (I do, We do, You do)

ASK

- Questions ①, ② and ③: *Can you explain when the 10s digit increases or decreases?*
- Question ②: *Which digit will change? How do you know?*
- Question ③: *Why does the 1s digit not change?*
- Question ③ *What number bonds will help you solve these calculations?*

IN FOCUS The focus for this section is for children to develop a fluency and understanding of why and how the 10s digit changes, by thinking about the place value and the number bonds they can use.

STRENGTHEN It may help some children to use a 100 square, to be able to visualise the effect of adding or subtracting 10s.

DEEPEN Ask children to think of several different solutions to the missing digits in 2☐4 – ☐0 = 204. Can they convince themselves or a partner that they have found all possible solutions?

ASSESSMENT CHECKPOINT Question ③ requires children to explain the calculations in terms of addition and subtraction of 10s.

ANSWERS

Question ① a): 325 + 60 = 385

Question ① b): 513 + 40 = 553

Question ②: 8 tens – 5 tens = 3 tens
582 – 50 = 532

Question ③ a): Abacus A:
5 tens – 4 tens = 1 ten
451 – 40 = 411
Abacus B
5 tens + 4 tens = 9 tens
451 + 40 = 491

Question ③ b): 414 + 70 = 484, 575 – 60 = 515,
124 + 60 = 184, 382 + 10 = 392,
280 – 10 = 270, 990 – 80 = 910

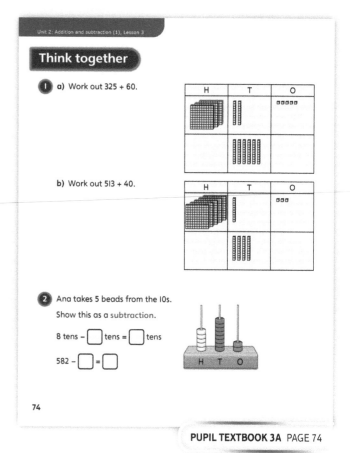

PUPIL TEXTBOOK 3A PAGE 74

PUPIL TEXTBOOK 3A PAGE 75

Practice

WAYS OF WORKING Independent thinking

IN FOCUS Here children further develop their understanding of adding 10s. In questions ❶ and ❷, children should link the number of 10s being added or subtracted to the number being added or subtracted. In question ❸, children should identify which digit will change and how they know. In question ❹, children should be encouraged to spot patterns and consider why they happen.

STRENGTHEN Provide children with a place value grid and base 10 equipment to represent the calculations and use them for support in finding the answers.

DEEPEN Ask children to explain the patterns that they notice in question ❹ and why this happens. In question ❻, encourage children to find all correct answers and explain how they know they have found them all.

THINK DIFFERENTLY In question ❺, children use their understanding of adding and subtracting 10s to find missing numbers within calculations.

ASSESSMENT CHECKPOINT Use questions ❶ to ❹ to assess whether children can add and subtract multiples of 10 to and from a 3-digit number where the calculations do not require an exchange. Use question ❺ to assess whether children can use this to find missing numbers within calculations.

ANSWERS Answers for the **Practice** part of the lesson can be found in the *Power Maths* online subscription.

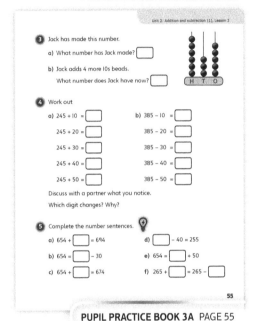

PUPIL PRACTICE BOOK 3A PAGE 54

PUPIL PRACTICE BOOK 3A PAGE 55

Reflect

WAYS OF WORKING Pair work

IN FOCUS Children should focus on convincing their partner what the 10s digits will be, without having to work out the full answer as proof. This way, they will have to give reasons about the place value in the calculation, rather than just the process.

ASSESSMENT CHECKPOINT Can children justify their decision using the relevant number bonds?

ANSWERS Answers for the **Reflect** part of the lesson can be found in the *Power Maths* online subscription.

PUPIL PRACTICE BOOK 3A PAGE 56

After the lesson ⏸

- Do children understand that the 10s digit increases or decreases depending on the number of 10s being added or subtracted?
- Can children use number bonds to calculate accurately?
- Can children also solve missing number problems?

Add/subtract 100s

Learning focus

In this lesson, children will add a multiple of 100 to a 3-digit number by using their knowledge of number bonds to add the 100s digits.

NATIONAL CURRICULUM LINKS

Number – addition and subtraction

Add and subtract numbers mentally, including: a 3-digit number and 1s, a 3-digit number and 10s, a 3-digit number and 100s.

ASSESSING MASTERY

Children can explain how to add a multiple of 100 to a 3-digit number in terms of the place value of the digits. Children can explain why, for example, when you add 300 to 654, the 100s digit increases by 3.

COMMON MISCONCEPTIONS

Children may resort to counting in 100s, rather than using knowledge of bonds and place value. Ask:
- *How do you solve 350 add 600?*

Children may struggle where there is a zero in the 1s or 10s place of the 3-digit number they are adding to. Ask:
- *What is the same and what is different about 204, 214 and 240?*

STRENGTHENING UNDERSTANDING

Encourage children to represent the calculations as jumps on a number line to illustrate whether the unknown number is greater or less than the number given. The number line need not be used as a counting strategy, but it can model the structure of calculations. Display or create part-whole models and use base 10 equipment or place value counters to support the use of number bonds.

GOING DEEPER

Challenge children to solve missing number problems, especially when worded as: *200 more than my number is 345. What is my number?* Encourage children to create and represent these missing number problems to help them solve them.

KEY LANGUAGE

In lesson: hundreds (100s), addition, subtraction, part-whole, zero (0), column, fewer

Other language to be used by the teacher: place value, multiple of 100, ones (1s), tens (10s), more, place value grids

STRUCTURES AND REPRESENTATIONS

Number line, part-whole models

RESOURCES

Mandatory: base 10 equipment, place value grids, place value counters

Optional: place value abacus

 In the eTextbook of this lesson, you will find interactive links to a selection of teaching tools.

Quick recap

Ask children questions such as 200 + 300, 700 – 200. What is the same about their answers? Can they find other examples that give the same answer?

Discover

Add/subtract 100s

Discover

WAYS OF WORKING Pair work

ASK

- Question ① a): *What is Jamilla's score? How do you know? What happens to her score when she hits one of the asteroids? Which asteroids will increase her score? Which will decrease her score? How do you know?*

IN FOCUS In question ① a), the key point is that children connect the idea of adding 300 with adding 3 hundreds. This will lead to a development in their understanding of more formal column methods over the next series of lessons. In question ① b), children should recognise that Jamilla's score has decreased.

PRACTICAL TIPS Children could represent Jamilla's score using base 10 equipment or place value counters and then use this to support them in building her scores before and after hitting asteroids and planets.

ANSWERS

Question ① a): 520 + 300 = 820 points

Question ① b): 820 − 420 = 400
820 − 400 = 420
The spaceship Jamilla shot was −400 points.

① a) Jamilla is playing a computer game.
She shoots the spaceship and scores 300 points.
What will her score be now?

b) Jamilla shoots another spaceship. Her score is now 420.
Which spaceship did she shoot?

76

PUPIL TEXTBOOK 3A PAGE 76

Share

WAYS OF WORKING Whole class teacher led

ASK

- Question ① a): *Which column will change? How do you know?*
- Question ① b): *Which column has changed? What does this mean about the planet Jamilla hit? What has her score gone down by?*

IN FOCUS The important point is for children to connect the addition and subtraction of 100s with the place value of the numbers being added or subtracted. Children should discuss why only the 100s digit changes.

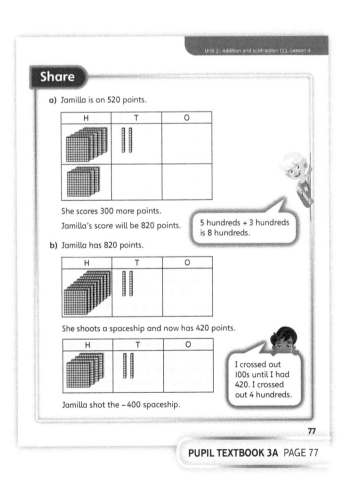

Share

a) Jamilla is on 520 points.

H	T	O

She scores 300 more points.
Jamilla's score will be 820 points.

5 hundreds + 3 hundreds is 8 hundreds.

b) Jamilla has 820 points.

H	T	O

She shoots a spaceship and now has 420 points.

H	T	O

Jamilla shot the −400 spaceship.

I crossed out 100s until I had 420. I crossed out 4 hundreds.

77

PUPIL TEXTBOOK 3A PAGE 77

Think together

WAYS OF WORKING Whole class teacher led (I do, We do, You do)

ASK

- Questions ❶: *Which column will change? How do you know? What digits will stay the same? What is 2 hundreds add 3 hundreds?*
- Question ❷: *Will Mo have more or fewer raisins now? How do you know? Which digit will change? Which will stay the same? What is 6 hundreds subtract 2 hundreds?*

IN FOCUS The focus of this section is for children to develop a fluency and understanding of why and how the 100s digit changes, by thinking about the place value and the number bonds they can use.

STRENGTHEN Provide children with a place value grid and either base 10 equipment or place value counters to represent the questions and support them in finding the answers.

DEEPEN In question ❸, encourage children to explain the pattern that they notice and why this happens. Can they create a similar pattern with a different starting number?

ASSESSMENT CHECKPOINT Use questions ❶ and ❷ to assess whether children can add and subtract multiples of 100 to and from a 3-digit number where no exchange is required.

ANSWERS

Question ❶ a): 268 + 300 = 568

Question ❶ b): 317 + 400 = 717

Question ❷: 629 – 200 = 429

Question ❸ a): 635

Question ❸ b): 635 + 200 = 835

Question ❸ c): 635 + 300 = 935

Question ❸ d): 635 – 200 = 435

Question ❸ e): 635 – 500 = 135

Only the hundreds digit changed each time.

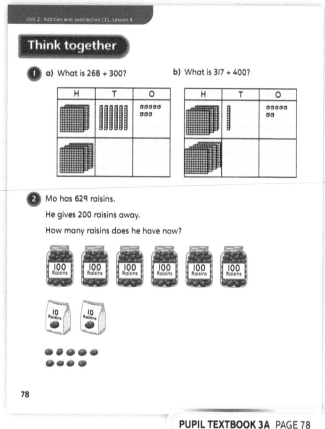

PUPIL TEXTBOOK 3A PAGE 78

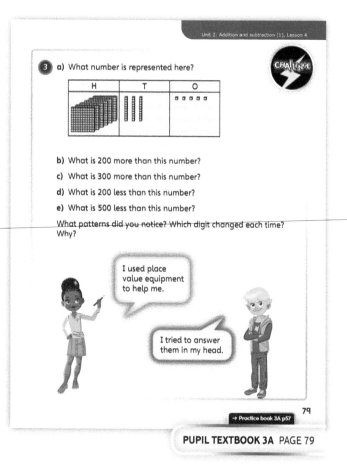

PUPIL TEXTBOOK 3A PAGE 79

Practice

WAYS OF WORKING Independent thinking

IN FOCUS The focus of this section is for children to develop their understanding and fluency in adding and subtracting multiples of 100. Questions ❶, ❷ and ❸ provide opportunity for children to consolidate this learning whilst question ❹ encourages them to spot patterns between calculations.

STRENGTHEN Provide children with a place value grid and either base 10 equipment or place value counters to represent the questions and support them in working out the answers.

DEEPEN In question ❹, encourage children to explain the pattern they notice and why this happens. Can they write the next calculation in each pattern? In question ❻, can they explain the order in which they approached the question and why they approached it in this way?

THINK DIFFERENTLY In question ❺, children use the patterns they have noticed to find missing numbers in word problems.

ASSESSMENT CHECKPOINT Use questions ❶ to ❹ to assess whether children can add and subtract multiples of 100 to and from 3-digit numbers. Use question ❺ to assess whether children can find missing numbers in questions.

ANSWERS Answers for the **Practice** part of the lesson can be found in the *Power Maths* online subscription.

PUPIL PRACTICE BOOK 3A PAGE 57

PUPIL PRACTICE BOOK 3A PAGE 58

Reflect

WAYS OF WORKING Independent thinking

IN FOCUS The focus of this section is to ensure that children understand why only the 100s digit changes when adding 100s to a 3-digit number.

ASSESSMENT CHECKPOINT Children should be able to explain that since the number of 10s and 1s will not change, neither will their digits.

ANSWERS Answers for the **Reflect** part of the lesson can be found in the *Power Maths* online subscription.

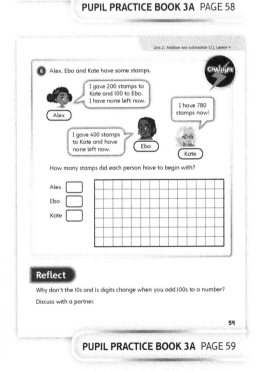

PUPIL PRACTICE BOOK 3A PAGE 59

After the lesson ⏸

- Do children understand that the 100s digit increases or decreases depending on the 100s being added or subtracted?
- Can children use number bonds to calculate accurately?
- Can children solve missing number problems?

Spot the pattern

Learning focus

In this lesson, children will explore patterns in addition and subtraction and the effect on different digits of adding or subtracting 1s, 10s or 100s. At this stage, they do not explore calculations requiring exchange.

Before you teach ⏸

- Can children identify which digits change when adding 5?
- Can they explain what is the same and what is different about these calculations: 11 + 2, 11 + 20, 20 + 11, 2 + 11?

NATIONAL CURRICULUM LINKS

Number – addition and subtraction

Add and subtract numbers with up to three digits, using formal written methods of columnar addition and subtraction.

Add and subtract numbers mentally, including: a 3-digit number and 1s, a 3-digit number and 10s, a 3-digit number and 100s.

ASSESSING MASTERY

Children can identify similarities and differences in the effect of adding or subtracting 1s, 10s or 100s, for example 2, 20 or 200.

COMMON MISCONCEPTIONS

Solving missing number problems can cause confusion as to which operation to use. Discuss whether children actually need to add or subtract in order to identify the missing number. What strategies do they use? Ask:
- *How could you solve* ☐ *+ 20 = 220?*
- *What if the missing number is in a different position, for example 200 +* ☐ *= 220?*

STRENGTHENING UNDERSTANDING

Using a part-whole model to represent the missing number problems can help support children as they develop their understanding of the calculations required.

GOING DEEPER

The function machines can be used for exploring missing number problems and for exploring hidden operations, too. Being able to investigate missing operations requires two steps of thinking.

KEY LANGUAGE

In lesson: subtraction, function machine, ones (1s), tens (10s), hundreds (100s), part-whole models, digit

Other language to be used by the teacher: operation

STRUCTURES AND REPRESENTATIONS

Function machines, part-whole model, place value grid

RESOURCES

Mandatory: base 10 equipment or place value equipment

Optional: part-whole models

 In the eTextbook of this lesson, you will find interactive links to a selection of teaching tools.

Quick recap 🔄

Draw a one-step function machine on the board. For example, a square, with an arrow pointing in and an arrow pointing out. In the square, write + 3. Give children an input number and ask them what the output number will be. Change the function, for example, to + 20. As an extension, you can give them the input and output numbers and see if they can work out the function.

Discover

Spot the pattern

Discover

WAYS OF WORKING Pair work

ASK

- Question ① a): *What is the same and what is different about the three function machines?*
- Question ① a): *What do you notice about the numbers being inputted into each function machine?*
- Question ① a): *Can you predict anything about how each number will change?*
- Question ① b): *What is different about how you work out the answer to question ① b) compared to ① a)?*

IN FOCUS The important aspect is to notice the similarities and differences in the functions in question ① a). The aim is for children to notice that they are linked, but that the place value of the digits may have a noticeable and predictable effect on how the input number is changed.

PRACTICAL TIPS The function machines could be role-played by the teacher or children. The children 'input' a number by handing the person a number written on a whiteboard, or by telling them the number. The function machine would then output a number in a similar way. (After having made some machine-like noises, say: *The output is 156.*)

ANSWERS

Question ① a): 154 + 2 = 156, 154 + 20 = 174,
154 + 200 = 354

Question ① b): 797 – 200 = 597,
597 + 200 = 797
Jamie's number was 597.

① a) Lee inputs 154 into each function machine.
What will the outputs be? Try to work them out in your head.

b) Jamie inputs a number into the + 200 machine. The output is 797.
What number did she put in?

80

PUPIL TEXTBOOK 3A PAGE 80

Share

WAYS OF WORKING Whole class teacher led

ASK

- Question ① a): *Which digit changes when you add 2, 20 or 200?*
- Question ① a): *Why do the other digits not change?*
- Question ① a): *Which number bonds help you solve each addition?*
- Question ① b): *Why is this a missing number problem?*
- Question ① b): *How does the part-whole model show how to find the missing number?*

IN FOCUS For question ① a), children should discuss and develop an awareness that the effect of adding 100s, 10s or 1s in the given examples is that only one digit of the input number changes. They need to understand that this is due to the place value of the numbers being added.

In question ① b), children should recognise that this is like a missing number problem. They need to be able to explain how the part-whole model represents this and support the reasons for choosing the operation required.

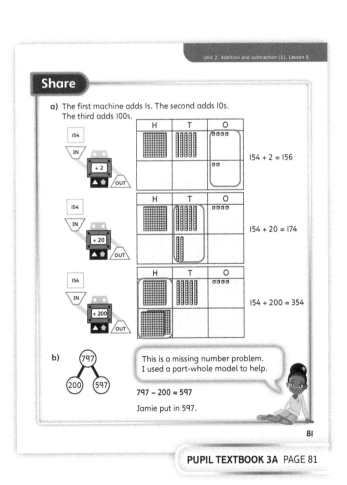

PUPIL TEXTBOOK 3A PAGE 81

Think together

Whole class teacher led (I do, We do, You do)

ASK

- Questions ❶, ❷ and ❸: *What operation will you need to perform?*
- Questions ❶ and ❷: *Can you predict how the digits will change?*
- Questions ❶, ❷ and ❸ b): *What do you notice about the effect on the digits?*

IN FOCUS In questions ❶ and ❷, children develop their expertise so that they understand whether they need to add or subtract and what digits change. They then extend this in question ❸ a) to reasoning about the effect on the different digits. By the end of the sequence, children should be able to explain how to reason based on the effect on the 1s, 10s or 100s digits.

STRENGTHEN Ensure that children represent their thinking using place value equipment and on a place value grid. Discuss how to use number bonds to work out the answers.

DEEPEN At this stage, we are not considering calculations which require exchange or that cross 10s or 100s. Deepen children's understanding by exploring the patterns in question ❸ b). Can children develop their own patterns to create a family of puzzles like these? Ask children to explain how their thinking has to change in order to invent the puzzles, as opposed to simply solving the question.

ASSESSMENT CHECKPOINT Can children explain the missing parts of the calculations in question ❸ b) by looking at which of the digits change?

ANSWERS

Question ❶: 321 + 5 = 326, 321 + 50 = 371, 321 + 500 = 821

Question ❷ a): 546 − 3 = 543

Question ❷ b): 546 − 30 = 516

Question ❷ c): 546 − 300 = 246

Question ❸ a): 259 − 253 = 6, 253 + 6 = 259, so + 6
 953 − 253 = 700, 253 + 700 = 953, so + 700
 253 − 203 = 50, 253 − 50 = 203, so − 50

Question ❸ b): 113 = 111 + 2   555 = 755 − 200
 131 = 111 + 20   555 = 557 − 2
 311 = 111 + 200   555 = 575 − 20

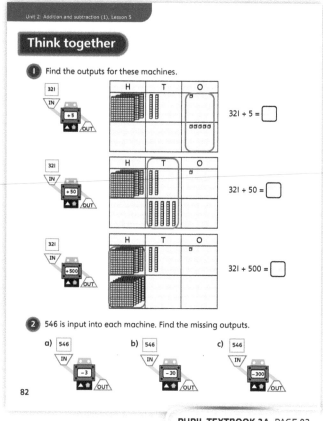

PUPIL TEXTBOOK 3A PAGE 82

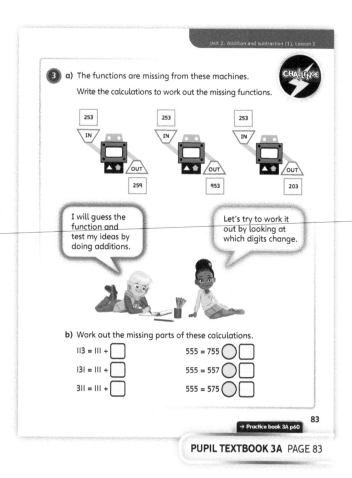

PUPIL TEXTBOOK 3A PAGE 83

Practice

WAYS OF WORKING Independent thinking

IN FOCUS The focus of question ❶ is on developing fluency in adding 1s, 10s or 100s. The numbers are carefully varied so that children add 4 ones, 4 tens and then 4 hundreds. This draws their attention to the effect of adding 1s, 10s and 100s. The missing number is also varied each time to help children develop greater fluency and flexibility. From question ❷ onward, children develop their reasoning around the problems. They apply these fluency and reasoning skills to word problems and to a variety of missing number problems presented in slightly different contexts. Question ❹ encourages children to reason about the effects of the digits in the function machine on the output number.

STRENGTHEN Base 10 equipment is used to support thinking in the earlier questions, but children should feel they can continue to use the equipment, even when not directly prompted. It may also help children to represent the calculations on part-whole models, in order to understand if they need to add to find the whole, or subtract to find a part.

DEEPEN The patterns of variation in question ❸ are designed to lead children to a deeper understanding. They will require a strong understanding of the structure of the patterns, in order to spot the different reasoning required. Question ❺ prompts children to notice that the technique they have been using does not apply to all situations. Children will start to consider which of these calculations require exchange.

ASSESSMENT CHECKPOINT Question ❹ requires a good understanding of how the calculations work, in order to identify which output matches which function.

ANSWERS Answers for the **Practice** part of the lesson can be found in the *Power Maths* online subscription.

Reflect

WAYS OF WORKING Independent thinking

IN FOCUS It is important that children understand which digits change when adding 1s, 10s and 100s. They should also be able to explain which digits do not change.

ASSESSMENT CHECKPOINT Can children explain their answers in terms of place value?

ANSWERS Answers for the **Reflect** part of the lesson can be found in the *Power Maths* online subscription.

After the lesson ⏸

- Do children know how to work out the calculations by considering which digits change?
- Can children solve missing number problems?
- Can children represent their thinking with place value equipment or models to justify their answers?

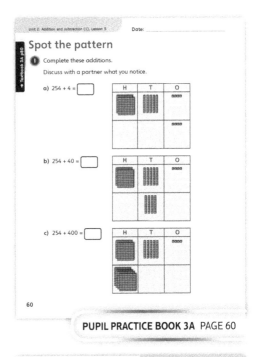

PUPIL PRACTICE BOOK 3A PAGE 60

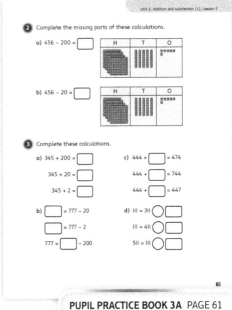

PUPIL PRACTICE BOOK 3A PAGE 61

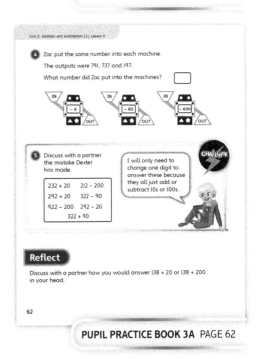

PUPIL PRACTICE BOOK 3A PAGE 62

121

Add 1s across 10

Learning focus

In this lesson, children will understand how to recognise additions where they will cross a 10, and know how to use exchange of 10 ones for 1 ten.

NATIONAL CURRICULUM LINKS

Number – addition and subtraction

Add and subtract numbers with up to three digits, using formal written methods of columnar addition and subtraction.

Add and subtract numbers mentally, including: a 3-digit number and 1s, a 3-digit number and 10s, a 3-digit number and 100s.

ASSESSING MASTERY

Children can add a 1-digit number to a 3-digit number and cross the 10 by exchanging 10 ones for 1 ten. Children can recognise when exchange of 10 ones for 1 ten is required.

COMMON MISCONCEPTIONS

Children may not understand that exchange is a different way of partitioning the number and think that they are changing the number. Ask:

- *What is the same and what is different about 100 + 50 + 5 and 100 + 40 + 15?*

The concept of exchange may cause difficulty for some children and they may assume you have to exchange for every addition. Ask:

- *Can you tell if the 10s digit will increase before working out the exact answer?*

STRENGTHENING UNDERSTANDING

Children will need to link this learning to knowledge of bonds within 20. It may help children to imagine or use a ten frame to support this thinking initially. Use of base 10 equipment to represent change from 1s to 10s in a concrete way will also support understanding.

GOING DEEPER

Encourage children to recognise whether the 10s digit will increase in an addition even before calculating the exact answer. Challenge children to recognise this as a checking strategy.

KEY LANGUAGE

In lesson: exchange, addition, subtraction, 10s digit, tens (10s), ones (1s), 10 ones, hundreds (100s), solutions, altogether, pattern

Other language to be used by the teacher: total, variation

STRUCTURES AND REPRESENTATIONS

Number line, part-whole model

RESOURCES

Mandatory: base 10 equipment, 0–9 digit cards

Optional: ten frame, blank laminated number lines, counters

 In the eTextbook of this lesson, you will find interactive links to a selection of teaching tools.

Quick recap

Ask children to work out 5 + 7. Discuss their methods. Encourage children to add 3 then add 2, or know 5 + 7 as a bond. Repeat for other examples that cross 10.

Discover

Unit 2: Addition and subtraction (1), Lesson 6

Add 1s across 10

Discover

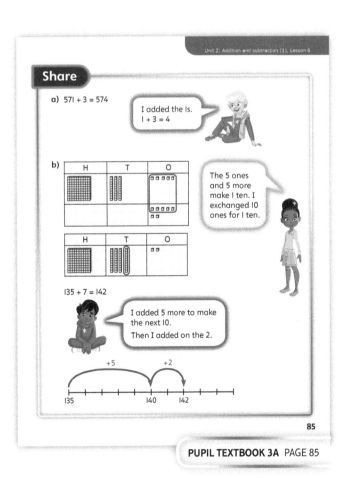

WAYS OF WORKING Pair work

ASK

- Questions ❶ a) and b): *What is the same and what is different about the calculations? What place value equipment could you use to represent the calculations?*

IN FOCUS In questions ❶ a) and b), the focus is to notice that children need to add the 1s, though there is a complication that had not been covered previously, as now the 1s digits may total to 10 or more.

PRACTICAL TIPS Children can replicate the context using digit cards. This will require them to think carefully about the place value of each digit.

ANSWERS

Question ❶ a): 571 + 3 = 574

Question ❶ b): 135 + 7 = 142

❶ a) Work out 571 + 3.

b) Work out 135 + 7.

84

PUPIL TEXTBOOK 3A PAGE 84

Share

WAYS OF WORKING Whole class teacher led

ASK

- Question ❶ b): *Can you explain why the number line uses a jump of 5 first? Why is there a jump of 2 next?*
- Question ❶ b): *How does the method showing base 10 equipment link to the number line method?*
- Question ❶ b): *How do you know the number of 1s to exchange?*

IN FOCUS In question ❶ b) the focus is first to recognise that the calculation requires exchange, or crossing of the next 10. Children should understand the decision about when to exchange and when not to exchange.

Ask children to look carefully at how the same concept is represented differently on the number line and with the base 10 equipment. Can children 'see' where the jump of 5 on the number line relates to the exchange of base 10 equipment? Children start to use the open number line for the calculations.

Share

a) 571 + 3 = 574

I added the 1s.
1 + 3 = 4

b)

H	T	O

The 5 ones and 5 more make 1 ten. I exchanged 10 ones for 1 ten.

H	T	O

135 + 7 = 142

I added 5 more to make the next 10.
Then I added on the 2.

+5   +2

135   140   142

85

PUPIL TEXTBOOK 3A PAGE 85

Think together

WAYS OF WORKING Whole class teacher led (I do, We do, You do)

ASK

- Question ❶ a) and b): *What number do you need to start from? How many jumps do you need to make? How do you know? What number will you land on?*
- Question ❷: *How many do you need to add to get to the next 10? How does this help you know whether the calculation will change the 10s digit?*

IN FOCUS The focus of this section is for children to understand when a calculation requires an exchange and why, and to be able to use a number line to support them in working out the answer.

STRENGTHEN Provide children with a blank laminated number line that they can label the start point on and then draw on jumps to help them calculate answers.

DEEPEN When considering whether a question will affect the 10s digit, ask children to think of the greatest and least possible numbers that will do this. Can they explain that when the 1s digit creates a bond to 10 or greater than 10, the 10s digit will change?

ASSESSMENT CHECKPOINT Use questions ❶ and ❷ to assess whether children can complete calculations which involve adding 1s and crossing a 10 boundary. Ensure they recognise why these calculations cross the 10 by looking at the 1s digit in each number.

ANSWERS

Question ❶ a): 316 + 5 = 321

Question ❶ b): 148 + 5 = 153

Question ❷ a): 248 + 6 = 254
　　　　　The 10s digit changes from 4 tens to 5 tens.

Question ❷ b): 842 + 6 = 848
　　　　　The 10s digit does not change.

Question ❷ c): 217 + 9 = 226
　　　　　The 10s digit changes from 1 ten to 2 tens.

Question ❷ d): 324 + 6 = 330
　　　　　The 10s digit changes from 2 tens to 3 tens.

Question ❸ a): Possible answers are: 442 + 9, 443 + 8, 444 + 7, 445 + 6, 446 + 5, 447 + 4, 448 + 3, 449 + 2

Question ❸ b): The hundreds digit will only change if the tens digit is 9 and the ones digits add to 10 or more.
　　　　　299 + 1 = 300, 198 + 4 = 202 are examples where this happens.

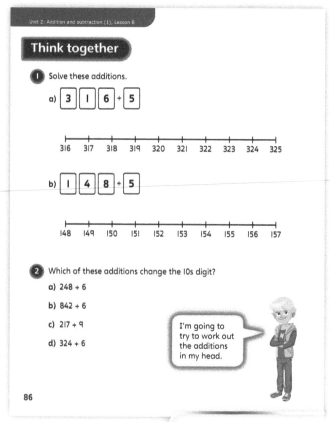

PUPIL TEXTBOOK 3A PAGE 86

PUPIL TEXTBOOK 3A PAGE 87

Practice

WAYS OF WORKING Independent thinking

IN FOCUS The focus of this section is for children to consolidate their understanding of adding 1s across a 10 boundary. In question ❶, they should recognise that, as the number they are adding increases by 1, so does the answer. In question ❸, they should recognise that as the total number of 1s is the same, so is the answer.

STRENGTHEN Provide children with base 10 equipment or place value counters that they can set out under their number lines to support them in finding the answers.

DEEPEN Encourage children to explain any patterns they notice in questions and why this happens. In question ❺, ask children to explain their answers and how they worked them out.

THINK DIFFERENTLY In question ❹, children should be able to identify calculations which will cross a 10 boundary and explain why this happens.

ASSESSMENT CHECKPOINT Use questions ❶ to ❸ to assess whether children can add 1-digit numbers to 3-digit numbers where the calculation involves an exchange. Use question ❹ to assess whether children can identify calculations which will involve an exchange without working out the answers. Use question ❺ to assess whether children can find missing numbers or digits in calculations.

ANSWERS Answers for the **Practice** part of the lesson can be found in the *Power Maths* online subscription.

Reflect

WAYS OF WORKING Independent thinking

IN FOCUS The focus here is for children to recognise the difference between calculations that involve an exchange and those that do not.

ASSESSMENT CHECKPOINT Children should be able to explain that as 5 + 3 is less than 10, this calculation will not change the 10s digit, but since 5 + 8 is greater than 10, this calculation will change the 10s digit.

ANSWERS Answers for the **Reflect** part of the lesson can be found in the *Power Maths* online subscription.

After the lesson ⏸

- Can children count beyond a multiple of 10 to find answers to a calculation?
- Can children recognise when an exchange is necessary?
- Can children solve missing number problems?

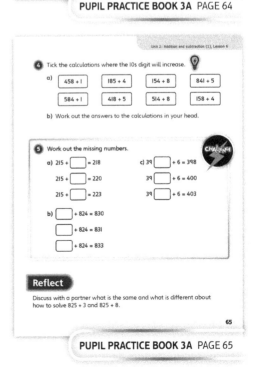

Date: _____

Unit 2: Addition and subtraction (1), Lesson 6

Add 1s across 10

❶ Work out the additions.

a) 318 + 5 = ☐

318 319 320 321 322 323 324 325 326 327

b) 318 + 6 = ☐

318 319 320 321 322 323 324 325 326 327

c) 318 + 7 = ☐

318 319 320 321 322 323 324 325 326 327

d) 318 + 9 = ☐

318 319 320 321 322 323 324 325 326 327

63

PUPIL PRACTICE BOOK 3A PAGE 63

Unit 2: Addition and subtraction (1), Lesson 6

❷ Work out the additions.

a) 215 + 7 = ☐

215 216 217 218 219 220 221 222 223 224

b) 566 + 7 = ☐

566 567 568 569 570 571 572 573 574 575

c) 629 + 7 = ☐

629 630 631 632 633 634 635 636 637 638

❸ Work out the additions.

a) 215 + 8 = ☐

b) 218 + 5 = ☐

c) Discuss with a partner what you notice about your answers to a) and b)

64

PUPIL PRACTICE BOOK 3A PAGE 64

Unit 2: Addition and subtraction (1), Lesson 6

❹ Tick the calculations where the 10s digit will increase.

a)
| 458 + 1 | 185 + 4 | 154 + 8 | 841 + 5 |
| 584 + 1 | 418 + 5 | 514 + 8 | 158 + 4 |

b) Work out the answers to the calculations in your head.

❺ Work out the missing numbers.

a) 215 + ☐ = 218
   215 + ☐ = 220
   215 + ☐ = 223

c) 39☐ + 6 = 398
   39☐ + 6 = 400
   39☐ + 6 = 403

b) ☐ + 824 = 830
   ☐ + 824 = 831
   ☐ + 824 = 833

Reflect

Discuss with a partner what is the same and what is different about how to solve 825 + 3 and 825 + 8.

65

PUPIL PRACTICE BOOK 3A PAGE 65

Add 10s across 100

Learning focus

In this lesson, children will develop their understanding of adding 10s to a 3-digit number, including examples which require exchange of 10 tens for 1 hundred.

Before you teach

- Can children answer questions such as: *What number is 17 tens?*
- Can children count in 10s, for example, from 65 up to 165?

NATIONAL CURRICULUM LINKS

Number – addition and subtraction

Add and subtract numbers with up to three digits, using formal written methods of columnar addition and subtraction.

Add and subtract numbers mentally, including: a 3-digit number and 1s, a 3-digit number and 10s, a 3-digit number and 100s.

ASSESSING MASTERY

Children can add multiples of 10 and recognise when they need to exchange 10 tens for 1 hundred.

COMMON MISCONCEPTIONS

Children may struggle with the flexible partitioning of the 10s required to cross the 100. Ask:
- *How many different ways can you partition 80 using 10s?*

Crossing the 100s can cause children to struggle with exchange. They may often forget to include the exchanged digit or, for example, misconstrue 13 tens as 103. Ask:
- *What number is 13 tens?*

STRENGTHENING UNDERSTANDING

Children should explore the concept of exchange using place value equipment. It may also help to represent the additions on a number line, and justify why the addition crosses into the next 100.

GOING DEEPER

Now that children will be able to add a multiple of 10 to any 3-digit number, challenge children to explore such conjectures as: *When you add 90 to a number, you always need to exchange 10 tens; When you add 50, you sometimes do not need to exchange.*

KEY LANGUAGE

In lesson: addition, tens (10s), number line, exchange, calculations, 3-digit number

Other language to be used by the teacher: hundreds (100s), ones (1s)

STRUCTURES AND REPRESENTATIONS

Number lines, place value grids

RESOURCES

Mandatory: base 10 equipment, place value counters

Optional: 0–9 digit cards

 In the eTextbook of this lesson, you will find interactive links to a selection of teaching tools.

Quick recap 🔁

Play a game of bingo using bonds within 20. Children pick ten numbers from 3 to 20. Say simple addition facts within 20. If they have got it on their grid, then they cross it off. The winner is the first person to cross off five numbers.

Discover

Add 10s across 100

Discover

WAYS OF WORKING Pair work

ASK

• Questions ❶ a) and b): *What is the same and what is different about the calculations? Which calculation requires more steps of thinking? How could you represent the additions?*

IN FOCUS The important point is for children to notice that they can employ the method of adding 10s in question ❶ a) but also recognise that they will need to adapt their thinking for the calculation in question ❶ b).

PRACTICAL TIPS Represent the age of the beech tree using place value equipment. Locate its age on a number line. Then represent the numbers to be added using equipment and jumps on a number line.

ANSWERS

Question ❶ a): 184 + 10 = 194.
The birch tree is 194 years old.

Question ❶ b): 184 + 20 = 204.
The horse chestnut tree is 204 years old.

Beech: 184 years old | Birch | Horse chestnut | Oak

❶ a) The birch tree is 10 years older than the beech tree.
How old is the birch tree?

b) The horse chestnut tree is 20 years older than the beech tree.
How old is the horse chestnut tree?

88

PUPIL TEXTBOOK 3A PAGE 88

Share

WAYS OF WORKING Whole class teacher led

ASK

• Question ❶ b): *How do you know that the age of the horse chestnut tree will be greater than 200? Are you exchanging 10 ones or 10 tens?*

• Question ❶ b): *Could you represent the calculation as 184 + 10 + 10? What would this look like?*

IN FOCUS The idea of exchange is key to this lesson. Here we see that one of the calculations crosses into the next 100, and children should discuss and explore how this is represented by the place value grids. Although children should gain developmental fluency, the concept of exchange will support later development of more formal written methods. Children should see that this is similar to the previous lessons where they were adding ones across 10. The main focus of this lesson is using base 10 and counters to show the exchange, however you may want to back this up with a number line.

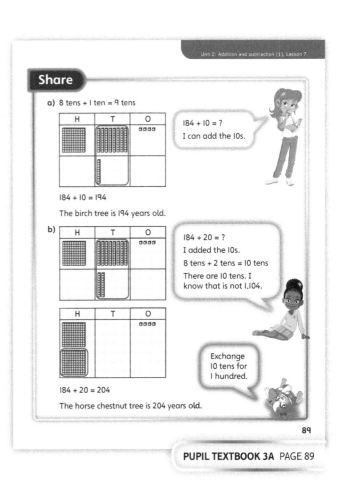

Share

a) 8 tens + 1 ten = 9 tens

184 + 10 = ?
I can add the 10s.

184 + 10 = 194
The birch tree is 194 years old.

b)

184 + 20 = ?
I added the 10s.
8 tens + 2 tens = 10 tens
There are 10 tens. I know that is not 1,104.

Exchange 10 tens for 1 hundred.

184 + 20 = 204
The horse chestnut tree is 204 years old.

89

PUPIL TEXTBOOK 3A PAGE 89

Think together

WAYS OF WORKING Whole class teacher led (I do, We do, You do)

ASK

- Questions ❶: *Which row represents the age of the beech tree? How do you know? Which columns are going to change? How do you know?*
- Questions ❷: *Which columns are going to change? How do you know? Will you need to make an exchange? How do you know?*

IN FOCUS The focus here is for children to make connections between the total number of 10s and how you can use this to decide whether an exchange will be necessary. They should be able to count in 10s beyond a 100 to help them find the answers. To help children understand the concept even more deeply you may want to explore this on a number line.

STRENGTHEN Provide children with base 10 equipment and a place value grid to support them in finding the answers. They could also draw jumps of 10 on a number line to support their counting.

DEEPEN In question ❸, encourage children to explain how they know the answer for each question and to make the decisions without performing the calculations. Can they summarise which digits always/sometimes/never change?

ASSESSMENT CHECKPOINT Use questions ❶ and ❷ to check that children can add multiples of 10 to a 3-digit number where the calculation requires an exchange.

ANSWERS

Question ❶: 184 + 50 = 234
The oak tree is 234 years old.

Question ❷: 263 + 70 = 333

Question ❸: Mia will need to do an exchange for c) and d).

Question ❸ a): 458 + 20 = 478

Question ❸ b): 458 + 30 = 488

Question ❸ c): 458 + 60 = 518

Question ❸ d): 458 + 80 = 538

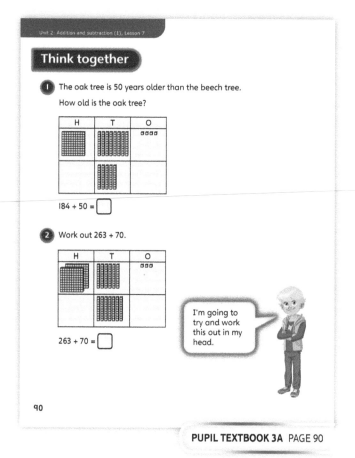

PUPIL TEXTBOOK 3A PAGE 90

PUPIL TEXTBOOK 3A PAGE 91

Practice

WAYS OF WORKING Independent thinking

IN FOCUS The focus of this section is for children to consolidate their learning on adding multiples of 10, crossing the 100. In question ❶, they perform these calculations where the base 10 equipment is represented for them. In question ❷, they work more abstractly and are encouraged to spot connections between questions and answers.

STRENGTHEN Provide children with a place value grid and base 10 equipment to represent the questions and support them in finding the answers.

DEEPEN In question ❻, encourage children to explain why Danny has chosen this method. Give them similar questions and ask them to adapt Danny's method to solve them. How many different ways can they find to answer Danny's question? Is one method always the most efficient?

THINK DIFFERENTLY In question ❹, children are exposed to a common misconception in adding 10s and encouraged to explain the mistake that has been made.

ASSESSMENT CHECKPOINT Use questions ❶ and ❷ to assess whether children can add 10s to a 3-digit number where the calculation requires an exchange. Use question ❸ to assess whether children can identify missing numbers in questions. Use question ❹ to check that children can identify the mistake that has been made and why it is incorrect.

ANSWERS Answers for the **Practice** part of the lesson can be found in the *Power Maths* online subscription.

Reflect

WAYS OF WORKING Independent thinking

IN FOCUS The sentence stem prompts children to really think about the relationship between the additions and the concept of place value. Encourage children to show examples of when exchange is necessary and when it is not necessary.

ASSESSMENT CHECKPOINT This section will show if children engage decision making processes, rather than simply performing rote calculations.

ANSWERS Answers for the **Reflect** part of the lesson can be found in the *Power Maths* online subscription.

After the lesson ▐▐

- Can children decide when exchange is necessary?
- Can children justify their answers to themselves and to others by invoking number bonds?
- Can children choose a representation to illustrate their mental methods?

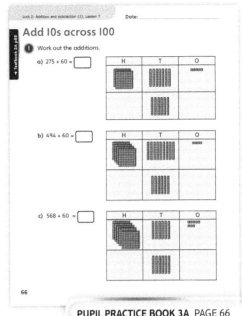

PUPIL PRACTICE BOOK 3A PAGE 66

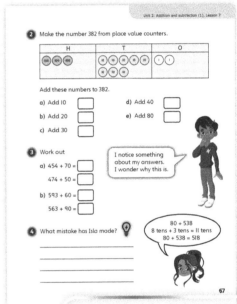

PUPIL PRACTICE BOOK 3A PAGE 67

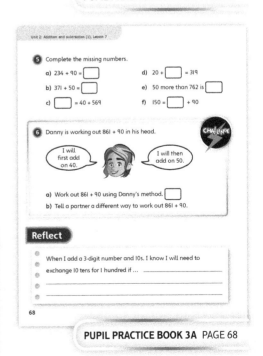

PUPIL PRACTICE BOOK 3A PAGE 68

Subtract 1s across 10

Learning focus

In this lesson, children will subtract a 1-digit number where the subtraction crosses a 10. Children understand how to exchange 1 ten for 10 ones.

Before you teach

- Can children answer questions such as: *Will the answer to 15 – 8 be greater than or less than 10?*

NATIONAL CURRICULUM LINKS

Number – addition and subtraction

Add and subtract numbers mentally, including: a 3-digit number and 1s, a 3-digit number and 10s, a 3-digit number and 100s.

Add and subtract numbers with up to three digits, using formal written methods of columnar addition and subtraction.

ASSESSING MASTERY

Children can explain how to use exchange of 1 ten for 10 ones to subtract a 1-digit number where they have to go across 10. Children will be able to represent their subtraction as two jumps on a number line. Children can justify their reasoning using their knowledge of number bonds within 20.

COMMON MISCONCEPTIONS

One of the most common misconceptions children make is to subtract the digits in the wrong order. For example, when calculating 234 – 7, children very often do 7 ones subtract 4 ones, as they think you have to subtract the smaller digit from the larger digit. This is one of the most common errors in learning to subtract. Ask:
- *What is the whole and what part are you subtracting?*

STRENGTHENING UNDERSTANDING

It may help children to represent the subtraction as two jumps on a number line, then explore the link with the exchange of place value equipment.

GOING DEEPER

It may be that imagining a number line to subtract is a more efficient mental method than writing a subtraction involving exchange. Challenge children to discuss this and decide which methods are most efficient and accurate for different kinds of subtraction.

KEY LANGUAGE

In lesson: subtraction, number line, exchange, tens (10s), ones (1s), calculations, zero (0), greater than (>), across

Other language to be used by the teacher: hundreds, less than (<)

STRUCTURES AND REPRESENTATIONS

Number line, part-whole model, place value grid

RESOURCES

Mandatory: base 10 equipment

Optional: bead strings, 0–9 digit cards

 In the eTextbook of this lesson, you will find interactive links to a selection of teaching tools.

Quick recap

Ask children to make the number 13 on two ten frames.

Ask children to demonstrate how they can subtract 7 from 13. Look for the method of subtracting 3 then 4. This is to help give children who do not know their bonds a method for subtracting when crossing 10.

Discover

WAYS OF WORKING Pair work

ASK

- Question ① a): *What is the whole?*
- Question ① a): *What is being taken away?*
- Question ① a): *Is it easy to subtract the 1s?*

IN FOCUS The focus of this question is for children to recognise that the number of 1s to be subtracted is greater than the 1s digit in the 3-digit number. This means they will require a different strategy from the one learnt previously. They will need to cross the 10.

PRACTICAL TIPS Represent the total number of parcels using base 10 equipment, and also locate it on a number line.

ANSWERS

Question ① a): $151 - 7 = 144$
There are 144 parcels left.

Question ① b):

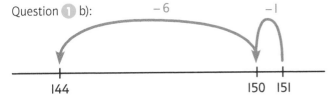

Subtract 1s across 10

Discover

① a) Make 151 using base 10 equipment.
Show how you can subtract 7 from 151.
How many parcels do they have left?

b) Show the subtraction on a number line.

92

PUPIL TEXTBOOK 3A PAGE 92

Share

WAYS OF WORKING Whole class teacher led

ASK

- Question ① a): *Why is Ash's suggestion wrong?*
- Question ① a): *Can you see two different ways of partitioning 151?*
- Question ①: *Can you use your knowledge of number bonds to work out how many 1s are left after subtracting 7?*

IN FOCUS Ash's comment voices the misconception about just subtracting the smaller digit from the larger digit.

The focus of the question is to understand that the exchange of 1 ten for 10 ones is necessary in order to be able to subtract the 7 ones from the 3-digit number.

Discuss how the exchange method in part a) and the number line in part b) show the same concept in a different way.

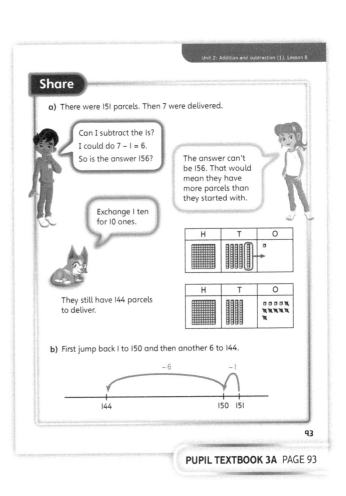

131

Think together

Whole class teacher led (I do, We do, You do)

ASK

- Question ❶: *What is happening in each place value grid? Why has a 10 been exchanged for 10 ones? How does this help? What is the answer?*
- Question ❷: *How does the number line help? How many jumps do you need to draw? When will you cross the 10? How do you know?*

IN FOCUS The focus of this section is for children to complete subtractions that cross a 10. In question ❶, they focus on doing this using base 10 equipment to make an exchange. In question ❷, they look at the number line method for counting back beyond a 10.

STRENGTHEN Provide children with a place value grid and base 10 equipment to represent the calculations and help them to solve them.

DEEPEN In question ❸, children use their knowledge of subtracting 1s to support them in working backwards and finding missing numbers in calculations.

ASSESSMENT CHECKPOINT Use questions ❶ and ❷ to assess whether children can subtract a 1-digit number from a 3-digit number where the calculations require an exchange.

ANSWERS

Question ❶: 144 − 8 = 136
There are 136 parcels left.

Question ❷ a): 143 − 2 = 141

Question ❷ b): 143 − 5 = 138

Question ❷ c): 143 − 7 = 136

Question ❷ d): 143 − 8 = 135

Question ❸ a): 236 − 229 = 7
236 − 7 = 229
Olivia subtracted 7.

Question ❸ b): 250 − 7 = 243
205 − 7 = 198
Max could use a number line, or a place value grid, or tens equipment.

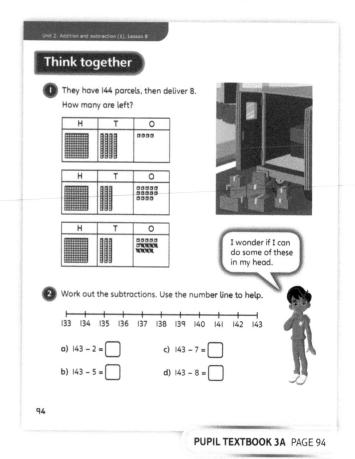

PUPIL TEXTBOOK 3A PAGE 94

PUPIL TEXTBOOK 3A PAGE 95

Practice

WAYS OF WORKING Independent thinking

IN FOCUS The focus of this section is for children to consolidate subtracting 1-digit numbers from 3-digit numbers where the calculation requires an exchange. They use number lines to count backwards to help them work out the answers in questions ① and ②. In question ③, they work abstractly or draw their own number line to calculate answers.

STRENGTHEN Provide children with base 10 equipment that they can use to build the start numbers on their number line and then make each jump to help them find the answers.

DEEPEN In question ⑤, children should be encouraged to explain how they found their answers and what patterns they noticed that supported them in finding the missing numbers.

THINK DIFFERENTLY In question ④, children are exposed to a common misconception and encouraged to explain the mistake that has been made.

ASSESSMENT CHECKPOINT Use questions ①, ② and ③ to assess whether children can subtract a 1-digit number from a 3-digit number where the calculation requires an exchange. Use question ④ to check that children can identify and explain common misconceptions.

ANSWERS Answers for the **Practice** part of the lesson can be found in the *Power Maths* online subscription.

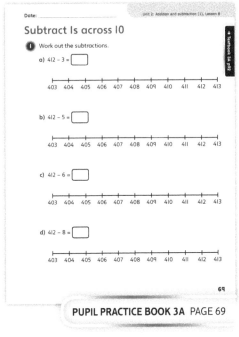

PUPIL PRACTICE BOOK 3A PAGE 69

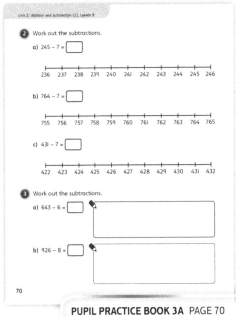

PUPIL PRACTICE BOOK 3A PAGE 70

Reflect

WAYS OF WORKING Group work

IN FOCUS Children should be able to explain how they know which subtractions require them to exchange a 10. Encourage each member of the group to use a different example subtraction to demonstrate their explanation.

ASSESSMENT CHECKPOINT Are children able to create their own examples to demonstrate their reasoning?

ANSWERS Answers for the **Reflect** part of the lesson can be found in the *Power Maths* online subscription.

After the lesson

- Can children explain when the 10s digit will decrease by 1?
- Can children solve the subtractions by using known number bonds?

PUPIL PRACTICE BOOK 3A PAGE 71

Subtract 10s across 100

Learning focus

In this lesson, children will subtract a multiple of 10 from a 3-digit number, including where they have to exchange 1 hundred for 10 tens.

Before you teach

- Can children identify what is the same and what is different about 5 – 4, 15 – 4, and 15 – 8?
- Can children partition 43 in three different ways? What about 143?

NATIONAL CURRICULUM LINKS

Number – addition and subtraction

Add and subtract numbers with up to three digits, using formal written methods of columnar addition and subtraction.

Add and subtract numbers mentally, including: a 3-digit number and 1s, a 3-digit number and 10s, a 3-digit number and 100s.

ASSESSING MASTERY

Children can recognise when they need to exchange 10s when subtracting a multiple of 10 from a 3-digit number. Children can perform the calculations accurately and fluently with mental methods, often supported by jottings.

COMMON MISCONCEPTIONS

A common misconception for children to have is thinking they can transpose the 10s digits. For example, in 230 – 70 they may perform 70 – 30 = 40 to give a 10s digit of 4. Ask:

- *How many 10s are being subtracted?*

STRENGTHENING UNDERSTANDING

Children need to see the number line and the exchange side by side, to recognise why the subtraction requires a part taken from a whole.

GOING DEEPER

This lesson models a number of different ways of representing the subtractions. Challenge children to explore and adapt the alternative representations. Can they think of a number of different ways of presenting the concept, and can they decide which shows the method required most effectively?

KEY LANGUAGE

In lesson: subtraction, exchange, difference, calculation, number line, part-whole, method, more, metres (m)

Other language to be used by the teacher: 3-digit number

STRUCTURES AND REPRESENTATIONS

Number lines, part-whole model, bar model, place value grid

RESOURCES

Mandatory: base 10 equipment

Optional: 0–20 digit cards, ball of string or wool

 In the eTextbook of this lesson, you will find interactive links to a selection of teaching tools.

Quick recap

Give the children digit cards/fans. Start with the number 15. Ask them to subtract 3 and to show the answer as fast as they can. Repeat by subtracting other numbers from 15. Choose a different start number. The aim is to develop increased fluency in bonds to 20.

Discover

Subtract 10s across 100

Discover

WAYS OF WORKING Pair work

ASK

• Question ❶ a): *What kind of problem is this? What calculation is required?*
• Question ❶ a): *What is the whole? What is the part to take away?*
• Question ❶ a): *How could you represent 210 m of fabric?*

IN FOCUS The focus of this lesson is for children to understand that the subtraction in question ❶ a) crosses a 100s boundary and so requires an exchange.

PRACTICAL TIPS This could be modelled with a ball of string or wool to represent the fabric. Children can represent the length using a number line, or using a bar model.

ANSWERS

Question ❶ a): 210 m – 20 m = 190 m
190 m are left.

Question ❶ b): 190 m – 140 m = 50 m
Jen sold 50 m.

❶ a) Jen has 210 m of dinosaur fabric to sell.
How much is left after she sells 20 m?

b) Jen sells some more dinosaur fabric. Now she has 140 m left.
How much did she sell?

96

PUPIL TEXTBOOK 3A PAGE 96

Share

WAYS OF WORKING Whole class teacher led

ASK

• Question ❶ a): *How does Astrid know that her first answer is incorrect?*
• Question ❶ a): *What mistake has Astrid made?*
• Question ❶ a): *What exchange has been made?*
• Question ❶ a): *Could you represent this question using a number line?*
• Question ❶ b): *How does the place value grid help solve this problem?*

IN FOCUS In question ❶ a), the important point for children to recognise is the need to exchange 1 hundred for 10 tens, so that the subtraction can be made. The calculations cross into the previous 100, and children should discuss and explore how this can be represented using part-whole models and number lines.

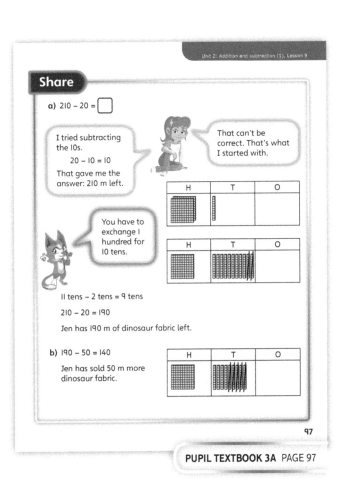

PUPIL TEXTBOOK 3A PAGE 97

Think together

WAYS OF WORKING Whole class teacher led (I do, We do, You do)

ASK

- Question **1**: *What is the whole? What is the part that you know?*
- Questions **1** and **2**: *How many 10s do you need to subtract?*
- Questions **1** and **2**: *What would this look like on a number line?*

IN FOCUS The important thing for children here is to develop fluency with the mental subtractions, alongside an understanding of the exchange required. In question **3**, children should be able to explain their thinking using the language of place value, and justify their answers based on their knowledge of number bonds within 20.

STRENGTHEN Ask children to represent the whole using place value equipment, and then enact the exchange using the equipment. They always need to remember to regroup and rename 1 hundred for 10 tens.

DEEPEN Question **3** presents the concept of exchange in a slightly alternative way. Ask children to discuss how this method is related to place value grids. Children should discuss the different methods and make judgements about which shows the concept most clearly, in their opinion, and ask them to justify with reasons why we use place value and how it helps develop mental methods of subtraction.

ASSESSMENT CHECKPOINT Use questions **1** and **2** to see whether children are able to use base 10 equipment in place value grids to help them subtract 2-digit numbers from 3-digit numbers, crossing the 100.

ANSWERS

Question **1**: 335 – 50 = 285
There is 285 m of space fabric left.

Question **2**: 213 – 80 = 133 m
Jen has 133 m of bee fabric more than Toshi.

Question **3** a): Rani has exchanged 1 hundred for 10 tens.
235 = 100 + 130 + 5
235 – 60 = 175

Question **3** b): 12 tens – 4 tens = 8 tens
328 – 40 = 288
12 tens – 5 tens = 7 tens
328 – 50 = 278
12 tens – 70 tens = 5 tens
328 – 70 = 258

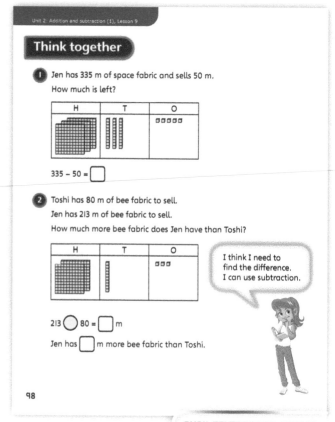

Think together

1 Jen has 335 m of space fabric and sells 50 m.
How much is left?

335 – 50 = ☐

2 Toshi has 80 m of bee fabric to sell.
Jen has 213 m of bee fabric to sell.
How much more bee fabric does Jen have than Toshi?

I think I need to find the difference. I can use subtraction.

213 ◯ 80 = ☐ m

Jen has ☐ m more bee fabric than Toshi.

98

PUPIL TEXTBOOK 3A PAGE 98

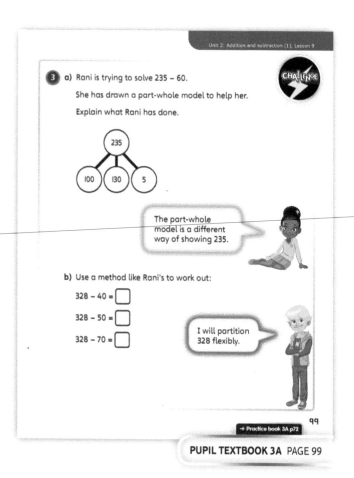

3 a) Rani is trying to solve 235 – 60.
She has drawn a part-whole model to help her.
Explain what Rani has done.

CHALLENGE

235
100 130 5

The part-whole model is a different way of showing 235.

b) Use a method like Rani's to work out:
328 – 40 = ☐
328 – 50 = ☐
328 – 70 = ☐

I will partition 328 flexibly.

→ Practice book 3A p72

PUPIL TEXTBOOK 3A PAGE 99

Practice

WAYS OF WORKING Independent thinking

IN FOCUS The focus of this section is for children to consolidate subtracting multiples of 10 from 3-digit numbers where the calculation requires an exchange. In question ❶, they use base 10 equipment to support them in finding the answers. In question ❷, they complete a series of connected calculations and should be encouraged to spot connections between the calculations and the answers.

STRENGTHEN Provide children with base 10 equipment and a place value grid to represent the calculations and support them in completing them. Encourage them to explain any exchanges they make, reinforcing the idea of 10 tens being equal to 100.

DEEPEN In question ❺, children should explain why the whole has been partitioned in such a way and how this helps find the answers. In question ❻, they should find different methods for working out the answer and compare these with a partner. Can they create a similar question for a partner?

THINK DIFFERENTLY In question ❸, ensure children use the correct starting number to find 30 more and 30 less.

ASSESSMENT CHECKPOINT Use questions ❶ and ❷ to assess whether children can subtract multiples of 10 from a 3-digit number where the calculations require an exchange. Use question ❸ to check that children can find different values given a starting number or a number 30 more. Use question ❹ to assess whether children subtract 10s where the calculation is not represented in a standard way.

ANSWERS Answers for the **Practice** part of the lesson can be found in the *Power Maths* online subscription.

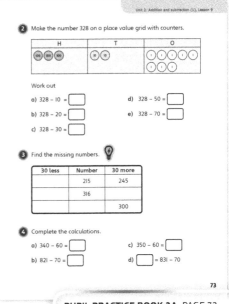

Reflect

WAYS OF WORKING Pair work

IN FOCUS Children use a given addition to derive related subtraction facts. They should observe how their knowledge of subtracting 1s across 10 can help them when crossing 10s across 100. They should be encouraged to explain each fact they find and how they know.

ASSESSMENT CHECKPOINT Children can find at least two related facts and explain how they have derived them.

ANSWERS Answers for the **Reflect** part of the lesson can be found in the *Power Maths* online subscription.

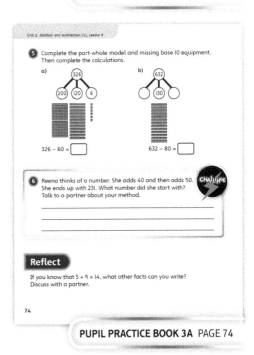

After the lesson ⏸

- Can children represent the exchange using place value equipment?
- Do children understand how to use number bonds to complete the subtractions?
- What different representations can children interpret?

Make connections

Learning focus

In this lesson, children focus on using simple calculations to find the answer to more complex calculations. They use their number bonds within 10 to break calculations down into more manageable steps and then use related facts to work out other calculations.

Before you teach

- Can children count in 10s from 0 to 200?
- Can children count in 100s from 0 to 1,000?
- Do children know common exchanges such as 1 ten = 10 ones?

NATIONAL CURRICULUM LINKS

Number – addition and subtraction

Solve problems, including missing number problems, using number facts, place value, and more complex addition and subtraction.

ASSESSING MASTERY

Children can use basic calculations to find related facts and solve calculations involving bigger numbers. They use their bonds to and within 10 to support them in working fluently with this.

COMMON MISCONCEPTIONS

Children may get confused with place value. For example, when using 5 + 7 = 12 to work out 50 + 70, children may write the answer as 1,200 as they think they need to add two zeros. Ask:
- *What is 5 tens add 7 tens? How do I write this in numerals?*

STRENGTHENING UNDERSTANDING

Provide children with base 10 equipment and a place value grid to represent calculations and support children in making connections.

GOING DEEPER

Encourage children to explain any connections they make and ask: *If I know this, what else can I find out?*

KEY LANGUAGE

In lesson: ones (1s), tens (10s), hundreds (100s)

Other language to be used by the teacher: related, facts, connections

STRUCTURES AND REPRESENTATIONS

Number lines, part-whole models, place value grid

RESOURCES

Mandatory: base 10 equipment, number lines, dice or dice simulator

Optional: laminated blank number line, laminated place value grids

 In the eTextbook of this lesson, you will find interactive links to a selection of teaching tools.

Quick recap

Ask children to roll four dice and to add the digits up as quickly as possible. You could do this as a whole class game using a dice simulator.

Discover

Make connections

Discover

WAYS OF WORKING Pair work

ASK

- Question ① a): *What is the starting number? Why does the number line not show one jump of 5? Why has the teacher chosen to split 5 into 3 and 2?*
- Question ① b): *How does knowing the answer to 5 add 7 help you to work out 50 add 70? What is the same about the calculations? What is different?*

IN FOCUS The focus of question ① b) is for children to make connections between calculations and recognise how they can use these connections to find answers to other calculations.

PRACTICAL TIPS Provide children with base 10 equipment to build the number on each number line and help them to make connections between the number line counting in 1s and the number line counting in 10s.

ANSWERS

Question ① a): 7 + 5 = 12

Question ① b): Using the basic fact 7 + 5 = 12
70 + 50 = 120

① a) What calculation is shown on the number line?

b) How can you use the calculation to work out the answer to Miss Hall's question?

100

PUPIL TEXTBOOK 3A PAGE 100

Share

WAYS OF WORKING Whole class teacher led

ASK

- Question ① a): *What is the bond to 10 for 7? How has this been used in the calculation? What is the bond to 5 for 3? How has this been used? How could you tell that the calculation was going to require an exchange?*
- Question ① b): *What is the same about the number lines in questions ① a) and b)? What is different? How does the part-whole model support the calculation?*

IN FOCUS Children look at different methods for answering the calculations and use bonds both to and within 10 to break a calculation down. They should recognise that if they know their bonds to 5, then they also know their bonds to 50 using 10s.

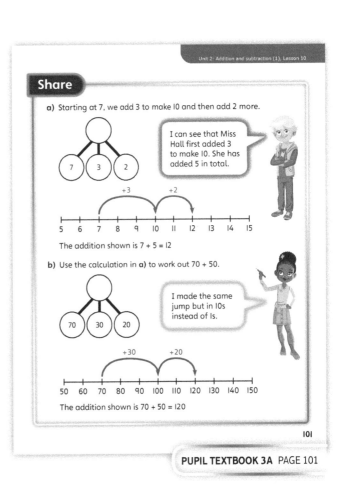

PUPIL TEXTBOOK 3A PAGE 101

Think together

Whole class teacher led (I do, We do, You do)

ASK

- Question **1** a): *What is the bond to 10 for 8? How many more do you need to jump? What is the answer?*
- Question **1** b): *What is the bond to 100 for 80? How many more do you need to jump? What is the answer?*
- Question **2** b): *How many do you need to subtract to get to 10 or 100? How many are left to subtract?*

IN FOCUS Here children use their bonds to 10 and 100 to find the answer to calculations that involve both addition and subtraction. In question **1**, they focus on adding, using their bonds to 10 to support them and then scaling these to find bonds to 100. In question **2**, they use their bonds to the number they are subtracting to support them in crossing the boundary.

STRENGTHEN Provide children with base 10 equipment to build the numbers under the number line to support them with their calculations.

DEEPEN In question **3**, encourage children to explain the connections and patterns they noticed and how they supported them in finding the answer to each calculation. Can they create a set of similar questions for a partner?

ASSESSMENT CHECKPOINT Use questions **1** and **2** to assess whether children can use patterns and connections to complete both additions and subtractions. Use question **3** to check how confident children are in using related facts to find the answers to calculations.

ANSWERS

Question **1** a): $8 + 6 = 14$

Question **1** b): $80 + 60 = 140$

Question **1** c): The same addition fact, $8 + 6 = 14$, is used in both calculations.
The first addition is adding ones, the second addition is adding tens.

Question **2** a): $11 - 6 = 5$

Question **2** b): $110 - 60 = 50$

Question **3**: $5 + 8 = 13$ $50 + 80 = 130$
           $80 + 50 = 130$ $130 - 80 = 50$

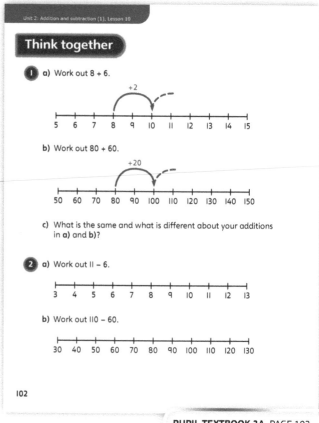

PUPIL TEXTBOOK 3A PAGE 102

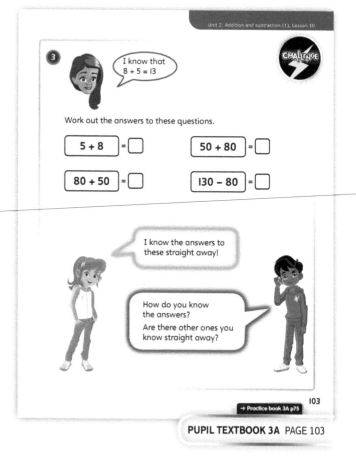

PUPIL TEXTBOOK 3A PAGE 103

Practice

WAYS OF WORKING Independent thinking

IN FOCUS In questions ❶ and ❷, children use their bonds to 10 and related bonds to 100 to find the answer to additions. In questions ❸ and ❹, children use their bonds within 10 and related bonds within 100 using 10s to find the answers to subtractions. They should be encouraged to make connections between parts a) and b) in each question. In question ❺, children use a given calculation to find the answer to related calculations.

STRENGTHEN Provide children with base 10 equipment and a place value grid to represent the calculation and support them in finding answers and spotting connections.

DEEPEN In questions ❺ and ❻, encourage children to explain the patterns and connections they spotted and explain how they then used these to find the answers. Can they create a similar set of questions for a partner?

ASSESSMENT CHECKPOINT Use questions ❶ and ❷ to assess whether children can make connections and use related facts to complete additions. Use questions ❸ and ❹ to assess whether children can make connections and use related facts to complete subtractions.

ANSWERS Answers for the **Practice** part of the lesson can be found in the *Power Maths* online subscription.

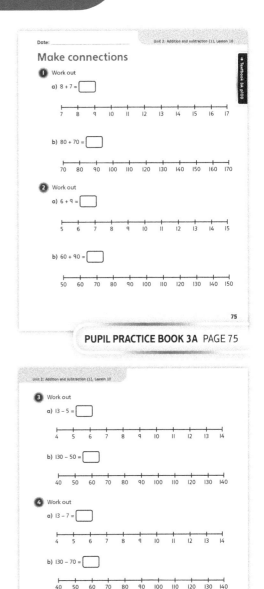

PUPIL PRACTICE BOOK 3A PAGE 75

PUPIL PRACTICE BOOK 3A PAGE 76

Reflect

WAYS OF WORKING Pair work

IN FOCUS Children should rehearse their answers and see if they have completed the sentence in terms of the reasoning required. Some children may just work out the answer, but the focus is on explaining the link between place value and exchange.

ASSESSMENT CHECKPOINT If children can explain their reasoning, then they will be able to understand if their own subtractions are accurate.

ANSWERS Answers for the **Reflect** part of the lesson can be found in the *Power Maths* online subscription.

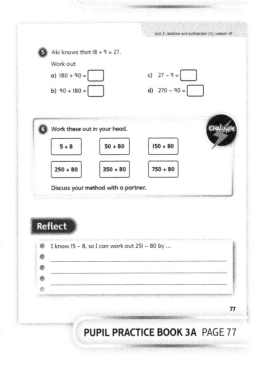

PUPIL PRACTICE BOOK 3A PAGE 77

After the lesson ⏸

- Can children use their bonds to 10 and related bonds to 100 to support them in working out related additions?
- Can children use their bonds within 10 and related bonds within 100 to support them in working out related subtractions?
- Can children explain any patterns and connections they spot?

End of unit check

> Don't forget the unit assessment grid in your *Power Maths* online subscription.

WAYS OF WORKING Group work adult led

IN FOCUS

- Question **1** focuses on children being able to calculate a subtraction of one 3-digit number from another.
- Question **2** focuses on children's understanding of where exchange of 1 ten for 10 ones is necessary.
- Questions **3**, **4** and **5** focus on children being able to calculate addition and subtraction involving an exchange accurately.
- Question **6** focuses on children's ability to solve missing digit calculations accurately.

ANSWERS AND COMMENTARY Children who have mastered this unit will be able to use mental methods, diagrams and place value grids to add and subtract 1-digit and 2-digit numbers to and from 3-digit numbers. Children will be able to explain where an exchange was necessary, and how this relates to crossing a 10 or 100.

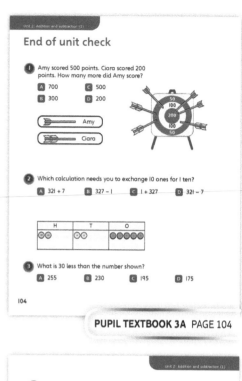

PUPIL TEXTBOOK 3A PAGE 104

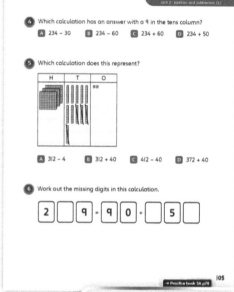

PUPIL TEXTBOOK 3A PAGE 105

Q	A	WRONG ANSWERS AND MISCONCEPTIONS	STRENGTHENING UNDERSTANDING
1	B	A suggests children have not understood parts and wholes.	Encourage children to draw number lines or manipulate place value equipment to justify or explore their own answers further, or to disprove errors.
2	D	A or B suggests children have not understood the different exchanges required for additions and subtractions.	
3	C	D suggests children have not accurately crossed the 100.	
4	C	D suggests children are not confident using the language of columns in preparation for the next unit.	
5	C	A suggests children have not developed a full understanding of the place value columns.	
6	249 = 95 + 154	Incorrect answers may suggest confusing place value columns.	

My journal

WAYS OF WORKING Independent thinking

ANSWERS AND COMMENTARY

Children should describe and explain the different methods represented, and highlight the similarities and differences. They will have to think very carefully about the action being shown, and the order of the different parts of the subtraction. Children will need to be able to use the language of partitioning and place value to explain what they notice.

If children need support in unpicking the methods, ask: *Which method would you use? What happens first in each method?*

If children are struggling to find similarities or differences, ask: *Do all the methods produce the same answer? How many 10s and 1s are being subtracted?*

Power check

WAYS OF WORKING Independent thinking

ASK

· *What numbers could you add and subtract before this unit?*
· *Do you know how to use place value equipment to support or explain your calculations?*
· *Have you built more confidence in any areas?*
· *Do you feel confident knowing when and how to use exchange?*
· *Can you show your answers in different ways?*

Power play

WAYS OF WORKING Pair work

IN FOCUS Use this **Power play** to encourage children to explore properties of calculations, and to make predictions and conjectures that they can then check.

ANSWERS AND COMMENTARY Children will most likely need to begin by using a trial and error method, then a trial and improve method, to explore the context. They should start to make conjectures about what digits could be involved, and should then test these and modify their predictions as appropriate.

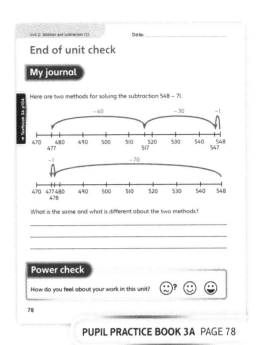

PUPIL PRACTICE BOOK 3A PAGE 78

PUPIL PRACTICE BOOK 3A PAGE 79

After the unit ⏸

· Can children use a mental method for different calculations?
· Can children use deep understanding to solve missing number and missing digit problems?

Strengthen and **Deepen** activities for this unit can be found in the *Power Maths* online subscription.

Unit 3
Addition and subtraction ②

Don't forget to watch the Unit 3 video!

WHY THIS UNIT IS IMPORTANT

This unit develops a depth of understanding of the key skills of formal addition and subtraction through place value, checking strategies and mental methods. Children will use their growing understanding to explore calculations which do or do not require exchange, developing fluency, accuracy and confidence in their ability to perform these calculations. They will be able to apply checking strategies to decide for themselves whether their answer is reasonable or likely to be an error.

WHERE THIS UNIT FITS

→ Unit 2: Addition and subtraction (1)
→ **Unit 3: Addition and subtraction (2)**
→ Unit 4: Multiplication and division (1)

In this unit, children build on previous learning to add and subtract numbers with up to three digits. They begin by using place value equipment and formal written methods of column addition and subtraction.

Before they start this unit, it is expected that children:
• know how to partition numbers to 1,000 flexibly
• understand the concept of exchange in addition and subtraction
• know how to represent additions and subtractions using place value equipment and a place value grid.

ASSESSING MASTERY

Children who have mastered this unit will be able to add or subtract numbers with up to three digits within 1,000. They will be able to justify whether or not an exchange was necessary and be able to explain the effect of doing an exchange in terms of place value. Children will also be able to justify an answer through checking strategies of approximation, estimation and the use of inverse operations.

COMMON MISCONCEPTIONS	STRENGTHENING UNDERSTANDING	GOING DEEPER
Children may resort to counting strategies rather than using known number bonds to add digits within an addition.	Use base 10 equipment to model an addition or subtraction, but encourage children to visualise the known number bond that finds the total or the part required to complete the parts of the calculation.	Explore the properties of numbers which result from adding odds and evens, or numbers that end in 5. Discuss problems with multiple solutions (for example, how many subtractions can you find that end in 1?).
Children may learn the process of column methods without fully understanding how it relates to place value and flexible partitioning.	Encourage children to use the language of 1s, 10s and 100s flexibly. Use a part-whole model alongside a column method to represent how a number has been partitioned.	There are many opportunities for all children to engage in missing number or missing digit problems, which dig deep into the mechanics of the methods. This requires children to explore how and why the process works.

UNIT STARTER PAGES

Use these pages to introduce the focus to children. You can use the characters to explore different ways of working.

STRUCTURES AND REPRESENTATIONS

Place value grid: Use this to help children to represent the partitions and exchanges required when adding and subtracting. It should support the understanding of column methods.

H	T	O

Base 10 equipment: This can be manipulated by children to model the differences between addition and subtraction (by taking away or by comparing). Base 10 equipment can be used in conjunction with place value grids to give structure.

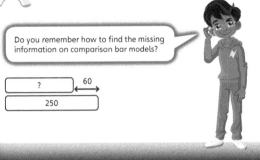

Column methods: Column methods will be used to present efficient and accurate addition and subtraction. Children should practise by using the scaffolded examples provided in the book, but should also experience writing their own calculations in columns.

	H	T	O
+			

KEY LANGUAGE

There is some key language that children will need to know as part of the learning in this unit.

- → add, addition
- → subtract, subtraction
- → total, altogether, sum
- → exchange
- → part-whole, whole, part
- → place value
- → hundreds (100s), tens (10s), ones (1s)
- → columns, column method
- → mental method, mentally
- → estimate, estimation
- → approximate, approx., approximation, approximately, about
- → fact family
- → bar model
- → digits, plus, minus
- → multiple
- → logically
- → 2-digit number, 3-digit number
- → calculation
- → zero (0)
- → order
- → number bond
- → how many more
- → difference

Unit 3
Addition and subtraction (2)

In this unit we will ...
- ⚡ Add and subtract 3-digit numbers
- ⚡ Decide if we need to exchange
- ⚡ Exchange across more than one column
- ⚡ Learn how to check our answers in different ways
- ⚡ Use bar models to solve 1- and 2-step problems

Do you remember how to find the missing information on comparison bar models?

| ? | → 60 ← |
| 250 | |

PUPIL TEXTBOOK 3A PAGE 106

We will need some maths words. Which words have you come across before? Which word means to find a rough answer?

exchange · column method · estimate · mental method · multiple · sum · digit · approximate · add · subtract · difference · plus · minus · total · place value

We need to remember about parts and wholes. Use this part-whole model to find a family of 8 facts.

240
150 · 90

PUPIL TEXTBOOK 3A PAGE 107

Add two numbers

Learning focus

In this lesson, children will learn to add two 3-digit numbers where no exchange is necessary. They will use a written column method and begin with the 1s, then the 10s and then the 100s.

Before you teach

- Can children add two 2-digit numbers?
- Do you have the necessary place value equipment ready to support learning in this lesson?
- Can children confidently use place value equipment in a place value grid?

NATIONAL CURRICULUM LINKS

Year 3 Number – addition and subtraction

Add and subtract numbers with up to three digits, using formal written methods of columnar addition and subtraction.

Add and subtract numbers mentally, including: a three-digit number and ones, a three-digit number and tens, a three-digit number and hundreds.

ASSESSING MASTERY

Children can explain how the written column method relates to the place value of an addition. They can write an addition accurately and use the columns to add two numbers accurately and efficiently.

COMMON MISCONCEPTIONS

Children may resort to the written column method, even when a mental method is more appropriate. Ask:
- *When you look at this addition, what method do you think is most useful? Why?*

No exchange is required in this lesson, which may lead children to assume that the method works in any order, perhaps by adding the 100s first. Ask:
- *Can you think of any reason why we might always start by adding the ones column?*

STRENGTHENING UNDERSTANDING

Ensure children have access to base 10 and place value equipment to support their understanding of place value. Encourage children to relate the numbers to the concrete elements of problems based around real-life contexts before building their confidence in abstract calculations.

GOING DEEPER

There are opportunities at various points through the lesson to explore additions which produce the same answers. Challenge children to explain how these are achieved through manipulating the place value of the additions and to describe how to find all the solutions systematically.

KEY LANGUAGE

In lesson: digit, ones (1s), tens (10s), hundreds (100s)

Other language to be used by the teacher: column

STRUCTURES AND REPRESENTATIONS

Place value grids, column addition

RESOURCES

Mandatory: base 10 equipment or other place value equipment

Optional: digit cards

 In the eTextbook of this lesson, you will find interactive links to a selection of teaching tools.

Quick recap

Recap with simple 2-digit + 2-digit addition examples where the answers do not cross the 10 or 100 (for example, 36 + 23).

Discover

Unit 3: Addition and subtraction (2), Lesson 1

Add two numbers

Discover

WAYS OF WORKING Pair work

ASK

- Question ① a): *How many 100s do you need for each number? How many 10s? How many 1s?*
- Question ① a): *What does each digit in 326 and 541 represent?*
- Question ① b): *How might Richard write the addition? Are there different ways?*
- Question ① b): *How can you use the base 10 equipment to help you?*

IN FOCUS In question ① a), ensure children use base 10 equipment to understand the value of each digit in the numbers. The key to question ① b) is encouraging children to explore how the base 10 representations support them in finding the total accurately. Ensure they understand that the total is just the combined base 10 cubes, rods and flats from each of the two numbers.

PRACTICAL TIPS Children could be provided with a place value grid and base 10 equipment to support them in representing both the numbers and the addition.

ANSWERS

Question ① a): Children make 326 and 541 using base 10 equipment.

Question ① b): 326 + 541 = 867

① a) Richard uses digit cards to make the numbers 3 2 6 and 5 4 1.

   Make Richard's numbers using base 10 equipment.

  b) Richard adds the numbers together.

   What is his total?

108

PUPIL TEXTBOOK 3A PAGE 108

Share

WAYS OF WORKING Whole class teacher led

ASK

- Question ① a): *How has each digit been represented?*
- Question ① b): *Do you need to count to find the total for each column?*

IN FOCUS The focus of question ① b) is on highlighting how the place value grid, combined with the base 10 equipment, is a very clear way of organising the addition. It helps clarify how to find the total by adding the digits in order. Children should use their knowledge of number bonds (rather than counting strategies) to find the total for each column.

Children should notice that they ought to start a column method by adding the numbers to the right first (i.e. the column with the least place value). This is important as they develop their understanding of place value in this way.

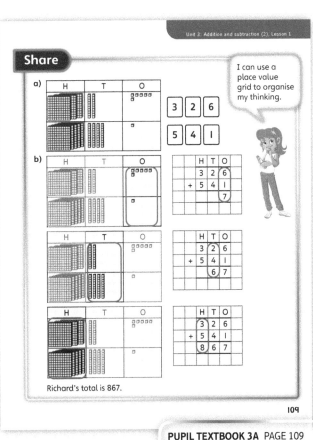

Share

I can use a place value grid to organise my thinking.

Richard's total is 867.

109

PUPIL TEXTBOOK 3A PAGE 109

Think together

Think together

WAYS OF WORKING Whole class teacher led (I do, We do, You do)

ASK

- Question **1**: *Which number bonds do you know that will help you find the totals of the 1s, 10s and 100s?*
- Question **1** b): *How many 1s, 10s and 100s will Jamilla need to show with the base 10 equipment?*
- Question **1** b): *Which order should she add the digits in? Does it matter?*
- Question **3**: *Where is the best place to start? Are there different ways of making a 9 or a 3 using the digit cards 1, 2, 3, 4, 5 and 6?*

IN FOCUS Develop children's confidence and fluency around the link between the place value representations and the column method. By the time children have completed question **3**, they may be confident enough to solve the additions without the support of place value equipment but they should still be encouraged to explain and justify their calculations using equipment and the language of 100s, 10s and 1s. In question **3**, children should explore different solutions by noticing the different pairs of digits that add up to 3 or 9.

STRENGTHEN Children should use digit cards alongside place value equipment to explore how moving the digits from one column to another changes the place value of the number. This should also change the way the number is represented using base 10 equipment. Remind them that they are working with the digit cards 1, 2, 3, 4, 5 and 6.

DEEPEN Question **3** has multiple solutions. Children should try to justify whether they have found all the possible options and to describe the method they have used to achieve this.

ASSESSMENT CHECKPOINT Can children use known number bonds to find the totals for each column in question **2** efficiently? Are children able to use the written column layout accurately?

ANSWERS

Question **1** a): 142 + 356 = 498

Question **1** b): 413 + 562 = 975

Question **2** a): 112 + 215 = 327

Question **2** b): 345 + 612 = 957

Question **2** c): 308 + 481 = 789

Question **2** d): 630 + 253 = 883

Question **3**: The 2 ones digits must be 1 and 2. Then the 10s and 100s digits will be any combination of the pairs 6 and 3, and 4 and 5. Examples include: 642 + 351, 431 + 562, 461 + 532.

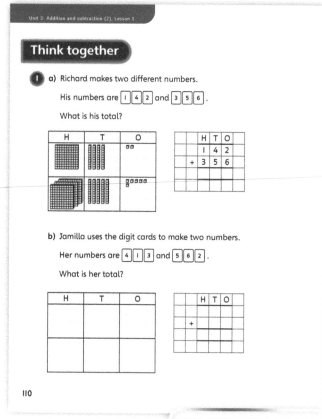

PUPIL TEXTBOOK 3A PAGE 110

PUPIL TEXTBOOK 3A PAGE 111

Practice

Independent thinking

In question **1** b), children should recognise that the lack of base 10 equipment in the tens column represents a '0'. Question **2** is only partly filled in, requiring children to realise that the total is given but some of the base 10 equipment is missing. They need to complete the missing base 10 equipment in the place value grid to find the addition.

Question **3** prompts children to solve calculations in a more abstract way, including using a mental method where appropriate rather than always reverting to the written column method. Question **5** encourages children to reason about the missing digits based on their knowledge of parts and wholes. Question **7** challenges children to use their understanding of variation to use the initial calculation to deduce the totals for related additions.

STRENGTHEN Although only question **1** directly prompts the use of base 10 equipment, children should have access to this to support their thinking. They should, however, use known bonds to find the totals and not rely on counting the base 10 equipment. Children may also find it helpful to have a column addition scaffold to support their layout.

DEEPEN Questions **6** and **7** are solved most effectively through deduction and a deep analysis of the numbers to be added. Encourage children to justify their answers through reasoning although they may find it helps to begin question **6** with a 'trial and improve' method. Similarly, some children may need to explore the additions in question **7** by using a column method for the first few examples, before pausing to look for a pattern to support their reasoning.

ASSESSMENT CHECKPOINT Solved accurately, question **5** demonstrates a sound understanding of how the column method relates to place value. Question **3** demonstrates fluency and accuracy and question **4** shows competency in applying the column method to real contexts.

ANSWERS Answers for the **Practice** part of the lesson can be found in the *Power Maths* online subscription.

Reflect

WAYS OF WORKING Pair work

IN FOCUS The important thing for children to notice is how the transcription error (writing 143 instead of 134) leads to an error. This highlights how children should check their working out against what the question is actually asking. The other mistake – incorrect addition of the hundreds column – prompts them to focus on accuracy.

ASSESSMENT CHECKPOINT Can children explain the errors that have been made in terms of place value?

ANSWERS Answers for the **Reflect** part of the lesson can be found in the *Power Maths* online subscription.

After the lesson ⏸

- Are children confident in representing the additions using base 10 and place value equipment on a place value grid?
- Do children use known number bonds to find column totals?
- Can children deduce missing digits and related answers, rather than relying on following a procedure without any deep understanding?

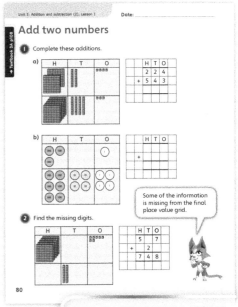

PUPIL PRACTICE BOOK 3A PAGE 80

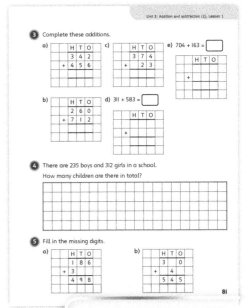

PUPIL PRACTICE BOOK 3A PAGE 81

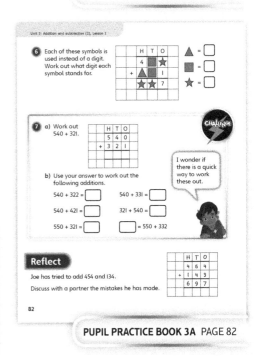

PUPIL PRACTICE BOOK 3A PAGE 82

149

Subtract two numbers

Learning focus

In this lesson, children will learn to subtract a 3-digit number from another 3-digit number where no exchange is necessary. They represent the subtraction as a written column subtraction.

Before you teach

- Can children successfully identify the wholes and parts in a calculation (for example, in $35 - 23 = 12$)?
- Are they able to use different methods to represent an addition or subtraction?

NATIONAL CURRICULUM LINKS

Year 3 Number – addition and subtraction

Add and subtract numbers with up to three digits, using formal written methods of columnar addition and subtraction.

Add and subtract numbers mentally, including: a 3-digit number and ones, a 3-digit number and tens, a 3-digit number and hundreds.

ASSESSING MASTERY

Children can subtract a 3-digit number from a 3-digit number using a written column method where no exchange is required. Children are increasingly confident in attempting to subtract a 3-digit number from a 3-digit number using a mental method, when appropriate.

COMMON MISCONCEPTIONS

Children may continue to find it confusing when subtraction is sometimes in the context of 'take away' and sometimes in the context of 'find the difference'. Ask:
- *What number do we need to subtract? What is the whole we are subtracting from?*

STRENGTHENING UNDERSTANDING

It may support some children to represent the numbers using base 10 equipment in order to explore and visually see subtraction as 'take away'.

GOING DEEPER

Explore the link between 'take away' and 'find the difference'. Encourage children to represent the whole number and then take away a part from the whole, by representing this with place value equipment. Then model the same subtraction as a find the difference. Ask: *How are the parts and the wholes related in each of these different ways of thinking about subtraction?*

KEY LANGUAGE

In lesson: subtract, subtraction, digit, mentally, mental method, logically, **multiple**

Other language to be used by the teacher: hundreds (100s), whole, part

STRUCTURES AND REPRESENTATIONS

Place value grids

RESOURCES

Mandatory: place value equipment (base 10 equipment, place value counters, cards and grids), dice

Optional: spinners, dice, paper clips, number lines

 In the eTextbook of this lesson, you will find interactive links to a selection of teaching tools.

Quick recap

Write the number 99 on the board. Ask children to roll two dice. Make a number with the result of the roll and ask them to subtract this number from 99.

Discover

Subtract two numbers

Discover

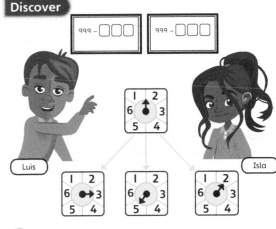

WAYS OF WORKING Pair work

ASK

- Question ❶ a): *What is the value of each of the digits in the number Luis creates with the spinner?*
- Question ❶ a): *How could Luis represent his subtraction?*
- Question ❶ a): *What is the whole that Luis is subtracting from?*
- Question ❶ b): *How many different numbers could Isla make using the digits 1, 6 and 6?*

IN FOCUS These questions are set in a mathematical game context. In question ❶ a), children may notice that they could make different numbers to subtract from 999, if they choose to reorder the digits for the 1s, 10s and 100s. For example, Luis chooses 352, but he could choose 523, 532, 253, 235 or 325. The important thing to notice is that they are subtracting from 999, which is the whole in each subtraction.

PRACTICAL TIPS Children could play this game themselves, using spinners or dice. If they do not have a spinner, they could make one using a paper clip as the spinning 'arrow'.

ANSWERS

Question ❶ a): 999 – 352 = 647
Luis scored 647.

Question ❶ b): 999 – 661 = 338, 999 – 616 = 383,
999 – 166 = 833
Isla could score 338, 383 or 833.

❶ a) Luis spins 3, 5 and 2.
He makes the subtraction 999 – 352.
What is his score?

b) Isla spins 1, 6 and 6.
Use 1, 6 and 6 in different combinations.
What could Isla's score be?

112

PUPIL TEXTBOOK 3A PAGE 112

Share

WAYS OF WORKING Whole class teacher led

ASK

- Question ❶ a): *Which number bonds help solve the subtraction for each digit?*
- Question ❶ a): *How has the whole been represented?*
- Question ❶ a): *Which part has been subtracted?*
- Question ❶ a): *Can you see the link between the column method and how it is shown on the number line?*
- Question ❶ b): *Why does the order in which you use the digits make a difference to the answer?*

IN FOCUS In question ❶ a), children should notice that the base 10 equipment is being used in a different way from when we model addition. In the additions, both numbers to be added are made and then the totals combined. However, in the subtraction, only the whole number is represented, and then the digits are subtracted from this (by crossing out or physically removing them). The number line is also used to represent another way of visualising the subtraction. Children should discuss what is the same and what is different about the two approaches. This is especially appropriate when thinking of subtraction as 'taking away'. Children may want to explore how this is related to finding the difference at a later stage, or at a relevant point in the lesson. To explore this, children may make the whole and then, to compare the whole with a part, you look at the part that is left once the other part has been taken away.

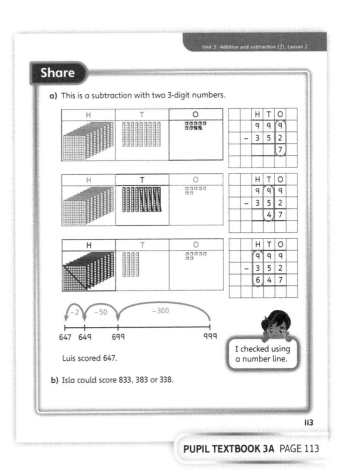

Share

a) This is a subtraction with two 3-digit numbers.

Luis scored 647.

b) Isla could score 833, 383 or 338.

I checked using a number line.

113

PUPIL TEXTBOOK 3A PAGE 113

Think together

WAYS OF WORKING Whole class teacher led (I do, We do, You do)

ASK

- Questions ❶, ❷ and ❸: *How many 1s, 10s or 100s are in the whole?*
- Questions ❶ and ❷: *How many 1s, 10s or 100s are being subtracted?*
- Questions ❶ and ❷: *How do the columns help represent the stages of the subtraction?*
- Question ❶: *What known number bonds help solve each step?*

IN FOCUS The important aspect of questions ❶, ❷ and ❸ is for children to gain fluency and accuracy in the use of column subtraction, alongside understanding how it relates to place value. Children should continue to recognise why and how the subtractions can be written in columns, and use the language of 1s, 10s and 100s to explain how known number bonds can be used to solve the stages of the calculation efficiently and accurately.

STRENGTHEN Represent each part of the subtraction separately, using the 1s base 10 equipment first, then the 10s and then the 100s. Give children the chance to use different types of place value equipment to develop a sense of the place value itself, rather than simply following a process without generating the deep understanding required.

For question ❸, some children may find a trial and improve approach is a useful way to begin the problem, rather than trying to solve it all in one step. Encourage children to try some different combinations and see if the resulting subtractions match any of the conditions as stated by the four children.

DEEPEN Question ❸ encourages children to make decisions about number properties. For example, challenge some children to explain how to choose the 1s digit to generate an even score or a multiple of 10, rather than simply listing all the possible solutions. Also encourage them to begin to subtract mentally and apply logic. For example, they may be able to work out mentally that 999 – 964 or 946 will produce an answer that is less than 100, meeting Reena's condition.

ASSESSMENT CHECKPOINT Can children find three different subtractions to satisfy question ❷ and solve them accurately? This includes one result where the answer is less than 100.

ANSWERS

Question ❶: 999 – 435 = 564
Jamilla's score is 564.

Question ❷: 678 – 446 = 232, 678 – 464 = 214,
678 – 644 = 34
Ebo could score 232, 214 or 34.

Question ❸: Mo needs to subtract a number with an odd number in the ones column. Reena needs to subtract a number with the 9 in the hundreds column. Ambika needs to subtract a number with a 9 in the ones column. Andy needs to subtract a number with a digit less than 5 in the hundreds column and with an even number in the ones column. Mo: 350 or 530. Reena: 35 or 53. Ambika: 350 or 530. Andy: 503.

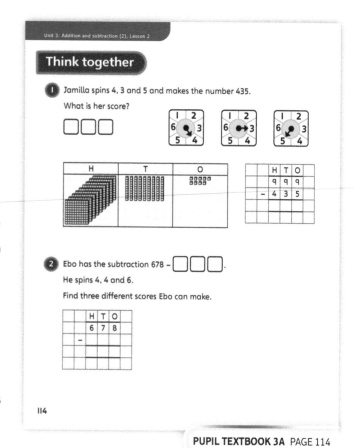

PUPIL TEXTBOOK 3A PAGE 114

PUPIL TEXTBOOK 3A PAGE 115

Practice

WAYS OF WORKING Independent thinking

IN FOCUS In questions ❶ and ❷, children generate understanding and fluency with continued scaffolding of place value equipment. In questions ❸ and ❺, children develop fluency in the abstract calculations and see how find the difference and take away are both related as subtractions, even attempting to mentally work out one or more of the answers. The important aspect is that children increase in fluency and accuracy as they move through the exercises. Questions ❹ and ❻ prompt them to apply mathematical thinking to probe the concept more deeply, through missing digits and by exploring the properties of the numbers generated through subtraction.

STRENGTHEN Make sure children have access to base 10 equipment, place value counters and place value grids to help scaffold their thinking. However, support children to become more fluent with the abstract calculations by asking them to visualise what the base 10 and place value equipment would look like. Encourage them to attempt to solve the calculations before checking using the equipment.

DEEPEN Challenge children to generate subtraction stories that match both 'take away' and 'find the difference' scenarios for some of the subtractions in these questions.

ASSESSMENT CHECKPOINT If children can complete the subtractions in question ❸ accurately and efficiently, then that should indicate sound fluency alongside an understanding of place value. This is especially the case for 688 – 34 and for questions ❸ c) and ❸ f) where they may attempt to use a mental method.

ANSWERS Answers for the **Practice** part of the lesson can be found in the *Power Maths* online subscription.

Reflect

WAYS OF WORKING Pair work

IN FOCUS Children should discuss the different methods of representing subtractions they have been using in the lesson, and then decide how to replicate them for the **Reflect** question.

ASSESSMENT CHECKPOINT Can children explain the place value of each part of the subtraction?

ANSWERS Answers for the **Reflect** part of the lesson can be found in the *Power Maths* online subscription.

After the lesson

- Do children have a good understanding of why writing subtractions in columns makes the process more efficient and accurate? Did children attempt to use a mental method to answer any of the questions?
- Do children use their knowledge of known number bonds to solve the steps of the calculations?
- Do children have a deep enough understanding of the approach to be able to solve missing number or missing digit problems?

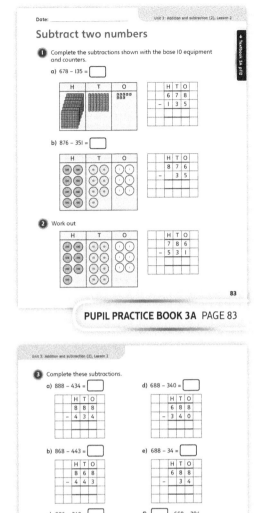

PUPIL PRACTICE BOOK 3A PAGE 83

PUPIL PRACTICE BOOK 3A PAGE 84

PUPIL PRACTICE BOOK 3A PAGE 85

Add two numbers (across 10)

Learning focus

In this lesson, children will learn to add two 3-digit numbers where exchange may be necessary, and to recognise when it is or is not necessary.

Before you teach ⏸

- Have children used exchange before?
- Do they know what it means to exchange 10 ones or 10 tens?

NATIONAL CURRICULUM LINKS

Year 3 Number – addition and subtraction

Add and subtract numbers with up to three digits, using formal written methods of columnar addition and subtraction.

Add and subtract numbers mentally, including: a 3-digit number and ones, a 3-digit number and tens, a 3-digit number and hundreds.

ASSESSING MASTERY

Children can add two 3-digit numbers and decide whether there is an exchange in none, one or two of the columns. They can explain how the exchange is presented in column written methods and demonstrate this using base 10 and place value equipment.

COMMON MISCONCEPTIONS

Some children may add mentally, starting with adding the 100s, then the 10s and then the 1s. If children do not begin by adding the 1s, then the 10s and then the 100s, adding where exchange is required can be difficult and prone to inaccuracy. Ask:

- Which part should we add first? Why do we add this part first?

STRENGTHENING UNDERSTANDING

Make sure children experience the concept of exchange through manipulating base 10 equipment on a place value grid.

GOING DEEPER

Completing missing digits in given column subtractions is a very good way to prompt a deeper understanding of the concept of exchange. Challenge children to develop their own missing digit calculations. Can they create examples where there are multiple solutions?

KEY LANGUAGE

In lesson: sum, exchange, ones (1s), tens (10s), hundreds (100s), addition

Other language to be used by the teacher: column method

STRUCTURES AND REPRESENTATIONS

Column addition, place value grids

RESOURCES

Mandatory: base 10 equipment and/or place value counters, place value grids

 In the eTextbook of this lesson, you will find interactive links to a selection of teaching tools.

Quick recap 🔁

Recall bonds to numbers within 20. Ask children to work out 6 + 7, 8 + 8, 9 + 4. Do they know them off by heart or do they need to work them out?

Discover

Add two numbers (across 10)

Discover

WAYS OF WORKING Pair work

ASK

- Question ①: *What number bonds can help us find the total of 1s, 10s and 100s?*
- Question ①: *Does your understanding of these number bonds suggest that you will need to exchange to answer this problem?*
- Question ①: *Is there anything we have learnt before that will help us to solve this accurately?*

IN FOCUS In question ① it is important that children recognise there is a need for exchange, as the sum of the 1s digits is greater than 9.

PRACTICAL TIPS Children should represent the numbers in the questions with place value equipment such as base 10 equipment, arranging the equipment in a way that demonstrates an efficient and accurate model of the addition.

ANSWERS

Question ① a): Children build the numbers 126 and 217 using base 10 equipment.

Question ① b): 126 + 217 = 343. Amal and Jen saw 343 birds in total.

① a) Make the numbers using base 10 equipment.

b) How many birds have Amal and Jen seen in total?

116

PUPIL TEXTBOOK 3A PAGE 116

Share

WAYS OF WORKING Whole class teacher led

ASK

- Question ①: *Can you see how the exchange is demonstrated using the base 10 equipment?*
- Question ①: *How is the exchange shown in the written column method?*
- Question ①: *Why do we not also exchange 10 tens?*

IN FOCUS Learning should focus on how the exchange is represented by the written method and understand how this links to the way it is represented using the base 10 equipment in the place value grids. Children should discuss why we add the ones column first and then continue from right to left, in terms of the effect on the exchange. The diagrams in **Share** will help you explain each of the steps. Use Sparks's comments to help too. It is useful to have children model this with their own equipment as you talk through it.

Share

a) and b) Amal and Jen saw 126 birds in the morning and 217 birds in the afternoon.

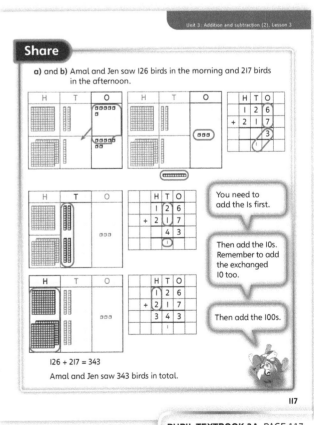

You need to add the 1s first.

Then add the 10s. Remember to add the exchanged 10 too.

Then add the 100s.

126 + 217 = 343
Amal and Jen saw 343 birds in total.

117

PUPIL TEXTBOOK 3A PAGE 117

Think together

Whole class teacher led (I do, We do, You do)

ASK

- Question ❶: *Did Amal and Jen see more birds in the morning or in the afternoon? How do you know?*
- Question ❶: *What can you use to represent the numbers? How can you use this to work out the total?*
- Question ❷: *What do you need to do when there are more than 10 ones?*

IN FOCUS The focus of this part of the lesson is on developing confidence and accuracy when completing additions that require an exchange in the ones column. Children should build the habit of working from right to left, and discuss how this accommodates any exchanges as the calculation builds.

STRENGTHEN Represent the calculations using base 10 equipment or place value counters in place value grids. This may help some children to visualise and understand any exchanges.

DEEPEN Question ❸ requires children to work out missing digits within a calculation that requires an exchange. At each step assess children's understanding by asking: *How did you work that out? Can you explain your reasoning?*

ASSESSMENT CHECKPOINT Use questions ❶ and ❷ to assess whether children can complete additions that require an exchange. Use question ❸ to assess whether children can find missing digits in questions.

ANSWERS

Question ❶:

	H	T	O
	2	2	6
+	2	1	5
	4	4	1
		1	

Question ❷ a):

	H	T	O
	3	7	8
+	2	1	7
	5	9	5
		1	

Question ❷ b): 126 + 239 = 365
348 + 348 = 696

Question ❸ a): The hidden digit is 5.

	H	T	O
	4	2	7
+	1	3	5
	5	6	2
		1	

Question ❸ b): 427 + 137 = 564
427 + 138 = 565

Multiple answers are possible, such as
427 + 135 = 562 or
420 + 132 + 552

PUPIL TEXTBOOK 3A PAGE 118

PUPIL TEXTBOOK 3A PAGE 119

Practice

WAYS OF WORKING Independent thinking

IN FOCUS These questions are designed to help children develop from using concrete or pictorial representations of place value, to become more confident and fluent in solving abstract calculations presented in columns. Question ③ ensures children are given an opportunity to write calculations in columns themselves, as this is essential and will ensure children can then complete the additions accurately.

STRENGTHEN Allow children to use concrete representations alongside the abstract representations, even in later questions, to support them in developing a deeper understanding of the addition.

DEEPEN Encourage children to reason around when an exchange is or is not necessary. This decision-making process should be made distinct from the process of performing the addition itself, so that children always maintain mathematical thinking alongside procedural fluency.

THINK DIFFERENTLY In question ④, children find missing digits where the total is given. Encourage them to explain each step in their reasoning. Ask: *Is it still important to work from right to left? Why?*

ASSESSMENT CHECKPOINT Use questions ① to ③ to assess whether children can accurately complete additions that require an exchange. Use question ⑤ to ensure children can apply this skill in a context. Use questions ④ and ⑥ to assess whether children can unpick their learning and apply it to slightly different contexts.

ANSWERS Answers for the **Practice** part of the lesson can be found in the *Power Maths* online subscription.

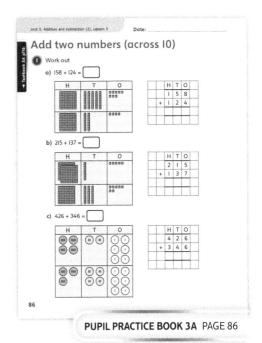

PUPIL PRACTICE BOOK 3A PAGE 86

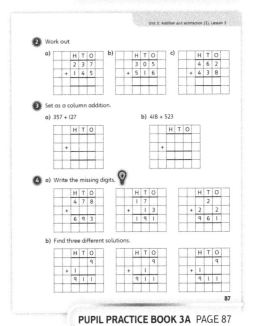

PUPIL PRACTICE BOOK 3A PAGE 87

Reflect

WAYS OF WORKING Pair work

IN FOCUS This question should allow children to use the language of place value and exchange in order to explain the mistake. They could discuss and rehearse their response in pairs before deciding how to write their answer as clearly and accurately as possible.

ASSESSMENT CHECKPOINT Can children explain the mistake using the language of place value rather than simply performing the calculation to find the 'right answer'?

ANSWERS Answers for the **Reflect** part of the lesson can be found in the *Power Maths* online subscription.

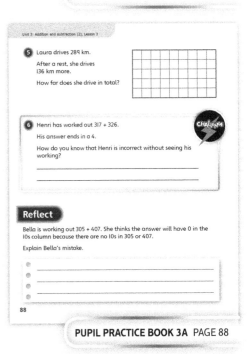

PUPIL PRACTICE BOOK 3A PAGE 88

After the lesson ⏸

- Are children confident representing an exchange in column addition?
- Can children spot when an exchange is or is not necessary?
- Do children look out for common mistakes like forgetting to add the exchanged digit?

Add two numbers (across 100)

Learning focus

In this lesson, children will build on the learning from the previous lesson to add 3-digit numbers where exchanges may be necessary in the 1s, 10s or both.

Before you teach

- Do children understand the equivalence between 10 ones and 1 ten and 10 tens and 1 hundred?
- Can children confidently complete addition calculations that require an exchange into the tens column only?

NATIONAL CURRICULUM LINKS

Year 3 Number – addition and subtraction

Add and subtract numbers with up to three digits, using formal written methods of columnar addition and subtraction.

Add and subtract numbers mentally, including: a 3-digit number and ones, a 3-digit number and tens, a 3-digit number and hundreds.

ASSESSING MASTERY

Children can add 3-digit numbers and decide whether there is an exchange in none, one or two of the columns. They can explain how the exchange is presented in column written methods and demonstrate this using place value equipment.

COMMON MISCONCEPTIONS

Children may start adding from the 100s, then the 10s and then the 1s. If children do not begin by adding the digit with the smallest place value, in this case the 1s, then exchanges can be difficult and prone to inaccuracy.

STRENGTHENING UNDERSTANDING

Make links back to the previous lesson where children exchanged into the tens column. Make sure children have an opportunity to represent calculations using place value equipment to ensure they make connections between exchanging into the 10s and exchanging into the 100s.

GOING DEEPER

Completing missing digits in given calculations is a good way to prompt a deeper understanding of the concept of exchange. Encourage children to decide how many exchanges are needed before completing the calculations as this will further develop their understanding.

KEY LANGUAGE

In lesson: exchange, ones (1s), tens (10s), hundreds (100s), addition

Other language to be used by the teacher: column method

STRUCTURES AND REPRESENTATIONS

Column addition, place value grids

RESOURCES

Mandatory: place value counters, place value grids

Optional: base 10 equipment

 In the eTextbook of this lesson, you will find interactive links to a selection of teaching tools.

Quick recap

Recall bonds to 20. Write down two numbers that add to make 12. Ask: *What can you tell me about two numbers that add together to make 13?*

Discover

Add two numbers (across 100)

WAYS OF WORKING Pair work

Discover

ASK

- Question ① a): *How many 100s, 10s and 1s are there in each number? How does this help to build the numbers?*
- Question ① b): *What do you need to do to add them together?*
- Question ① b): *Do you need to make an exchange in the ones column? How do you know?*
- Question ① b): *What happens when there are more than 9 tens?*

IN FOCUS In question ① b), it is important that children recognise there is a need for an exchange from the tens column into the hundreds column as the sum of the 10s digit is greater than 9. Children should recognise this from similar learning in the previous lesson.

PRACTICAL TIPS Provide children with place value counters and grids to represent the numbers and support them in adding them together.

> Add together 185 and 341.

① a) Help Mr Jones make the numbers in the place value grid.

b) Add the two numbers together to find the total.

120

ANSWERS

Question ① a): Children build the numbers 185 and 341 using place value counters or base 10 equipment in a place value grid.

Question ① b):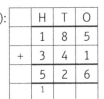

	H	T	O
	1	8	5
+	3	4	1
	5	2	6
	1		

Share

WAYS OF WORKING Whole class teacher led

ASK

- Question ① a): *How does the number of each counter link to the digits in the numbers?*
- Question ① b): *Is there an exchange in the 1s? How do you know?*
- Question ① b): *What do you do when there are more than 9 tens? What is 10 tens the same as?*

IN FOCUS In question ① b), learning should focus on how the exchange is represented by the written method and understand how this links to the way it is represented using the place value equipment. Children should discuss why we add the ones column first and then continue from right to left, in terms of the effect on the exchange.

In this lesson, children are encouraged to use place value counters instead of base 10 equipment. This shows flexibility. Some children may prefer to use base 10 equipment and this is fine. Work through each of the steps with the children, they could model it along with you. Relate adding across 100 by exchanging 10s to the previous lesson where they exchanged 1s.

Share

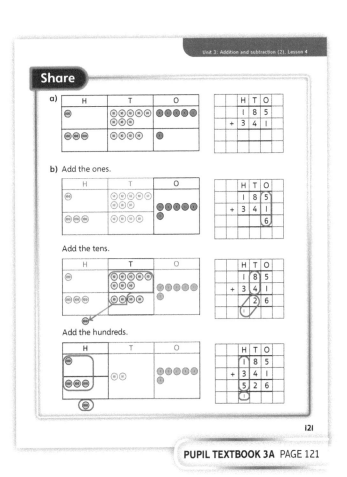

a)

b) Add the ones.

Add the tens.

Add the hundreds.

121

159

Think together

Whole class teacher led (I do, We do, You do)

ASK

- Question **1** a): *Do you need to make an exchange? How do you know?*
- Question **1** c): *What is the same and what is different about adding two numbers and adding three numbers?*

IN FOCUS The focus of this part of the lesson is on developing confidence and accuracy when completing additions that require an exchange in either the 1s, 10s or both. Question **1** c) gives children the opportunity to apply what they have learnt about exchanging to adding three 3-digit numbers. Children should build the habit of working from right to left, and discuss how this accommodates any exchanges as the calculation builds.

STRENGTHEN Represent the calculations using place value equipment. It may help some children to visualise and understand any exchanges.

DEEPEN Question **3** requires children to work out the number of exchanges without necessarily completing the calculations. Ensure they reflect on the mistake made by Max in question **2** to avoid this misconception occurring. Encourage children to explain their reasoning

ASSESSMENT CHECKPOINT Use question **1** to assess whether children can complete additions that require an exchange. Use question **2** to assess whether children can identify a common misconception.

ANSWERS

Question **1** a):

H	T	O
4	9	5
+ 3	8	4
8	7	9
1		

b):

H	T	O
2	5	3
+ 1	7	4
4	2	7
1		

Question **1** c):

H	T	O
1	2	1
2	7	3
+ 1	4	3
5	3	7
1		

Question **2**: Max is incorrect. He will exchange the 1s but this will mean that the 10s will add up to 10 and so another exchange is needed.

H	T	O
1	8	4
+ 2	1	7
4	0	1
1	1	

Question **3**:

No exchange	1 exchange	2 exchanges
253 + 123	253 + 174	253 + 279
		253 + 188
		253 + 149

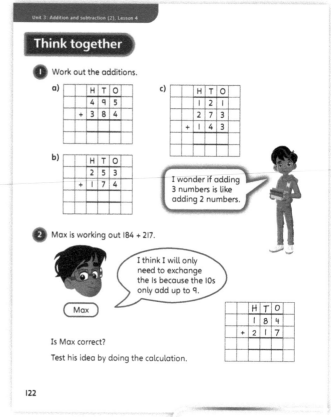

Think together

1 Work out the additions.

a)
H	T	O
4	9	5
+ 3	8	4

b)
H	T	O
2	5	3
+ 1	7	4

c)
H	T	O
1	2	1
2	7	3
+ 1	4	3

I wonder if adding 3 numbers is like adding 2 numbers.

2 Max is working out 184 + 217.

I think I will only need to exchange the 1s because the 10s only add up to 9.

Max

H	T	O
1	8	4
+ 2	1	7

Is Max correct?

Test his idea by doing the calculation.

122

PUPIL TEXTBOOK 3A PAGE 122

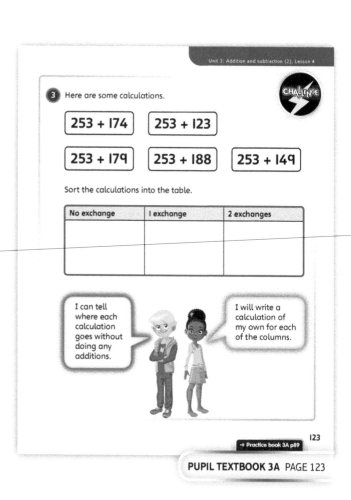

3 Here are some calculations.

CHALLENGE

253 + 174	253 + 123

253 + 179	253 + 188	253 + 149

Sort the calculations into the table.

No exchange	1 exchange	2 exchanges

I can tell where each calculation goes without doing any additions.

I will write a calculation of my own for each of the columns.

123

→ Practice book 3A p89

PUPIL TEXTBOOK 3A PAGE 123

Practice

WAYS OF WORKING Independent thinking

IN FOCUS These questions are designed to help children develop from using concrete or pictorial representations of place value in questions ① and ②, to become more confident and fluent in solving abstract calculations presented in columns in questions ⑤ and ⑥. Children will also develop their reasoning around when an exchange is or is not necessary.

STRENGTHEN Provide children with place value equipment to use alongside the abstract calculations to support them in making connections between the two representations.

DEEPEN Before tackling a question such as question ⑦, prompt children to consider how many exchanges are necessary by asking: *What exchanges are needed? How do you know?*

ASSESSMENT CHECKPOINT Use questions ① and ② to assess whether children can accurately complete additions that require an exchange in either the 1s, 10s or both. Use questions ③ and ④ to assess whether children can apply this in context. Use questions ⑤ to ⑥ to assess children's level of understanding and how they apply this.

ANSWERS Answers for the **Practice** part of the lesson can be found in the *Power Maths* online subscription.

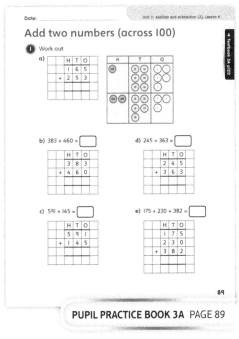

PUPIL PRACTICE BOOK 3A PAGE 89

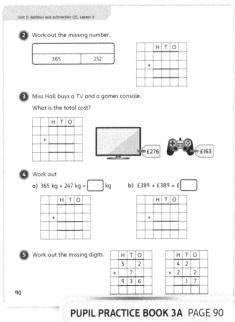

PUPIL PRACTICE BOOK 3A PAGE 90

Reflect

WAYS OF WORKING Pair work

IN FOCUS The focus here is for children to explain in their own words how to add two 3-digit numbers.

ASSESSMENT CHECKPOINT Ensure children mention the need to start from the place value column with the lowest value, which in this case is the 1s, and why this is the case. Ensure they use accurate language when talking about exchanges.

ANSWERS Answers for the **Reflect** part of the lesson can be found in the *Power Maths* online subscription.

After the lesson

- Are children confident representing an exchange in column addition?
- Can children spot when an exchange is and is not necessary?
- Can children accurately complete additions that require an exchange in either the 1s, 10s or both?

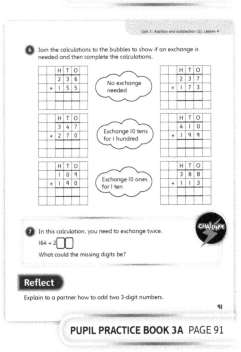

PUPIL PRACTICE BOOK 3A PAGE 91

Subtract two numbers (across 10)

Learning focus

In this lesson, children will develop their fluency with column subtraction of 3-digit numbers to include calculations where exchange is necessary across one or two columns.

Before you teach

- Do children know what is different between the kinds of exchange in addition and in subtraction? Do they understand this?
- Can they recognise which of these requires an exchange: 157 – 72 or 175 – 72?

NATIONAL CURRICULUM LINKS

Year 3 Number – addition and subtraction

Add and subtract numbers with up to three digits, using formal written methods of columnar addition and subtraction.

Add and subtract numbers mentally, including: a 3-digit number and ones, a 3-digit number and tens, a 3-digit number and hundreds.

ASSESSING MASTERY

Children can represent column subtractions involving exchange across 10 and can explain when and why it is necessary.

COMMON MISCONCEPTIONS

There is a danger that children assume that they always subtract the smaller digit from the larger digit, which can lead to a misconception where the subtractions require exchange. In this lesson, it is important that children understand they are subtracting from a whole. Ask:

- *What is the number to subtract? What is the whole we are subtracting from?*

STRENGTHENING UNDERSTANDING

Use a part-whole model to represent the subtractions so that children can clearly identify the whole from which the part is subtracted.

GOING DEEPER

There are numerous opportunities to deepen fluency through solving missing number and missing digit problems. Further depth may also be gained through challenging children to create their own subtraction problems. Where suitable, they could also try to work without the aid of equipment or apply mental methods.

KEY LANGUAGE

In lesson: digit, exchange, ones (1s), tens (10s), hundreds (100s), subtract, mentally

Other language to be used by the teacher: column subtraction

STRUCTURES AND REPRESENTATIONS

Place value grid, number line, column subtraction

RESOURCES

Mandatory: place value equipment (such as base 10 equipment and place value counters)

Optional: number lines, part-whole models

 In the eTextbook of this lesson, you will find interactive links to a selection of teaching tools.

Quick recap

Play number bonds bingo. Ask children to pick six numbers from 11 to 20. They may pick a number more than once. Ask number bond questions involving answers from 11 to 20; children cross off these numbers from their grids when they hear a question with that number as the answer.

Discover

Subtract two numbers (across 10)

Discover

WAYS OF WORKING Pair work

ASK

- Question ① a): *How does the base 10 equipment link to the digits in the number?*
- Question ① b): *What is the part that is being subtracted?*

IN FOCUS In question ① b), the vital point is that children recognise the context requires a subtraction and then that the numbers involved will require an exchange.

PRACTICAL TIPS It may help children to sketch their own version of the staircase and draw a stick figure on to the sketch. Alternatively, they could role-play a dramatic sketch. They may find it lends some sense of reality to the context if they discuss times they may have had to walk a long way or climb a large number of steps. The sense of 'are we there yet?' is familiar to many children and may help them to understand the context of the problem.

ANSWERS

Question ① a): Children build 361 using base 10 equipment using 3 flats, 6 rods and 1 cube.

Question ① b): 361 − 147 = 214
Aki has 214 steps left to climb.

① a) Make the number 361 using base 10 equipment.

b) How many steps does Aki have left to climb?

124

PUPIL TEXTBOOK 3A PAGE 124

Share

WAYS OF WORKING Whole class teacher led

ASK

- Question ①: *Why do we need to make an exchange?*
- Question ①: *Which part of the written column method shows that we have exchanged 1 ten for 10 ones?*
- Question ①: *Why has the 6 been crossed out and a 5 been written instead? What do the 6 and the 5 represent?*
- Question ①: *Why do we need to start by subtracting the 1s?*

IN FOCUS The focus of question ① b) is to recognise the need for an exchange and choose the appropriate exchange. Children need to represent the exchange in the written method and perform the subtraction beginning with the 1s and working through the 10s and then the 100s.

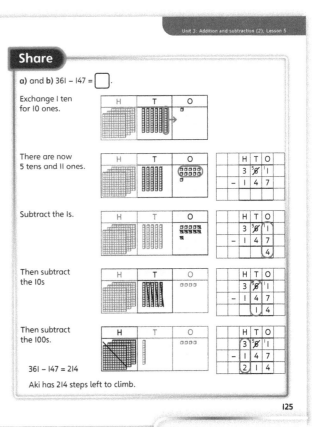

PUPIL TEXTBOOK 3A PAGE 125

Think together

Unit 3: Addition and subtraction (2), Lesson 5

Think together

WAYS OF WORKING Whole class teacher led (I do, We do, You do)

ASK

- Question ❶: *What is the whole? What is the part being subtracted?*
- Question ❷: *How can you tell a mistake has been made? What is the mistake?*

IN FOCUS The important aspect of question ❶ (and indeed with all the questions in this part of the lesson) is for children to recognise where the exchange is needed, and how to represent it in written methods. Children should build confidence and fluency, alongside understanding, by using the language of place value and base 10 equipment to represent the process.

STRENGTHEN Children should explore the exchanges necessary using a range of place value equipment such as place value counters, base 10 equipment or any other suitable equipment.

DEEPEN Question ❷ requires children to recognise and explain some common mistakes. Challenge children to think of some word problems that would match the two subtractions.

ASSESSMENT CHECKPOINT Can children correct the subtractions for question ❷? This will indicate whether they have sufficient understanding of both the need to identify and subtract from the whole and also of the exchange process; the order in which the exchange needs to be completed and how to accurately show the exchange when using a written method.

ANSWERS

Question ❶:

H	T	O
5	$^5\cancel{6}$	$^1 1$
− 3	2	5
2	3	6

Question ❷ a): In the ones column the 1 has been subtracted from 5, but this is the wrong way round. The 5 should have been subtracted from the 1 requiring an exchange.

$341 - 235 = 106$

Question ❷ b): In the ones column an exchange has been made so that $13 - 5 = 8$ but this exchange has not been taken into account in the calculation in the tens column where 5 tens has been subtracted from 8 tens instead of from 7 tens.

Question ❸ a): In the ones column the digit (7) being subtracted from the original number (2) is greater; hence an exchange is needed.

Question ❸ b): $482 - 13\mathbf{5} = 34\mathbf{7}$;
$482 - 13\mathbf{6} = 34\mathbf{6}$;
$482 - 13\mathbf{3} = 34\mathbf{9}$

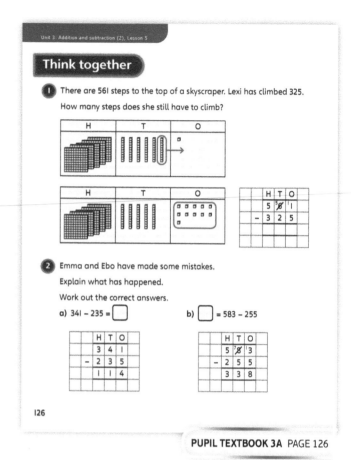

PUPIL TEXTBOOK 3A PAGE 126

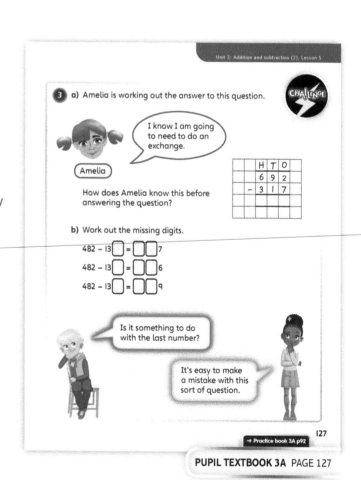

PUPIL TEXTBOOK 3A PAGE 127

Practice

WAYS OF WORKING Independent thinking

IN FOCUS The focus of this section is on children being able to accurately complete subtractions that cross a 10. They should move from using concrete and pictorial representations in questions ❶ and ❷ to using the abstract representations in questions ❸, ❹ and ❺, whilst also understanding this abstract procedure.

STRENGTHEN Ensure children have access to place value equipment to support them in making links between the concrete, pictorial and abstract representations.

DEEPEN Before starting the calculations, ask: *Is an exchange needed? Can you explain how you know?* This can be built upon and developed further when calculating missing digits in question ❺.

THINK DIFFERENTLY In question ❻, children use the pictorial representation to identify the question being answered. They should be able to explain their reasoning for each part.

ASSESSMENT CHECKPOINT Use questions ❶, ❸ and ❹ to assess whether children can accurately complete subtractions where the calculation crosses the 10. Use question ❷ to assess whether children can identify common misconceptions.

ANSWERS Answers for the **Practice** part of the lesson can be found in the *Power Maths* online subscription.

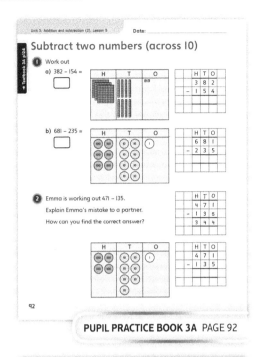

PUPIL PRACTICE BOOK 3A PAGE 92

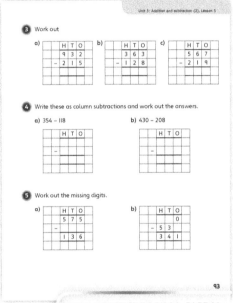

PUPIL PRACTICE BOOK 3A PAGE 93

Reflect

WAYS OF WORKING Pair work

IN FOCUS Here children explain in their own words how to complete a subtraction where an exchange is necessary.

ASSESSMENT CHECKPOINT Ensure children explain each step of the calculation including how they know that an exchange is necessary. They should use correct mathematical language when explaining their reasoning.

ANSWERS Answers for the **Reflect** part of the lesson can be found in the *Power Maths* online subscription.

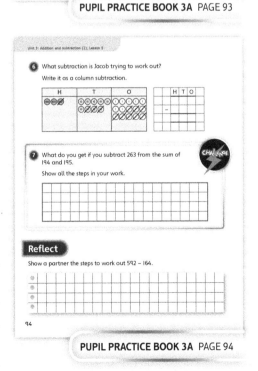

After the lesson

- Can children explain how to represent an exchange in written column subtraction?
- Are children able to use bonds within 20 to solve the subtractions in each place value position?
- Can children exchange across more than one column when necessary?

PUPIL PRACTICE BOOK 3A PAGE 94

Subtract two numbers (across 100)

Learning focus

In this lesson, children will further develop their fluency with column subtraction of 3-digit numbers, to include calculations where an exchange is necessary across one or two columns.

Before you teach

- Do children understand the difference between exchanges in an addition and in a subtraction?
- Can children recognise where a subtraction requires an exchange?
- Can children complete subtractions that cross a 10?

NATIONAL CURRICULUM LINKS

Year 3 Number – addition and subtraction

Add and subtract numbers with up to three digits, using formal written methods of columnar addition and subtraction.

Add and subtract numbers mentally, including: a 3-digit number and ones, a 3-digit number and tens, a 3-digit number and hundreds.

ASSESSING MASTERY

Children can represent column subtractions involving exchange across one or two columns and can explain when and why it is necessary.

COMMON MISCONCEPTIONS

There is a danger that children assume that they always subtract the smaller digit from the greater digit, which can lead to a misconception where the subtractions require an exchange. In this lesson, it is important that children understand they are subtracting from a whole. Ask:

- *Which number is the whole? Which number is being subtracted?*

STRENGTHENING UNDERSTANDING

Use a part-whole model to represent the subtractions so that children can clearly identify the whole from which the part is being subtracted. Use place value equipment to support the understanding of any exchanges

GOING DEEPER

There are numerous opportunities to deepen fluency through solving missing number and missing digit problems. Further depth may also be gained through challenging children to create their own subtraction problems that require a given number of exchanges.

KEY LANGUAGE

In lesson: digit, exchange, ones (1s), tens (10s), hundreds (100s), subtract

Other language to be used by the teacher: column subtraction

STRUCTURES AND REPRESENTATIONS

Place value grids, part-whole models, number lines, column subtractions

RESOURCES

Mandatory: place value equipment (such as base 10 equipment and place value counters)

Optional: number lines, part-whole models

 In the eTextbook of this lesson, you will find interactive links to a selection of teaching tools.

Quick recap

Give children base 10 equipment or place value counters. Ask children to work in pairs and show you 453 – 127 using the place value equipment.

Discover

Subtract two numbers (across 100)

Discover

Last year I took you to football practice on 184 days.

WAYS OF WORKING Pair work

ASK

- Question ① b): *Why will this calculation work out the number of days that Alex did not go to football practice?*
- Question ① b): *Do you need to make an exchange in the 1s? What about the 10s? How do you know?*

IN FOCUS The focus of question ① b) is for children to understand the steps needed to complete a subtraction where an exchange is needed from the 100s to the 10s.

PRACTICAL TIPS Provide children with place value equipment and a place value grid to represent the calculation and support them in understanding the steps needed to complete it.

ANSWERS

Question ① a): Children build 365 using place value counters.

Question ① b):

	H	T	O
	²3	¹6	5
−	1̶	8	4
	1	8	1

① a) There are 365 days in a year.
   Make the number 365 using place value counters.

b) Subtract 184 from 365 to work out the number of days that Alex did not go to football practice.

128

PUPIL TEXTBOOK 3A PAGE 128

Share

WAYS OF WORKING Whole class teacher led

ASK

- Question ① a): *Why do you think you need to make the number 365? What is significant about that number?*
- Question ① b): *Why do you not need to make both numbers?*
- Question ① b): *Why do you not need to make an exchange to the ones column?*
- Question ① b): *How do you know you need to make an exchange from the hundreds to the tens column? How do you do this?*

IN FOCUS In question ① b), ensure that children can represent, understand and explain every step that is needed when completing the column subtraction. Make links to previous learning where children completed subtractions that crossed a 10.

To help understanding, ask children to work through each step in pairs, and then together as a class. Each time, link what is happening with the place value counters to the abstract column calculations.

Share

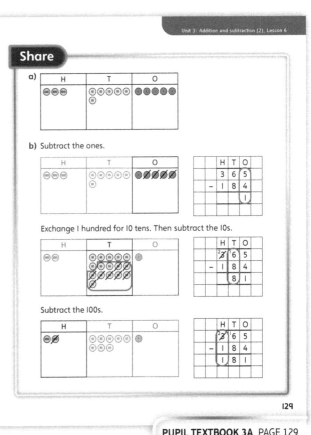

a)

b) Subtract the ones.

Exchange 1 hundred for 10 tens. Then subtract the 10s.

Subtract the 100s.

129

PUPIL TEXTBOOK 3A PAGE 129

Think together

Think together

WAYS OF WORKING Whole class teacher led (I do, We do, You do)

ASK

- Question **①**: *Do you need to make an exchange? How do you know?*
- Question **②**: *What is the same about the calculations? What is different? How can you see this in the answers?*

IN FOCUS The focal point here is on completing subtractions where an exchange is needed from the hundreds column to the tens column. In question **②**, children should recognise that the number of 1s in each answer is the same and explain why this happens.

STRENGTHEN Provide children with place value equipment to represent the calculations and support them with understanding the abstract column procedure.

DEEPEN Ask children to make connections between questions and answers, specifically when working through question **②**. In question **③**, children should be able to explain how to complete the calculations where they need to exchange across more than one column.

ASSESSMENT CHECKPOINT Use questions **①** and **②** to assess whether children can accurately complete subtractions that require exchanges.

ANSWERS

Question **①**:

H	T	O
⁴5̸ ¹1		9
− 1	4	5
3	7	4

Question **②**: The 100s digits and the 1s digits stay the same and there is always an exchange from the hundreds column to the tens column. The 10s digit varies each time.

Question **②** a):

H	T	O
⁷8̸ ¹2		7
− 1	4	4
6	8	3

c):

H	T	O
⁷8̸ ¹2		7
− 1	7	4
6	5	3

Question **②** b):

H	T	O
⁷8̸ ¹2		7
− 1	5	4
6	7	3

d):

H	T	O
⁷8̸ ¹2		7
− 1	8	4
6	4	3

Question **③** a):

H	T	O
³4̸ ¹¹2̸		¹3
− 1	4	7
2	7	6

Question **③** b): Dexter needs to make an exchange in the hundreds column first, this will give him 4 hundreds and 10 tens, he can then exchange 1 ten into 10 ones to make the first subtraction.

H	T	O
⁴5̸ ⁹0̸		¹6
− 3	2	8
1	7	8

Think together

① Work out 519 − 145.

H	T	O

	H	T	O
	5	1	9
−	1	4	5

② Work out the following.
Do you notice any patterns?
What stays the same? What is different?

a)

H	T	O
8	2	7
− 1	4	4

c)

H	T	O
8	2	7
− 1	7	4

b)

H	T	O
8	2	7
− 1	5	4

d)

H	T	O
8	2	7
− 1	8	4

130

PUPIL TEXTBOOK 3A PAGE 130

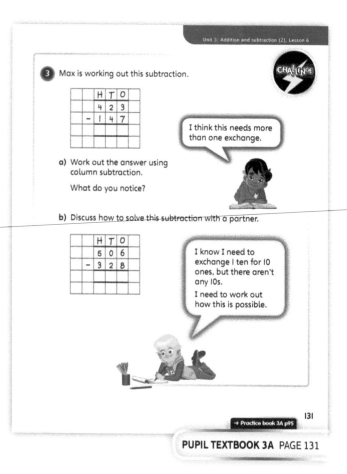

③ Max is working out this subtraction.

H	T	O
4	2	3
− 1	4	7

I think this needs more than one exchange.

a) Work out the answer using column subtraction.

What do you notice?

b) Discuss how to solve this subtraction with a partner.

H	T	O
5	0	6
− 3	2	8

I know I need to exchange 1 ten for 10 ones, but there aren't any 10s.

I need to work out how this is possible.

→ Practice book 3A p95

131

PUPIL TEXTBOOK 3A PAGE 131

Practice

WAYS OF WORKING Independent thinking

IN FOCUS The focus of this learning is for children to be able to complete subtractions using concrete and pictorial representations in questions ❶ and ❷, as well as just using the abstract procedure in question ❸. Children should use the concrete and pictorial representations to support their understanding of the abstract concept.

STRENGTHEN Provide children with place value equipment to help them make connections between this and the abstract representation and support them in understanding the steps they need to take.

DEEPEN Ask children to explain why a calculation does or does not require an exchange, how many are needed before completing it and how they know.

ASSESSMENT CHECKPOINT Use questions ❶ to ❸ to assess whether children can confidently and accurately complete subtractions that require an exchange across one column. Use question ❹ to assess whether they can then apply this in context. Use question ❺ to assess whether children can accurately complete subtractions that require an exchange across two columns.

ANSWERS Answers for the **Practice** part of the lesson can be found in the *Power Maths* online subscription.

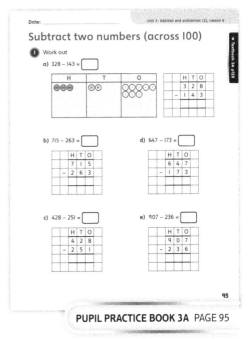

PUPIL PRACTICE BOOK 3A PAGE 95

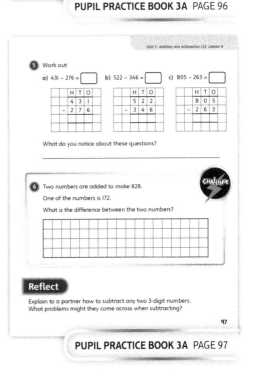

PUPIL PRACTICE BOOK 3A PAGE 96

Reflect

WAYS OF WORKING Pair work

IN FOCUS The focus of this activity is for children to explain in their own words how to subtract any pair of 3-digit numbers.

ASSESSMENT CHECKPOINT Ensure children use correct mathematical language when explaining their steps. They should be able to explain how to identify the need for an exchange, as well as the steps needed to complete this exchange.

ANSWERS Answers for the **Reflect** part of the lesson can be found in the *Power Maths* online subscription.

PUPIL PRACTICE BOOK 3A PAGE 97

After the lesson

- Can children complete subtractions that require an exchange across one column?
- Can children complete subtractions that require more than one exchange?

Add a 3-digit and a 2-digit number

Learning focus

In this lesson, children will develop written methods for addition, including exchange of 10s and 1s.

NATIONAL CURRICULUM LINKS

Year 3 Number – addition and subtraction

Add and subtract numbers with up to three digits, using formal written methods of columnar addition and subtraction.

Add and subtract numbers mentally, including: a 3-digit number and ones, a 3-digit number and tens, a 3-digit number and hundreds.

ASSESSING MASTERY

Children can add a 3-digit and a 2-digit number accurately using a written column method.

COMMON MISCONCEPTIONS

Children may find it confusing where an exchange has a 'knock-on' effect, such as in 128 + 73, where the exchange of 1s also causes an exchange of 10s. Ask:

- *Will this need an exchange of 1s, 10s, or both?*

Children set out the calculation incorrectly and align numbers from the left. Ask:

- *What is the value of each digit? How do we align the 1s, 10s etc.? How do we have to align our digits?*

STRENGTHENING UNDERSTANDING

Encourage children to use a range of place value equipment to explore the idea of exchanging 10 ones for 1 ten and 10 tens for 1 hundred. Allow children to experience the exchange by physically combining and regrouping the equipment.

GOING DEEPER

Challenge children to create different additions with the same total. Ask them to explore the relationship between calculations that create the same totals.

KEY LANGUAGE

In lesson: addition, 3-digit number, 2-digit number, altogether, total, exchange, ones (1s), tens (10s), hundreds (100s), digit, columns, calculation, 10 ones, 10 tens

Other language to be used by the teacher: column method, column additions

STRUCTURES AND REPRESENTATIONS

Place value equipment, column additions

RESOURCES

Mandatory: place value grids, base 10 equipment, place value counters, 0–9 digit cards

 In the eTextbook of this lesson, you will find interactive links to a selection of teaching tools.

Quick recap

Say two 3-digit numbers and ask children to add them together. Use a mix of questions from previous lessons, including ones where they cross a 10 or 100 or both.

Discover

Add a 3-digit and a 2-digit number

Discover

WAYS OF WORKING Pair work

ASK

- Question ①: *Can you represent the prices using place value equipment?*
- Question ①: *How many 1s are there in the parts? How many 10s?*
- Question ①: *What is the best way to represent the additions?*
- Question ①: *Which order should you add the parts?*

IN FOCUS The important aspect for children to recognise in both parts of question ① is that the calculations require exchange.

PRACTICAL TIPS Although the context uses money, it is best not to represent the values using coins or notes. Instead, use place value equipment to represent the different amounts shown in the shop.

ANSWERS

Question ① a): 275 + 16 = £291

Question ① b): 45 + 61 = £106

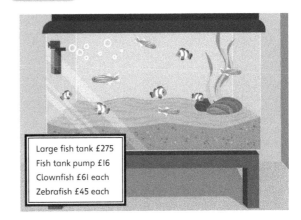

Large fish tank £275
Fish tank pump £16
Clownfish £61 each
Zebrafish £45 each

① a) Zoe buys a large fish tank and a pump.
How much does Zoe spend altogether?

b) Aaron buys a zebrafish and a clownfish.
How much does this cost in total?

132

PUPIL TEXTBOOK 3A PAGE 132

Share

WAYS OF WORKING Whole class teacher led

ASK

- Question ① a): *Can you see the exchange of 10 ones?*
- Question ① a): *How is this represented in the column method?*
- Question ① a): *Where is the exchanged 10 shown?*
- Question ① a): *Why do we add the 10s before we add the 100s?*

IN FOCUS These examples represent the first time that children have been asked to add numbers with a different number of digits. Discuss with children the correct way of aligning them (1s under 1s etc.).

By making the numbers in a place value table, this should help children understand why the digits are aligned as they are.

Work through each of the steps with the children, linking the exchanges and other parts of the calculation to the column method.

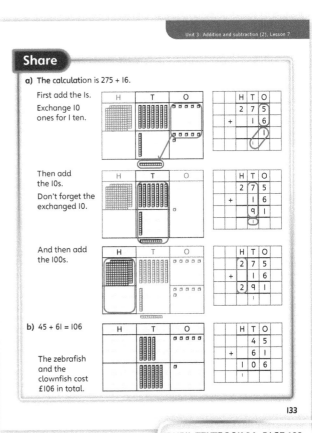

PUPIL TEXTBOOK 3A PAGE 133

Think together

Think together

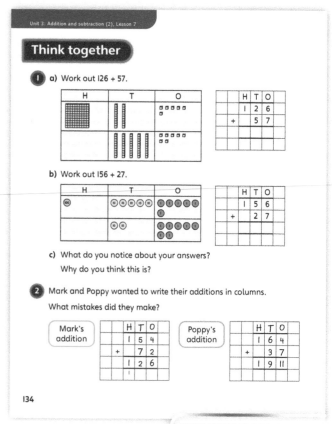

PUPIL TEXTBOOK 3A PAGE 134

WAYS OF WORKING Whole class teacher led (I do, We do, You do)

ASK

- Question ❶: *Can you spot if exchanges will be needed?*
- Question ❶: *How do you represent the exchange in the column method?*
- Question ❶: *Can you explain why the additions have the same answer?*
- Question ❷: *What mistakes should we look out for?*

IN FOCUS The focus of this lesson is for children to develop confidence and fluency in the column methods involving exchanges, while continuing to deepen their understanding of exchange and adding numbers that have different numbers of digits. In questions ❶ and ❷, it is vital that children are able to explain the process of the calculation method in terms of the exchanges needed and known number bonds.

STRENGTHEN Children may find it very useful to use place value equipment to model each calculation alongside writing the column method. They should complete the calculation in stages and use the place value equipment to model the process, without relying on a counting strategy. For example, if there are 3 ones and 9 ones, children should use their knowledge of number bonds to work out $3 + 9 = 12$, rather than counting the place value equipment.

DEEPEN Questions ❶ and ❸ contain prompts to consider different additions that produce the same answers. Challenge children to explore the effect of transposing digits.

ASSESSMENT CHECKPOINT If children can find and explain the mistakes in question ❷, then they have developed a good understanding of the column method.

ANSWERS

Question ❶ a): $126 + 57 = 183$

Question ❶ b): $156 + 27 = 183$

Question ❶ c): The number of 1s added is the same and the number of 10s added is the same.

Question ❷: Mark has not exchanged 10 tens for 1 hundred: $154 + 72$ is not 126 but 226. Poppy has shown 2 digits in the 1s column of the answer, rather than exchanged the 10 ones for 1 ten. $164 + 37$ is not 1,911 but instead 201.

Question ❸: There are 10 possible combinations.
$338 + 83 = 383 + 38 = 421$ because both involve $38 + 83$, with 3 as the hundreds digit.
$333 + 88$ and $388 + 33$ also $= 421$ as $33 + 88 = 38 + 83 = 121$, because addition can be done in any order.
$883 + 33 = 833 + 83 = 916$ because $83 + 33 = 33 + 83$
$838 + 33 = 871 = 833 + 38 = 871$ because $33 + 38 = 38 + 33$
The totals that are different involve adding 2 threes or 2 eights in either the ones or tens or both.
$338 + 38 = 376$, $383 + 83 = 466$

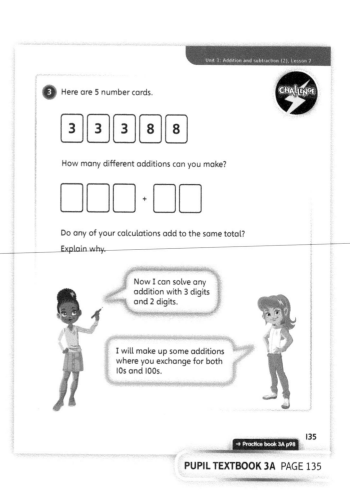

PUPIL TEXTBOOK 3A PAGE 135

172

Practice

WAYS OF WORKING Independent thinking

IN FOCUS Children develop their fluency and also make decisions about which parts need exchanging. They will need to understand how to use column addition accurately in order to solve calculations, but in question ⑥, they will also need to reason some answers based on their understanding of inverse operations for missing number and missing digit problems.

STRENGTHEN Support children to write their own column additions by using place value grids or column addition scaffolds.

DEEPEN Question ⑦ requires children to reason about additions which have to exchange both 10s and 1s. They will need to work logically to find the greatest and smallest totals that Dana could make.

THINK DIFFERENTLY Question ⑥ prompts children to reason to find the missing digits, rather than simply to rely on following the calculation method by rote.

ASSESSMENT CHECKPOINT Accurate completion of question ③ will demonstrate fluency in children's understanding of how to add a 3-digit number and a 2-digit number.

ANSWERS Answers for the **Practice** part of the lesson can be found in the *Power Maths* online subscription.

Reflect

WAYS OF WORKING Independent thinking

IN FOCUS The focus of this activity is to ensure children are able to identify, explain and correct common misconceptions that arise within column additions.

ASSESSMENT CHECKPOINT Children should be able to explain why the mistake has been made as well as overcome it. They should also be able to work out the correct answer.

ANSWERS Answers for the **Reflect** part of the lesson can be found in the *Power Maths* online subscription.

After the lesson

- Are children able to represent the additions using place value equipment?
- Can children explain how to represent an exchanged 10 or 100 in a written column method?
- Do children use their knowledge of bonds to complete the addition of digits?

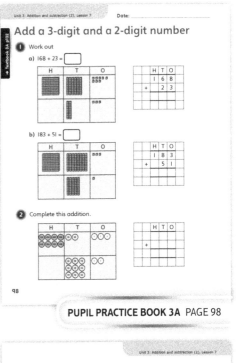

PUPIL PRACTICE BOOK 3A PAGE 98

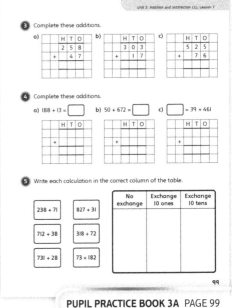

PUPIL PRACTICE BOOK 3A PAGE 99

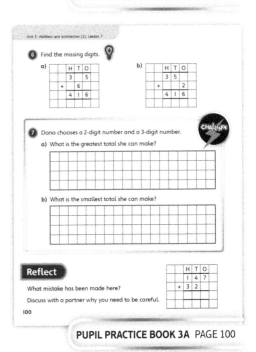

PUPIL PRACTICE BOOK 3A PAGE 100

Subtract a 2-digit number from a 3-digit number

Learning focus

In this lesson, children will subtract using column methods with exchange where necessary.

Before you teach

- What is the same and what is different about the methods used for addition and subtraction?

NATIONAL CURRICULUM LINKS

Year 3 Number – addition and subtraction

Add and subtract numbers with up to three digits, using formal written methods of columnar addition and subtraction.

Add and subtract numbers mentally, including: a 3-digit number and ones, a 3-digit number and tens, a 3-digit number and hundreds.

ASSESSING MASTERY

Children can use column subtraction to complete subtractions of a 2-digit number from a 3-digit number, where exchange is needed.

COMMON MISCONCEPTIONS

Representing the exchange in written column methods can be challenging for children, as they may struggle to understand why certain digits are crossed out and small digits are written next to other digits. Ask:
- *Why is this number crossed out? What does the '1' mean that has been added next to the 5 ones?*

Children may struggle where there is an exchange required but there is a 0 in the tens column. Ask:
- *Can you exchange 10s in this calculation: 305 – 58?*

STRENGTHENING UNDERSTANDING

Children will need plenty of experience of regrouping the place value equipment and discussion of how this exchange and regrouping is connected to the way column subtractions are written. Make sure there are plenty of opportunities for children to explore this method alongside the place value equipment.

GOING DEEPER

Encourage children to think about whether a written column method is the most efficient way of solving each subtraction. It may be that certain subtractions are better suited to a mental method, or one represented on a number line. Challenge children to scan the questions for where alternative methods may be more appropriate.

KEY LANGUAGE

In lesson: subtraction, 2-digit number, 3-digit number, ones (1s), tens (10s), hundreds (100s), order, exchange, columns, zero (0), addition

Other language to be used by the teacher: regroup, column method, column subtraction, partition

STRUCTURES AND REPRESENTATIONS

Number line, column subtractions

RESOURCES

Mandatory: base 10 equipment, place value grids, 0–9 digit cards

Optional: place value counters

 In the eTextbook of this lesson, you will find interactive links to a selection of teaching tools.

Quick recap

Say two 3-digit numbers and ask children to subtract the first number from the second number. Use examples where there are no exchanges and examples where there is more than one exchange.

Discover

Pair work

ASK

- Question 1 a): *How do you know what operation is required? What is it about the problem that suggests subtraction?*
- Question 1 a): *What do you notice about the digits in the two numbers?*
- Question 1 a): *Would you need to represent the 38 as well as the 175?*

IN FOCUS The important aspect of question 1 a) is that children recognise that an exchange is required, as the 8 ones needs to be subtracted, but 175 has 5 ones.

PRACTICAL TIPS Represent the number of trees using place value equipment.

ANSWERS

Question 1 a): Luis subtracted the ones in the wrong order.

Question 1 b): 175 – 38 = 137

Subtract a 2-digit number from a 3-digit number

Discover

1 a) Luis worked out how many new trees survived. What mistake did he make?

```
  H  T  O
  1  7  5
-    3  8
  1  4  3
```

Luis

7 tens – 3 tens = 4 tens
8 ones – 5 ones = 3 ones.
So 175 – 38 = 143.

b) What is the correct answer?

136

PUPIL TEXTBOOK 3A PAGE 136

Share

Whole class teacher led

ASK

- Question 1 a): *Can you explain Luis's mistake?*
- Question 1 a): *Can you see how the exchange is represented by the base 10 equipment?*
- Question 1 b): *What does this exchange look like in the written column method?*

IN FOCUS The most important aspect of teaching this question is to help children to understand how to represent the exchange in written columns. In question 1 b), they will need to understand why the digit 7 is crossed out and replaced by a 6, and what the 1 written next to the 5 represents. The danger is that children simply learn this as a process, rather than understanding why these changes are made.

Remind children, as in the previous lesson, that these numbers have a different number of digits. Therefore, it is important that the digits are aligned correctly first.

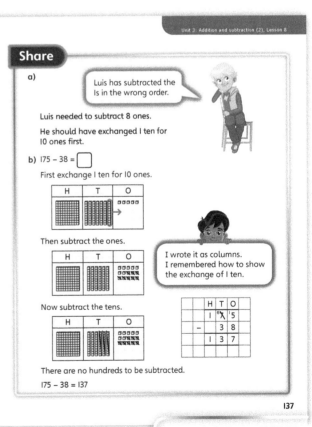

Share

a)
Luis has subtracted the 1s in the wrong order.

Luis needed to subtract 8 ones.

He should have exchanged 1 ten for 10 ones first.

b) 175 – 38 = ☐

First exchange 1 ten for 10 ones.

Then subtract the ones.

I wrote it as columns. I remembered how to show the exchange of 1 ten.

Now subtract the tens.

There are no hundreds to be subtracted.
175 – 38 = 137

137

PUPIL TEXTBOOK 3A PAGE 137

Think together

Unit 3: Addition and subtraction (2), Lesson 8

WAYS OF WORKING Whole class teacher led (I do, We do, You do)

ASK

- Question ❶: *What is the whole? What is being subtracted? How do you know?*
- Question ❶: *Do you need to make an exchange? How do you know?*
- Question ❷: *What calculation do you need to complete? How do you know?*

IN FOCUS The focus of question ❷ is for children to be able to decide which subtractions they need to do from the information they are given. They then need to complete the subtractions.

STRENGTHEN Provide children with place value equipment to support them in making connections between concrete and abstract representations. Encourage them to draw diagrams such as part-whole models to support them in deciding what calculations they need to complete.

DEEPEN Children could be stretched to consider what the question was if, for example, in question ❷ the calculation was 85 – 43. When working through the challenge question, children should explain their thinking and reasoning at each point.

ASSESSMENT CHECKPOINT Use question ❶ to assess whether children can confidently and accurately subtract a 2-digit number from a 3-digit number. Use question ❷ to assess whether children can interpret questions and decipher what is being asked in order to form and answer the correct calculation.

ANSWERS

Question ❶ a): 183

Question ❶ b): 308

Question ❶ c): 239

Question ❷ a): 271 – 43 = 228

Question ❷ b): 271 – 85 = 186

Question ❸ a): First, exchange 1 hundred for 10 tens, then one of those tens for 10 ones:
$39^{1}5 - 7 = 398$
$39^{1}5 - 17 = 388$
$39^{1}5 - 217 = 188$

Question ❸ b): 42

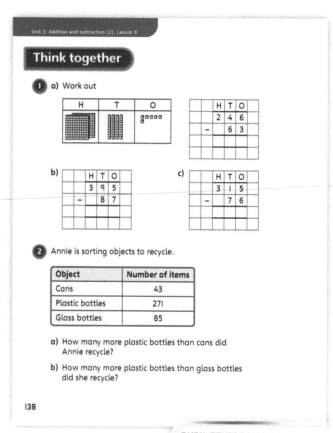

Think together

❶ a) Work out

② Annie is sorting objects to recycle.

Object	Number of items
Cans	43
Plastic bottles	271
Glass bottles	85

a) How many more plastic bottles than cans did Annie recycle?

b) How many more plastic bottles than glass bottles did she recycle?

138

PUPIL TEXTBOOK 3A PAGE 138

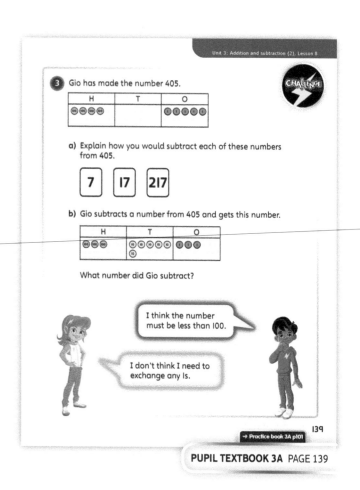

③ Gio has made the number 405.

a) Explain how you would subtract each of these numbers from 405.

7   17   217

b) Gio subtracts a number from 405 and gets this number.

What number did Gio subtract?

I think the number must be less than 100.

I don't think I need to exchange any 1s.

→ Practice book 3A p101

139

PUPIL TEXTBOOK 3A PAGE 139

Practice

WAYS OF WORKING Independent thinking

IN FOCUS The key point of questions ❶ and ❷ is to allow children the chance to build confidence and fluency alongside an understanding of place value. In question ❸, children are required to correctly align the digits. Look to ensure that children line up the 1s under the 1s and 10s under 10s etc. Ask children to look back at early examples for help.

STRENGTHEN Build confidence by working on subtractions that only require exchange of one column at a time, until children are confident with this aspect. Begin with examples where the 1s require an exchange of 1 ten, and then move to where the 10s require an exchange of 1 hundred.

DEEPEN The challenge in question ❻ requires significant logical deduction. Children will need to reason from the fact that the 100s digit changes, and so the subtraction of the 10s digit *does* require an exchange. This is a very logical puzzle and children may have to explore it using some trial and improve approaches first.

ASSESSMENT CHECKPOINT Question ❸ requires a good level of understanding of column subtraction and of how to exchange in one or two columns.

ANSWERS Answers for the **Practice** part of the lesson can be found in the *Power Maths* online subscription.

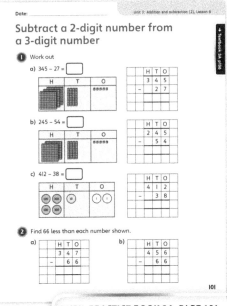

PUPIL PRACTICE BOOK 3A PAGE 101

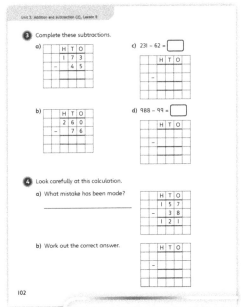

PUPIL PRACTICE BOOK 3A PAGE 102

Reflect

WAYS OF WORKING Pair work

IN FOCUS In pairs, children should look back over the previous lessons in this unit and discuss what they have learnt.

ASSESSMENT CHECKPOINT Have children retained the learning from the previous lessons in this unit?

ANSWERS Answers for the **Reflect** part of the lesson can be found in the *Power Maths* online subscription.

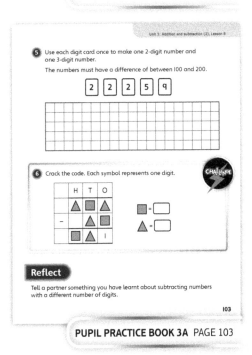

PUPIL PRACTICE BOOK 3A PAGE 103

After the lesson ⏸

- Can children represent the exchange in subtractions?
- Can children understand how to write the exchange using written column methods?

Complements to 100

Learning focus

In this lesson, children make number bonds to 100 using a 100 square.

Before you teach

- Are children secure with number bonds to 10?
- How will you ensure children do not make both 10s and 1s total 10?

NATIONAL CURRICULUM LINKS

Year 3 Number – addition and subtraction

Add and subtract numbers with up to three digits, using formal written methods of columnar addition and subtraction.

ASSESSING MASTERY

Children can make links to other known facts, such as number bonds to 10, and work out unknown quantities using these facts rather than simply calculating the unknown value. Children can make number bonds to 100 using a 100 square.

COMMON MISCONCEPTIONS

Children may miscount how many squares there are until 100 when using the 100 square to calculate. Ask:
- *Can you use your knowledge of number bonds to 10 to check the number of 1s that you have added?*

Children may confuse the game board used in the lesson with a 100 square. Ask:
- *What is the same and what is different between this board and a 100 square?*

STRENGTHENING UNDERSTANDING

Give children a list of number bonds to 10 and multiples of 10 that make 100. This will encourage them to use these known facts rather than simply counting the number of squares to 100 each time. You could ask children to identify which facts helped them calculate the answer to each question.

A bead string (with tens in alternating colours) may help children see how many 1s are needed to make the next multiple of 10 and subsequently how many 10s there are until 100.

GOING DEEPER

Ask: *Can you complete the fact family for each calculation? Which known facts have you used for each calculation? What other calculations in the fact family can you find?*

KEY LANGUAGE

In lesson: number bond, addition, number sentence, tens (10s), ones (1s), how many more

Other language to be used by the teacher: 100 square, multiples of 10, count on, add

STRUCTURES AND REPRESENTATIONS

Part-whole model, 100 square

RESOURCES

Mandatory: 100 squares, counters

Optional: Base 10 equipment, bead strings, blank part-whole models, blank addition scaffolds

 In the eTextbook of this lesson, you will find interactive links to a selection of teaching tools.

Quick recap

Write a multiple of 10 on the board. Ask: *What do you need to add to this number to make 100? How quickly can you find the answer?*

Discover

WAYS OF WORKING Pair work

ASK

- Questions a) and b): *What numbers are the counters on?*
- Questions a) and b): *How many more squares before the next multiple of 10?*
- Questions a) and b): *How many 10s until 100?*

IN FOCUS For each part of question , encourage children to use known facts to calculate how many more squares there are to 100. They should first consider what square each counter is on, then how many 1s are needed to make the next multiple of 10, and then finally how many 10s there are to 100.

PRACTICAL TIPS Children could use a bead string or base 10 equipment to model the number bonds to 100.

ANSWERS

Question a): 87 + 13 = 100

Zac has to move 13 squares.

Question b): 51 + 49 = 100

Emma has to move 49 squares.

Share

WAYS OF WORKING Whole class teacher led

ASK

- Questions a) and b): *Who counted in 1s to get to 100? Was this an efficient strategy to use?*
- Questions a) and b): *Who calculated the answer in a more efficient way?*
- Questions a) and b): *Can you write your own number sentence to make 100?*
- Question a): *How do you know 87 + 12 is not correct?*

IN FOCUS Use question b) to highlight that counting in 1s is not an efficient strategy, even though doing so can lead to the right answer – it is a slow method and mistakes may result from miscounting. Agree that the 'make 10' strategy – using number bonds to 10 to reach the next multiple of 10, and then counting the number of rows of 10 to 100 – is more efficient.

STRENGTHEN Strengthen understanding by using base 10 equipment to make the representations in this part of the lesson. This will help children see how many more are needed to make 100.

Complements to 100

Discover

 a) How many squares does Zac have to move to get to 100?

Complete the number sentence and the part-whole model.

87 + ☐ = 100

b) How many squares does Emma have to move to get to 100?

Complete the number sentence and the part-whole model.

51 + ☐ = 100

140

PUPIL TEXTBOOK 3A PAGE 140

PUPIL TEXTBOOK 3A PAGE 141

Think together

WAYS OF WORKING Whole class teacher led (I do, We do, You do)

ASK

- Question **1**: *What is the whole? What are the parts? What calculation do you need to do? How do you know?*
- Question **1**: *How does the 100 square represent the question?*
- Question **2**: *What calculations do you need to do to work out the missing parts? How do you know?*

IN FOCUS The focus of questions **1** and **2** is on children being able to form calculations in order to work out complements to 100. They should be able to explain why they are completing the calculation and what method they are using and why.

STRENGTHEN Provide children with blank 100 squares to colour in to support them in working out the complements to 100.

DEEPEN Encourage children to consider common mistakes that could be made and how these could be overcome. For example, when using part-whole models and 100 squares, the 'whole' in the part-whole model should always be 100. If they have entered a different number in there, they know they have made a mistake.

ASSESSMENT CHECKPOINT Use questions **1** and **2** to assess whether children can confidently and accurately find the complement to 100 for a given number.

ANSWERS

Question **1**:

Question **2** a):

100	
72	28

Question **2** b):

Question **3** a): The missing number is 38.

Question **3** b): 37 + 63 = 100; 43 + 57 = 100; 22 + 78 = 100

Question **3** c): 44 + 66 = 110. Both digits have been added to make 10 rather than adding 6 ones to 44 and then adding 5 more tens to 50 to make 100.

Think together

1 Complete the part-whole model.

2 Find the missing numbers.

a)

b)

I used a 100 square to help me.

142

PUPIL TEXTBOOK 3A PAGE 142

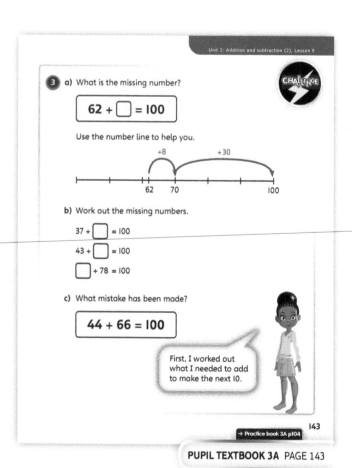

3 a) What is the missing number?

$$62 + \boxed{} = 100$$

Use the number line to help you.

+8   +30

62   70   100

b) Work out the missing numbers.

$37 + \boxed{} = 100$

$43 + \boxed{} = 100$

$\boxed{} + 78 = 100$

c) What mistake has been made?

$$44 + 66 = 100$$

First, I worked out what I needed to add to make the next 10.

143

→ Practice book 3A p104

PUPIL TEXTBOOK 3A PAGE 143

Practice

WAYS OF WORKING Independent thinking

IN FOCUS Question ❶ asks children to work out complements to 100, based on a concrete board game. Children move on to pictorial representations in questions ❷ and ❸, before moving towards abstract representations in question ❹.

Question ❻ will challenge children as they are not given the starting number, meaning they cannot simply count on to 100. They will need to use their understanding of place value, recognising, for example, that in part a) the 3 represents 3 tens and the 4 represents 4 ones.

STRENGTHEN Children should use a bead string or base 10 equipment to strengthen their understanding of the strategies used to make 100. They could work with a partner and comment on what they are doing and which facts they are using as they work through the calculation.

DEEPEN Challenge children to complete the calculation ☐ + ☐ = 100 (where the two boxes have different values). Encourage children to work systematically to begin with and then predict how many different possible answers they think there are. Can they justify their prediction?

THINK DIFFERENTLY In question ❺, check children understand that both circles in the number sentence represent the same number; they cannot find any two numbers that add to 100.

ASSESSMENT CHECKPOINT Children should be able to explain the steps that they have completed mentally and which known number facts they have used.

ANSWERS Answers for the **Practice** part of the lesson can be found in the *Power Maths* online subscription.

Reflect

WAYS OF WORKING Independent thinking

IN FOCUS The **Reflect** part of the lesson requires children to combine what they have learnt in the lesson with known facts in order to find different ways to complete the calculation. Challenge children to work systematically and find all the possibilities.

ASSESSMENT CHECKPOINT Assess whether children are working systematically. Do they understand the concept of both parts having 0 ones? Ask: *What do the 2 zeros represent? How many 1s are in both numbers?*

ANSWERS Answers for the **Reflect** part of the lesson can be found in the *Power Maths* online subscription.

After the lesson ⏸

- Are children confident at counting in 10s from any number?
- How many children were able to use their knowledge of number bonds to 10 to count to the nearest multiple of 10?

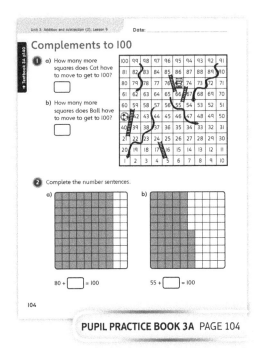

PUPIL PRACTICE BOOK 3A PAGE 104

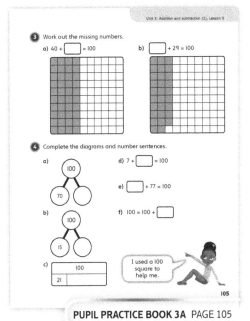

PUPIL PRACTICE BOOK 3A PAGE 105

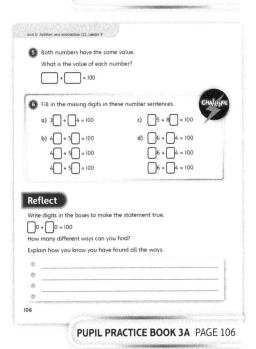

PUPIL PRACTICE BOOK 3A PAGE 106

Estimate answers

Learning focus

In this lesson, children will develop skills of estimation and approximation to allow them to make simple checks of additions and subtractions. They have not yet learnt rounding, so the approach builds on number sense and approximate position on a number line.

Before you teach

- Can children confidently suggest numbers near 100? How about far from 100?
- Do they understand the meaning of 'about' and 'roughly' in the context of estimating?

NATIONAL CURRICULUM LINKS

Year 3 Number – addition and subtraction

Estimate the answer to a calculation and use inverse operations to check answers.

ASSESSING MASTERY

Children can use a rough approximation to estimate answers to calculations by adding 100s mentally.

COMMON MISCONCEPTIONS

There may be some confusion over the difference in meaning between estimation (making a justified or educated 'guess' based on the information available) and approximation (choosing a rough value to stand in where a precise value is not known). Ask:

- *Is 101 + 201 approximately 200 or 300?*
- *Estimate how many children are in our classroom today.*

STRENGTHENING UNDERSTANDING

Many children find it confusing to estimate or approximate, thinking that a precise answer is always necessary or preferable. They may find it difficult to accept that a rough answer can sometimes be more useful or appropriate.

Use a washing line with 100s pegged at regular intervals. Give children the opportunity to discuss which numbers would be placed near to the 100s and to explore the relative locations of numbers to either side of a 100 mark.

GOING DEEPER

There are some contexts where either an over-estimate or an under-estimate is always preferable. For example, if you need at least 22 water bottles for a school trip, then having 30 is better than having 20, even though 20 is the closer approximation. Discuss this with children and ask them if they can see where this consideration might come into some of the problems in the lesson.

KEY LANGUAGE

In lesson: estimate, estimation, approximate, approximation, approx., **approximately**, about

Other language to be used by the teacher: roughly

STRUCTURES AND REPRESENTATIONS

Number line, column method

RESOURCES

Mandatory: number lines, number or digit cards

Optional: washing line, pegs, 100s numbers to 1,000, paper clips, cubes, concrete objects in the classroom such as counters

 In the eTextbook of this lesson, you will find interactive links to a selection of teaching tools.

Quick recap

Take number cards from 1 to 100 (or use digit cards). Turn over a card and ask children to quickly work out the bond to 100. How many can they do in 2 minutes?

Discover

Estimate answers

WAYS OF WORKING Pair work

ASK

- Question ➊: *What does approx. mean?*
- Question ➊: *How many matchsticks could be in each bag? Do you know the exact amount? Can you work it out? Does it matter?*

IN FOCUS Here children consider the idea of approximate values for calculations. They should recognise that, whilst they do not know the exact number of matchsticks in any of the bags, they can still answer questions ➊ a) and ➊ b) using the information they are given.

PRACTICAL TIPS Model a similar situation using smaller numbers of concrete objects in the classroom. For example, tell children you have approximately 10 counters in your hand. What are their guesses as to how many you actually have? How can they apply this logic when working with bigger numbers?

ANSWERS

Question ➊ a): 211 is approximately 200.
Ebo counted the bag that has approximately 200 matchsticks.

Question ➊ b): 595 is approximately 600. 600 − 200 = 400.
Ebo should use the bag with approximately 400 matchsticks.

Discover

➊ a) Ebo counts all of the matchsticks from one bag.
There are exactly 211. Which bag did he count from?

b) Ebo needs to use 2 bags to make his model.
Which other bag should he use?

144

PUPIL TEXTBOOK 3A PAGE 144

Share

WAYS OF WORKING Whole class teacher led

ASK

- Question ➊ a): *How does the number line help work out the answer?*
- Question ➊ b): *Approximately how many more matchsticks does Ebo need? Which bag is the best for this?*
- Question ➊ b): *Does Ebo definitely have enough matchsticks? How do you know?*

IN FOCUS Here children focus on the idea of using approximate values within a calculation. In question ➊ a), they should use the number line to support them in identifying how many matchsticks could be in each bag and help them in choosing the correct one they need.

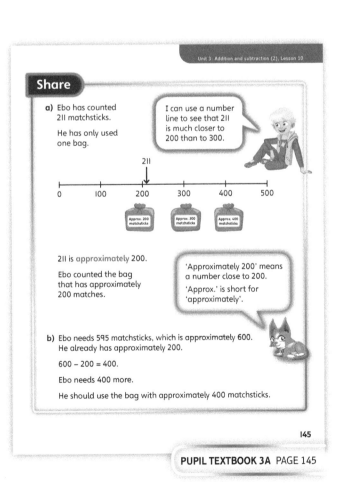

145

PUPIL TEXTBOOK 3A PAGE 145

Think together

WAYS OF WORKING Whole class teacher led (I do, We do, You do)

ASK

- Question **1**: *Why does each character think this? Whose approximation is more accurate? Why?*
- Question **2**: *How can approximations be useful for checking answers?*

IN FOCUS Here children look further at approximate calculations and consider how they can be useful as a method for checking calculations. They should be encouraged to discuss in detail the two approximations in question **1** and the rationale behind each, and also recognise and be able to explain why Dexter's approximation will be more accurate.

STRENGTHEN Provide children with a number line counting in 100s that they can use to support them in finding estimations and approximations.

DEEPEN In question **3**, children should be able to explain the steps and reasoning behind their thinking and how they came to their answer. They could be stretched to consider what the calculations could be for other approximations.

ASSESSMENT CHECKPOINT Use question **1** to assess whether children understand which approximation is more accurate and why. Use question **2** to assess whether children understand how to use approximations as a method for checking answers.

ANSWERS

Question **1**: 381 + 398 is approximately 400 + 400 = 800. Dexter is correct.

Question **2**: 512 − 280 is approximately 500 − 300 = 200. Alex should try the calculation again.

Question **3**: 105 g is approximately 100 g, 407 g is approximately 400 g, 196 g is approximately 200 g and 612 g is approximately 600 g. 400 + 200 = 600 and 400 − 200 = 200. The two bags of flour are 407 g and 196 g.

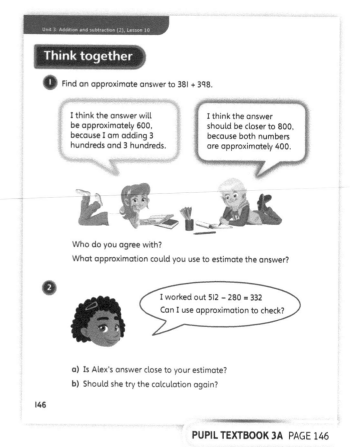

PUPIL TEXTBOOK 3A PAGE 146

PUPIL TEXTBOOK 3A PAGE 147

Practice

WAYS OF WORKING Independent thinking

IN FOCUS Questions ❶ and ❷ prompt children to use their skills of estimation and approximation to find approximate answers to additions and subtractions. Question ❸ requires children to use the estimations as a checking strategy.

STRENGTHEN Ask children to represent the numbers on 0–1,000 number lines and use the representations to support their approximations.

DEEPEN Question ❺ starts to challenge children to think about under- and over-estimates. Although it seems that Andy's idea (that the answer might be a bit more or a bit less than 500) is reasonable, given that the numbers are less than 200 and 300, the answer has to be less than 500.

THINK DIFFERENTLY Question ❸ prompts children to find the likely mistake by using estimation as a checking strategy.

ASSESSMENT CHECKPOINT Answers to questions ❸ and ❹ will demonstrate if children have understood how the concept of estimation can be used in relation to calculations.

ANSWERS Answers for the **Practice** part of the lesson can be found in the *Power Maths* online subscription.

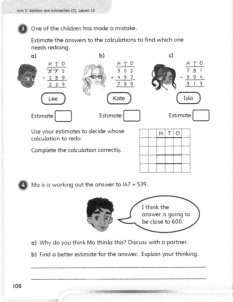

PUPIL PRACTICE BOOK 3A PAGE 107

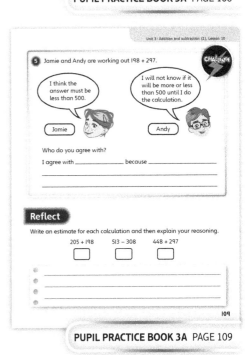

PUPIL PRACTICE BOOK 3A PAGE 108

Reflect

WAYS OF WORKING Pair work

IN FOCUS Children should discuss and agree reasonable estimates for each calculation and then discuss mental approaches to the calculations.

ASSESSMENT CHECKPOINT Can children make estimates without immediately resorting to the process of column methods?

ANSWERS Answers for the **Reflect** part of the lesson can be found in the *Power Maths* online subscription.

After the lesson

- Can children justify their approximations and estimates?
- Are children able to estimate answers to calculations?
- Can children use estimates to decide if a reasonable answer has been given for a calculation?

PUPIL PRACTICE BOOK 3A PAGE 109

Inverse operations

Learning focus

In this lesson, children will learn to use inverse operations and fact families as checking strategies. They will also use them to help make appropriate calculations more efficient as mental strategies.

NATIONAL CURRICULUM LINKS

Year 3 Number – addition and subtraction

Estimate the answer to a calculation and use inverse operations to check answers.

ASSESSING MASTERY

Children can use the part-whole concept to find an inverse operation to check an addition or a subtraction. Also, they are able to decide when it is more efficient to use mental strategies over written methods.

COMMON MISCONCEPTIONS

Once children learn written methods for calculations, they may revert to the algorithm rather than first thinking about the most efficient or accurate way to solve a calculation. The use of fact families in this lesson prompts children to make decisions about when a mental method may be more appropriate than the written algorithm. Ask:

- *What fact family could we use to work out $198 - 10 = \boxed{}$? What if we try $198 - \boxed{} = 10$; or $10 + \boxed{} = 198$; or $198 = \boxed{} + 10$? Can we answer this mentally now, with the aid of the fact family?*

STRENGTHENING UNDERSTANDING

The part-whole model is used throughout the lesson, to help children visualise the families of facts. Ask: *Can you think of the way to solve this that is least likely to produce a mistake?*

GOING DEEPER

Challenge children to make the link with the learning in the previous lesson about estimates and approximations. Children could employ and adjust approximations in calculations such as $399 + \boxed{} = 601$, by working out $400 + \boxed{} = 600$ mentally, and then adjusting to find the precise answer.

KEY LANGUAGE

In lesson: subtraction, part-whole, approximately, fact family

Other language to be used by the teacher: addition, approximate, estimate, inverse operation

STRUCTURES AND REPRESENTATIONS

Part-whole model, number lines

RESOURCES

Optional: part-whole models, number lines, scrap paper

 In the eTextbook of this lesson, you will find interactive links to a selection of teaching tools.

Quick recap

Draw a part-whole model on the board, for example, 12 as the whole and 7 and 5 as the parts. Ask: *Can you write down the fact family? Can you write four facts? Eight facts?*

Discover

Inverse operations

Discover

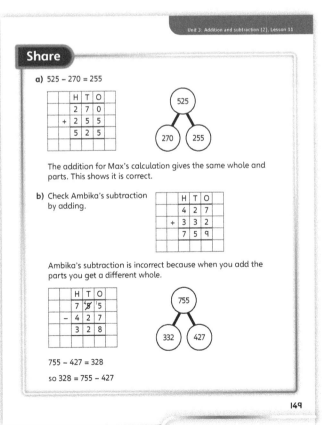

WAYS OF WORKING Pair work

ASK

- Questions ① a) and b): *What is the whole and what are the parts represented in each calculation?*
- Questions ① a) and b): *Can you think of related additions based on these parts and wholes?*
- Questions ① a) and b): *How can you decide if a calculation is correct?*

IN FOCUS For questions ① a) and b), children should discuss whether the calculations seem to have reasonable answers. They may employ a strategy from the previous lesson, which used estimation and approximation. They should also identify the parts and the whole from each calculation and then represent them in a part-whole model. They can then use that structure to find the additions to check the calculations.

PRACTICAL TIPS Children should use part-whole models to replicate the calculations, perhaps by writing the numbers on scraps of paper so that they do not appear permanent as this is a lesson about checking strategies where calculations may or may not be correct.

ANSWERS

Question ① a):

$270 + 255 = 525$

b)

$427 + 332 = 759$
The correct calculation is
$755 - 427 = 328$

① a) Use an addition to check Max's subtraction.

   b) Use an addition to check Ambika's subtraction.

148

PUPIL TEXTBOOK 3A PAGE 148

Share

WAYS OF WORKING Whole class teacher led

ASK

- Question ① a): *Can you spot how the addition matches the parts of the original subtraction?*
- Questions ① a) and b): *Which calculation should be checked carefully?*
- Question ① b): *Can you see what the likely mistake was?*
- Question ① b): *Can you draw a new part-whole model to check that 755 – 427 = 328 is correct?*

IN FOCUS Questions ① a) and b) show how to use the part-whole model to identify an addition that can be used to check the subtraction. Children should discuss how the addition in question ① b) produces a different whole from the parts in the original answer. This means that it should be checked again. Where the parts add to produce the same whole as in the original calculation, this should give confidence that the original was correct. Children should note that this is not conclusive proof of being correct or incorrect but it is a helpful way of spotting likely errors.

Share

a) $525 - 270 = 255$

	H	T	O
	2	7	0
+	2	5	5
	5	2	5

The addition for Max's calculation gives the same whole and parts. This shows it is correct.

b) Check Ambika's subtraction by adding.

	H	T	O
	4	2	7
+	3	3	2
	7	5	9

Ambika's subtraction is incorrect because when you add the parts you get a different whole.

	H	T	O
	7	8̷⁴	5
–	4	2	7
	3	2	8

$755 - 427 = 328$
so $328 = 755 - 427$

149

PUPIL TEXTBOOK 3A PAGE 149

Think together

WAYS OF WORKING Whole class teacher led (I do, We do, You do)

ASK

- Question **1**: *Why should an inverse operation be used to check the answer rather than just completing the calculation again?*
- Question **2**: *Why has Emma made these mistakes? What could she do in future to avoid them?*

IN FOCUS The focus here is on children's understanding of the benefit of using inverse operations as a method for checking calculations. They should recognise that, if they just did the calculations again, they may make the same mistakes. Children should be able to identify and explain any mistakes that have been made.

STRENGTHEN Encourage children to draw representations such as part-whole models to support them in identifying what the inverse calculation is.

DEEPEN Encourage children to find multiple ways of checking answers, drawing on all of their learning from this unit. Ask: *Which ways are the most efficient? Which ways are the most accurate?* Encourage children to discuss their reasoning with a partner.

ASSESSMENT CHECKPOINT Use questions **1** and **2** to assess whether children can confidently and accurately use inverse operations as a method for checking calculations.

ANSWERS

Question **1** a): Subtraction 612 − 371 = 341 needs checking again. 612 − 371 = 241 is correct.

Question **1** b): 334 + 477 = 811, so the addition is incorrect.

Question **2** a): Mistake 1: Emma has exchanged incorrectly. Mistake 2: 0 − 9 is not 9.

Question **2** b): 501 = 499 + 2, so 501 − 499 = 2

Question **3**: Children should discuss using the inverse (i.e. subtraction) to check. They could also approximate each number and use estimation to check.

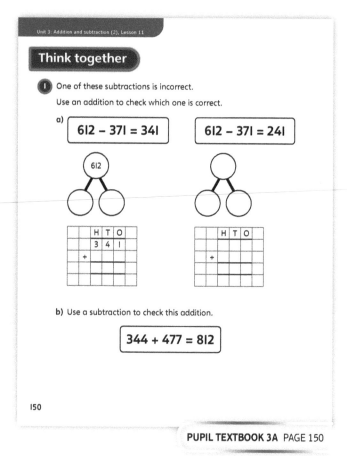

Unit 3: Addition and subtraction (2), Lesson 11

Think together

1 One of these subtractions is incorrect.

Use an addition to check which one is correct.

a)

612 − 371 = 341   612 − 371 = 241

b) Use a subtraction to check this addition.

344 + 477 = 812

150

PUPIL TEXTBOOK 3A PAGE 150

Unit 3: Addition and subtraction (2), Lesson 11

2 Emma has worked out 501 − 499 = 91.

That doesn't look right. 499 is approximately 500, so the answer should be close to 0.

a) What two mistakes has she made?

b) How could you work this out in your head?

CHALLENGE

3 Reena is working out this addition.

176 + 741

She gets the answer 917.

Discuss with a partner different ways you can check Reena's answer. Find as many ways as you can.

I will look at what each number is close to and approximate.

I will do a subtraction.

→ Practice book 3A p110

151

PUPIL TEXTBOOK 3A PAGE 151

Practice

WAYS OF WORKING Independent thinking

IN FOCUS Questions ❶, ❸ and ❹ encourage children to use inverse operations and make decisions about which calculations are likely to be correct. Question ❷ prompts children to use fact families to make decisions about how to use mental calculation methods to solve calculations efficiently.

STRENGTHEN The part-whole model is the ideal way to represent the structure needed and to inform which operation to use as a checking strategy. Build children's confidence first with some simpler calculations within 100.

DEEPEN In question ❺, children should explain their rationale behind where the answer will fall on the number line. They should be encouraged to find different methods for checking answers and discuss the advantages and disadvantages of each, including efficiency and accuracy.

ASSESSMENT CHECKPOINT Answers to questions ❸ and ❹ will demonstrate a good level of understanding of the main objective for the lesson. It encourages children to recognise the parts and whole in a calculation and then successfully employ an inverse operation to check their answer.

ANSWERS Answers for the **Practice** part of the lesson can be found in the *Power Maths* online subscription.

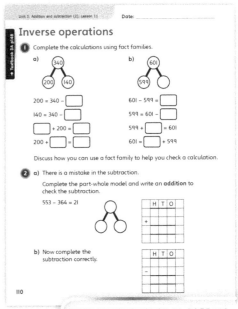

PUPIL PRACTICE BOOK 3A PAGE 110

PUPIL PRACTICE BOOK 3A PAGE 111

Reflect

WAYS OF WORKING Pair work

IN FOCUS Children should consider the methods from the previous lesson and the current lesson. The prompt is open-ended, allowing children plenty of freedom to demonstrate the depth of their understanding of the different checking strategies.

ASSESSMENT CHECKPOINT Can children justify their reasoning using the language of parts and wholes?

ANSWERS Answers for the **Reflect** part of the lesson can be found in the *Power Maths* online subscription.

After the lesson

- Do children understand how to use fact families to find related calculations?
- Can children decide whether a checking calculation confirms or casts doubt on a calculation?
- Can children use fact families, where appropriate, to inform mental calculation strategies?

PUPIL PRACTICE BOOK 3A PAGE 112

Problem solving

Learning focus

In this lesson, children will use single bar model to represent word problems that require addition or subtraction.

Before you teach

- Are children able to represent a simple addition as a bar model?
- Are they able to represent a simple subtraction as a bar model?
- Can they confidently identify whether it is the whole or a part that is missing in a word problem?

NATIONAL CURRICULUM LINKS

Year 3 Number – addition and subtraction

Solve problems, including missing number problems, using number facts, place value, and more complex addition and subtraction.

ASSESSING MASTERY

Children can use a bar model to represent the context of a word problem and to identify the operation required based on whether the missing information is the whole or a part.

COMMON MISCONCEPTIONS

Children can struggle to identify the missing information in word problems and so find it difficult to know which operation to use. The bar model can represent the structure of a problem to support decisions about which calculation to use. However, the model itself does not yield the numerical answer and so children will need to know that they are required to complete the calculation using a method they have learnt previously. Try posing a word problem, such as: *Farmer Beth has 310 animals on her farm. 200 are sheep and the rest are cows. How many cows does she have?* Then ask:

- *What information are you told in this word problem? What information is missing?*
- *If we model this in a bar model, is it the whole or a part that we are missing? Do we need to add or subtract in order to find the missing information?*

STRENGTHENING UNDERSTANDING

Children can find it difficult to draw the bars and the braces of a bar model to their own satisfaction. Using strips of paper, which children can fold, tear or cut to an appropriate length, can support children in creating their own bar models. The process of creating the bar models can help to cement the idea of how the model works.

GOING DEEPER

More than one bar model can be used to represent a problem successfully. Ask children to identify which problems could be represented in different ways and make decisions about which bar models they think best represent a given context.

KEY LANGUAGE

In lesson: subtraction, approximation, bar model

Other language to be used by the teacher: operation, approximate, raise, raised

STRUCTURES AND REPRESENTATIONS

Bar model

RESOURCES

Optional: strips of paper

 In the eTextbook of this lesson, you will find interactive links to a selection of teaching tools.

Quick recap

Say a number from 1 to 100. Ask children to quickly work out the bond to 100. How many can they do in 2 minutes?

Discover

WAYS OF WORKING Pair work

ASK

- Question ① a): *Can you identify the prices you need to use?*
- Question ① a): *How can you prove that an addition is the appropriate operation to use?*
- Question ① a): *What sort of diagram could you use?*
- Question ① b): *What does 'approximation' or 'approximate' mean?*

IN FOCUS In question ① a), it is important that children recognise what information they need to find in the picture in order to find the solution. Children may recognise that the operation required is addition, but some children may not be able to justify why this requires addition rather than subtraction.

PRACTICAL TIPS Spend some time discussing the items in the shop context. Children could create price labels for the items identified in each problem. When representing the bars, children could cut strips of paper to represent the different parts of the bar model.

ANSWERS

Question ① a): 275 + 99 = 374. She spent £374 in total.

Question ① b): 275 + 100 = 375. £375 is very close to £374.

Problem solving (1)

Discover

① a) Holly bought a racing bike and paid to have a service. How much did she spend in total?

   b) Write an approximation to check the calculation.

152

PUPIL TEXTBOOK 3A PAGE 152

Share

WAYS OF WORKING Whole class teacher led

ASK

- Question ① a): *What does the '?' represent on the bar model?*
- Question ① a): *How does the bar model show we need to use the operation of addition?*

IN FOCUS In question ① a), the bar model very clearly represents the concept of the parts and wholes. The brace with the question mark identifies that it is the whole that needs to be calculated. The most important aspect of this phase of the lesson is recognising that the bar model supports the relationship between the different aspects of a story problem and can justify which calculation needs to be used to find the solution.

PUPIL TEXTBOOK 3A PAGE 153

Think together

WAYS OF WORKING Whole class teacher led (I do, We do, You do)

ASK

- Question **1**: *Which information do you know – the parts or the whole? Where do you need to look to find information about the missing part?*
- Question **1**: *What operation do you need to do to find the whole?*
- Question **2**: *Do you need to add or subtract to find the answer? How can you use the bar model to help you?*
- Question **3** a): *Whose idea do you think best represents the word problem? Do both methods work? Can you draw a bar model with three parts?*

IN FOCUS Question **2** is important because it prompts children to recognise that the missing information is one of the parts, rather than the whole. Therefore, they need to identify that the appropriate operation to use to solve this question is a subtraction.

STRENGTHEN Use strips of paper for children to fold or cut and then write on to represent the different parts of a bar model. Using concrete items to represent the bar model may help children to visualise more clearly the relationship between the parts and the whole and understand which operation they need to perform in order to solve a given problem.

DEEPEN Question **3** requires the addition of three numbers. Some children may use one bar to represent the addition of the first two parts and then draw a second bar for the addition of the total from the first bar model with the remaining part. Other children may combine all three parts into one bar model, and represent the steps of the calculation by showing a brace for two parts, and then adding the third part to the total shown by that brace.

ASSESSMENT CHECKPOINT Can children confidently explain how the different elements of the bar models represent aspects from the word problems?

ANSWERS

Question **1**: 159 + 25 = 184: Holly's dad spent £184.

Question **2**: 468 – 349 = 119: The child's bike cost £119.

Question **3** a): 275 + 25 + 50 = 350: Toshi spent £350.

Question **3** b): ☐ + 25 + 25 = 399, 399 – 25 – 25 = 349: The bike cost £349, so it must have been a tandem. Children may use a part-whole model or a bar model to show this problem.

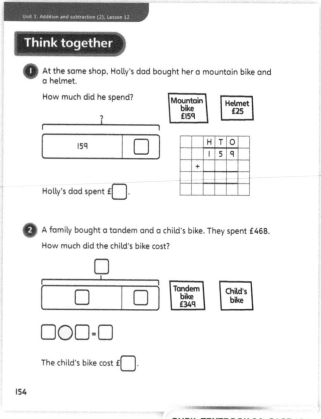

PUPIL TEXTBOOK 3A PAGE 154

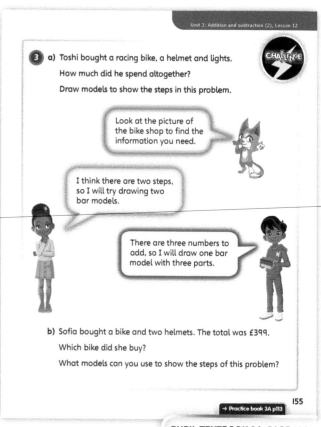

PUPIL TEXTBOOK 3A PAGE 155

Practice

WAYS OF WORKING Independent thinking

IN FOCUS Questions **2** and **3** challenge children to create their own bar models. This focuses their attention on which elements of the bar model they already know, which information they need to find and, therefore, which operation is needed to solve the question.

STRENGTHEN Children may find it boosts their confidence to draw their bar models on a whiteboard first. They can then adapt it to fit once they have started to explore the problems through the bar model itself.

THINK DIFFERENTLY Question **4** prompts children to consider and recognise where an incorrect operation has been used. Alex has jumped to a conclusion because she is trying to find the answer too quickly.

ASSESSMENT CHECKPOINT If children can draw their own bar models correctly in questions **2** and **3**, then they should be demonstrating a good grasp of how a bar model represents word problems.

ANSWERS Answers for the **Practice** part of the lesson can be found in the *Power Maths* online subscription.

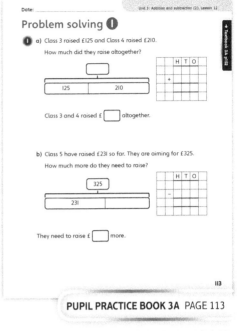

PUPIL PRACTICE BOOK 3A PAGE 113

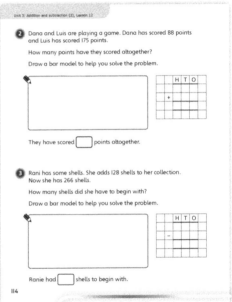

PUPIL PRACTICE BOOK 3A PAGE 114

Reflect

WAYS OF WORKING Independent thinking

IN FOCUS Rather than focusing on children's ability to draw bar models, this activity challenges children to think about how the different elements of the bar model really relate to the context of a word problem.

ASSESSMENT CHECKPOINT Have children recognised that the bar model represents a problem where a whole is known and a part is known, and so written a word problem that requires a subtraction?

ANSWERS Answers for the **Reflect** part of the lesson can be found in the *Power Maths* online subscription.

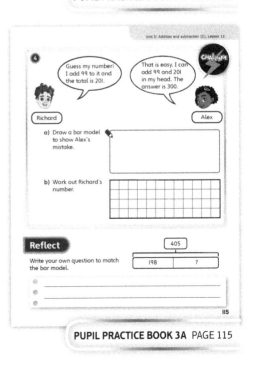

PUPIL PRACTICE BOOK 3A PAGE 115

After the lesson ⏸

- Can children explain how the bar model relates to the particular context of a word problem?
- Can children identify the missing information required from a word problem, and whether it is a whole or a part?
- Have children developed some confidence in drawing their own bar models to solve problems?

Problem solving ②

Learning focus

In this lesson, children will develop their use of the bar model to include two bars to represent comparison and to tackle problems with two or more steps.

NATIONAL CURRICULUM LINKS

Year 3 Number – addition and subtraction

Solve problems, including missing number problems, using number facts, place value, and more complex addition and subtraction.

ASSESSING MASTERY

Children can represent and justify problem-solving decisions through the use of comparison bar models.

COMMON MISCONCEPTIONS

Misconceptions can arise when children encounter comparison bar models, as opposed to single bars. Children may find it difficult to know how to represent the whole and how to reason between the two bars. Ask:
- *Which part of this bar model represents the whole?*
- *What are the parts that make up the whole?*

STRENGTHENING UNDERSTANDING

Allow children to represent the bars using strips of paper or resources such as coloured rods. The use of concrete materials will aid visualisation and strengthen their understanding of the relationship between the parts and the whole.

GOING DEEPER

Challenge children to create their own word problems or alternative word problems to match the bars.

KEY LANGUAGE

In lesson: subtract, bar model, part

Other language to be used by the teacher: runs (in a cricket context), operation, column addition, column subtraction, whole, missing information

STRUCTURES AND REPRESENTATIONS

Bar model, column addition and subtraction

RESOURCES

Mandatory: bar models

Optional: strips of paper, coloured rods, lengths of string, place value equipment

 In the eTextbook of this lesson, you will find interactive links to a selection of teaching tools.

Discover

Problem solving (2)

Discover

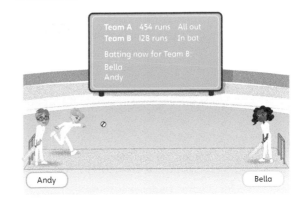

WAYS OF WORKING Pair work

ASK

- Question ① a) and b): *How can you show which operation is needed?*
- Question ① a) and b): *What numbers are relevant in solving this problem?*

IN FOCUS The important aspect of question ① a) is to recognise that it is a comparison problem, and so children need to understand that it can be solved using a find the difference subtraction. Are children able to represent this using a bar model?

PRACTICAL TIPS Some children may want to represent the amounts using place value equipment or strips of paper to create a bar model. In question ① a), they should arrange the place value equipment in a way that allows for the direct comparison this problem requires.

ANSWERS

Question ① a): 454 – 128 = 326. Team A has scored 326 more runs than Team B.

Question ① b): 128 + 105 + 83 = 316. Team B has now scored 316 runs.

① a) How many more runs has Team A scored than Team B?

b) Bella and Andy start batting for Team B.
Bella scores 105 and Andy scores 83.
How many runs has Team B scored now?

156

PUPIL TEXTBOOK 3A PAGE 156

Share

WAYS OF WORKING Whole class teacher led

ASK

- Question ① a): *Can you explain how the bar model represents the comparison clearly?*
- Question ① a): *Which part of the bar model shows the information we need to work out?*
- Question ① b): *Can you think of an alternative way to solve this?*

IN FOCUS Question ① a) demonstrates the power of a comparison bar model for problems which involve finding the difference. The important point is that children notice how to represent the missing information on the bar model.

For question ① b), children may discuss how to find the solution in a different order. For instance, they may find a way of adding the two new scores first, then adding that onto the 128 runs Team B had already.

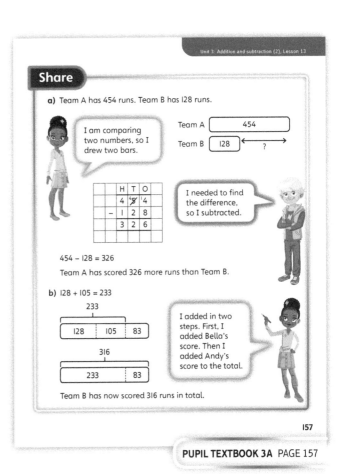

PUPIL TEXTBOOK 3A PAGE 157

Think together

WAYS OF WORKING Whole class teacher led (I do, We do, You do)

ASK

- Question **1**: *What operation do you need to perform to find the difference between the two scores?*
- Question **2**: *What information is missing?*
- Question **2**: *What do we need to subtract in the second part of this calculation? What is the whole and what do we need to subtract from the whole?*

IN FOCUS Question **2** requires two steps of calculation, including a single bar addition and then a comparison bar, to find the difference. The important aspect of this phase of the lesson is for children to develop some confidence with representing the missing information on a comparison bar model.

STRENGTHEN Encourage children to use strips of paper, coloured rods or lengths of string to represent the two bars required. This will help them to visualise the parts and whole of the bar model.

DEEPEN Question **3** presents two different bar models to represent a problem. Challenge children to explore how the different bar models show the same information but in different ways. Ask them to justify which approach they think better represents the story context in terms of how clear it makes the path to the solution.

ASSESSMENT CHECKPOINT Can children explain how the two steps of the calculation in question **2** are represented in different aspects of the bar model?

ANSWERS

Question **1**: 451 – 317 = 134
Isla's team scored 134 more runs.

Question **2**: 165 + 56 = 221 and then 320 – 221 = 99;
Jamilla and Emma need 99 more runs to get the same score as Mo and Lexi.

Question **3**: 188 + 188 + 56 = 432

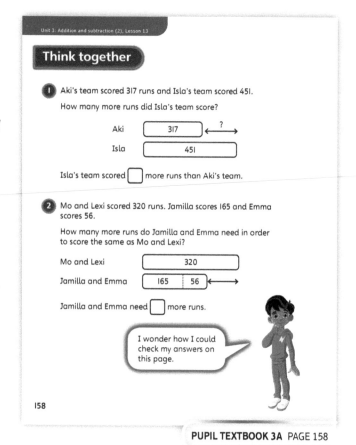

PUPIL TEXTBOOK 3A PAGE 158

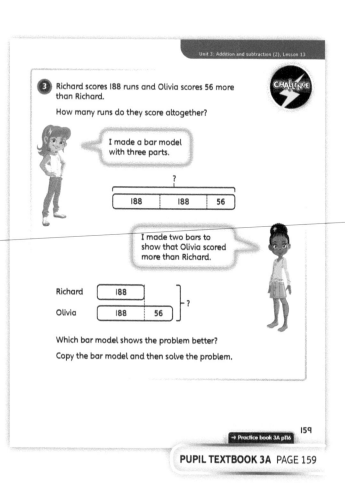

PUPIL TEXTBOOK 3A PAGE 159

Practice

WAYS OF WORKING Independent thinking

IN FOCUS Questions ❶ and ❷ provide children with the bar models; they have to add the numbers and identify the calculations required. Questions ❸ and ❹ extend this by challenging children to draw and label their own bar models to represent the problems.

STRENGTHEN Many of the problems require more than one step and children may struggle to identify how to begin. Encourage them to use the bar model concept to sketch the elements of the problem without trying to solve it at all. The first stage is simply to correctly represent the given information and then the second stage is to identify and organise the calculation by looking carefully at the bar model.

DEEPEN Question ❺ requires children to recognise how to represent the whole. They need to find the difference and start with Ebo's number in order to find Zac's number and accurately complete the bar model. Ensure children understand that it is only the shaded parts of the bar model that add up to 801.

THINK DIFFERENTLY Not every question is best represented using a comparison bar. For example, question ❹ b) is best represented as a single bar. Question ❸ uses a comparison, but the difference is known, and children have to use this information to find one of the parts.

ASSESSMENT CHECKPOINT Confident solutions to question ❹ will represent a good level of understanding.

ANSWERS Answers for the **Practice** part of the lesson can be found in the *Power Maths* online subscription.

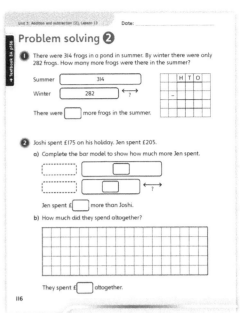

PUPIL PRACTICE BOOK 3A PAGE 116

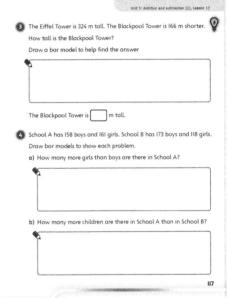

PUPIL PRACTICE BOOK 3A PAGE 117

Reflect

WAYS OF WORKING Group work

IN FOCUS This focuses not on the use of a bar model to get an answer, but on the decisions about how to represent different problems using different bar models.

ASSESSMENT CHECKPOINT Can children explain when the comparison bar is appropriate and when it is not?

ANSWERS Answers for the **Reflect** part of the lesson can be found in the *Power Maths* online subscription.

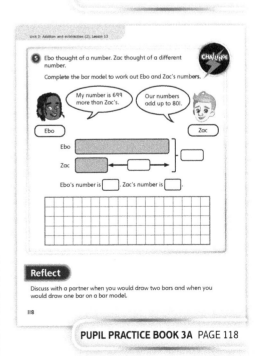

PUPIL PRACTICE BOOK 3A PAGE 118

After the lesson ⏸

- Can children represent a 'find the difference' problem as a comparison bar model?
- Do children know how to decide which bar models to draw?
- Can children identify the required missing information in a bar model?

End of unit check

> Don't forget the unit assessment grid in your *Power Maths* online subscription.

WAYS OF WORKING Group work teacher led

IN FOCUS

- Questions ❶ to ❸ focus on accuracy and efficiency using the column methods of addition and subtraction and exchange.
- Questions ❹ and ❺ focus on the checking strategies of approximation and using an inverse operation (in this case, an addition to check a subtraction).
- Question ❻ focuses on problem solving. This is a SATs-style question.

ANSWERS AND COMMENTARY

Children who have mastered this unit will be able to add or subtract numbers with up to 3 digits within 1,000. They will be able to justify whether or not an exchange is necessary and be able to explain the effect of doing an exchange in terms of place value. Children will also be able to justify an answer through checking strategies of approximation, estimation and the use of inverse operations.

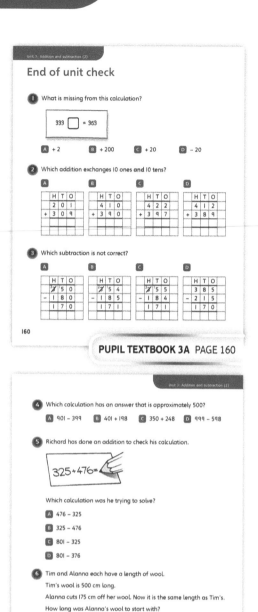

PUPIL TEXTBOOK 3A PAGE 160

PUPIL TEXTBOOK 3A PAGE 161

Q	A	WRONG ANSWERS AND MISCONCEPTIONS	STRENGTHENING UNDERSTANDING
1	C	A and B suggest a lack of understanding of place value. D indicates children misinterpreted the missing number problem as a subtraction.	Place value: allow children to model their reasoning and justify decisions using place value equipment.
2	D	A may suggest they are confused when 0 is a place holder.	Use the equipment to discuss the decisions and the calculations, but not as a counting strategy to explain the concepts.
3	B	A, C or D suggest children are not confident using exchange.	
4	A	B and C suggest children just added the 100s. D indicates children may find it difficult to approximate to 1,000.	
5	C	A, B or D suggest they are not confident with the relationship between wholes and parts in additions and subtractions.	Children should also be encouraged to use any other models or images that support their reasoning, such as part-whole models or number lines.
6	675 cm	An answer of 325 cm suggests children have not understood parts and wholes in this context.	

My journal

WAYS OF WORKING Independent thinking

ANSWERS AND COMMENTARY Answers will vary. Children do not have to write the answers, although they may choose to explore the best methods to find the answers as a way of deciding which are more complex. Instead, children should make and justify decisions about the different levels of complexity or difficulty in solving the different calculations.

If children are struggling to form an opinion or find it difficult to move beyond simply finding the answers, ask:
- Which question would you choose to attempt first?
- Are there any you would solve in a different way from the others?
- What is the same and what is different about the calculations?
- Which do you think would be most likely to cause you to make a mistake?

If children are struggling to explain their reasoning, ask:
- Could you explain how many steps each option takes?
- What do people tend to get wrong in class?
- Have you made any mistakes in this unit that you are reminded of here?

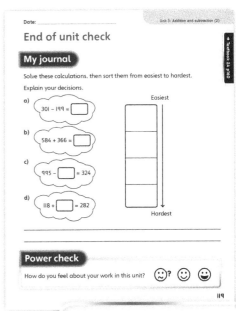

PUPIL PRACTICE BOOK 3A PAGE 119

Power check

WAYS OF WORKING Independent thinking

ASK
- What numbers could you add and subtract before this unit?
- Do you have any new skills?
- Have you built more confidence in any areas?
- Do you feel confident knowing when and how to use exchange?
- Can you check your own answers now?

Power play

WAYS OF WORKING Pair work

IN FOCUS Use this **Power play** to give children the opportunity to explore calculations in the context of a strategic game. The game can be adapted, as per Sparks's suggestion, to deepen the challenge by attempting to create an answer close to a specified number. Allow children to use the numbers generated by their spins in whatever position or order they like in their addition. This will encourage strategic thinking and challenge their understanding of place value.

ANSWERS AND COMMENTARY Supply pairs of children with a 0–5 spinner or a dice. Children will be choosing their digits to try to block their opponent, although the spinner provides a random element to the game. This random element will mean that children will aim for and 'hope' for certain numbers to come up on the spinner (or dice), meaning they will be using the mathematical skill of predicting, with a particular focus on place value in the context of addition.

PUPIL PRACTICE BOOK 3A PAGE 120

After the unit ⏸

- Can children explain why exchange is required for some calculations and how it relates to place value?
- Can children justify and explain their answers using checking strategies and sound understanding of different methods?

Strengthen and **Deepen** activities for this unit can be found in the *Power Maths* online subscription.

Unit 4
Multiplication and division ①

Mastery Expert tip! 'In Year 2, children only get a quick understanding of multiplication and division. It is vital that children thoroughly understand the basics for their work in Year 3. For example, do they know what equal groups and simple arrays look like? Do they know the difference between grouping and sharing? If not, I make time to focus on these concepts!'

Don't forget to watch the Unit 4 video!

WHY THIS UNIT IS IMPORTANT

This unit is essential to recap the key messages around multiplication and division that children focussed on in Key Stage 1. It begins by asking children to show their understanding of equal groups. They then take time to look at arrays and how they can show two different multiplications. This unit also provides the opportunity to recap multiples of 2, 5 and 10 and explore if a given number is a multiple of one of these numbers. Finally, there is an opportunity to recap work on sharing and grouping and ensure that children know the difference between the two concepts. By the end of this unit, it is hoped that children have the knowledge they need to tackle other work on multiplication and division when they start exploring other times-tables. This knowledge is fundamental to future learning and so, if necessary, take extra time to explore the concepts in detail.

WHERE THIS UNIT FITS

→ Unit 3: Addition and subtraction (2)
→ **Unit 4: Multiplication and division (1)**
→ Unit 5: Multiplication and division (2)

This unit acts as a recap of children's work in Year 2, where multiplication and division were introduced and equal and unequal groups were explored. It reminds children of the difference between equal sharing and equal grouping, as it is possible some children spent only a little amount of time on this. This unit provides essential preparation for their work on developing and securing multiplication knowledge.

Before they start this unit, it is expected that children:
• know how to make equal groups, for example, using counters
• can count in 2s, 5s and 10s.

ASSESSING MASTERY

Children who have mastered this unit will be able to count in multiples of 2, 5 and 10 and be able to recall some associated multiplication facts. As well as this, children should know when groups are equal and unequal and know what multiplication fact is shown. They will be able to make and draw arrays and identify what two facts are shown by the arrays. Finally, they should start to know the difference between sharing and grouping situations.

COMMON MISCONCEPTIONS	STRENGTHENING UNDERSTANDING	GOING DEEPER
Children think that equal groups have to look the same.	Show children examples of equal groups that do not look the same, for example, using strawberries on a plate or counters on a ten frame. Ask children what is the same. Usually, it is the fact there is the same amount of items in each group.	Ask children to draw their own equal groups. Ask them to make equal groups for a particular multiplication fact.
Children do not know that 2 × 5 is the same as 5 × 2. Knowledge of commutativity helps children work out more complicated multiplication facts quicker.	Draw a 2 × 5 array and ask children if they can see the two groups of five and also the five groups of two. Use this to explain that 5 × 2 is the same as 2 × 5. It is fine if they find working out two 5s easier than five 2s. They can apply this technique going forward.	As children work through the 2, 4 and 8 times-tables in the next unit, ask them if they can see how they are linked.

UNIT STARTER PAGES

Introduce the unit using teacher-led discussion. Ask children to think back to their work in Year 2 and identify which images show equal groups. Ask them what else they remember from Year 2 on this topic.

STRUCTURES AND REPRESENTATIONS

Arrays: This model shows the total of a multiplication and reinforces commutativity. It can also be used to demonstrate sharing and grouping.

$5 \times 2 = 10$

Bar model: This model represents the situation in multiplication and division word problems and shows the link between multiplication and repeated addition.

$5 \times 2 = 10$

2	2	2	2	2

KEY LANGUAGE

There is some key language that children will need to know as part of the learning in this unit.

→ times-tables
→ equal, unequal
→ repeated addition
→ array, bar model
→ commutative
→ multiple
→ group, groups, grouping
→ share, sharing
→ multiply, multiplication sentence
→ divide, division statement, division facts
→ remainder
→ even, odd
→ columns, rows

PUPIL TEXTBOOK 3A PAGE 162

PUPIL TEXTBOOK 3A PAGE 163

Multiplication – equal groups

Learning focus

In this lesson, children will recap their knowledge from Year 2 and be able to recognise equal groups. For any equal groups, children should be able to write down the associated multiplication fact, know how it links to repeated addition and know how to find the answer.

Before you teach

- Do children know some multiplication facts for 2, 5 and 10?
- Can they recognise equal groups and unequal groups?
- Can they write a multiplication statement for any equal groups?

NATIONAL CURRICULUM LINKS

Year 3 Number – multiplication and division

Write and calculate mathematical statements for multiplication and division using the multiplication tables that they know, including for two-digit numbers times one-digit numbers, using mental and progressing to formal written methods.

Recall and use multiplication and division facts for the 3, 4 and 8 multiplication tables.

ASSESSING MASTERY

Children can recognise equal groups and write down the correct multiplication sentence for each group. Children can recognise groups that are not equal and say why they are not equal.

COMMON MISCONCEPTIONS

Children may think that groups that do not look the same are not equal. For example, if strawberries on several plates have a different arrangement on each plate, children may think the groups are not equal, even though there are the same number of strawberries on each plate. Ask:
- *How many are in each group? Are there the same number in each group? Does this mean they are equal?*

Children may over rely on repeated addition to work out multiplication sentences. Encourage children to start to know multiplication facts from the times-tables to help instant recall of the answers. Ask:
- *Can you say the multiplication fact that matches the repeated addition?*

STRENGTHENING UNDERSTANDING

Children who are struggling with the concepts in this lesson should use cubes or counters to represent the objects. For example, the number of towers of cubes will represent the number of groups. Children should then be able to compare whether there are the same number in each group by comparing the heights or lengths of the towers.

GOING DEEPER

Ask children to show their own equal groups and unequal groups. For example, ask: *Make four groups that are equal. Can you write a multiplication sentence for each one? Draw a story to represent 3 × 5. Explain your thinking.*

KEY LANGUAGE

In lesson: multiplication (×), equal groups, multiplication sentence, grouping, groups, equal, total, repeated addition

Other language to be used by the teacher: not equal, times-table, multiply

STRUCTURES AND REPRESENTATIONS

Arrays

RESOURCES

Mandatory: cubes and counters

Optional: number lines

 In the eTextbook of this lesson, you will find interactive links to a selection of teaching tools.

Quick recap 🔎

Give children ten counters and ask them to share them equally between five people. How have they shown equal groups? Now ask them to share the counters between three people. Ask: *Have you made equal groups? Why or why not?*

Discover

Multiplication – equal groups

Discover

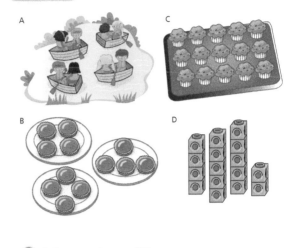

WAYS OF WORKING Pair work

ASK

- Question ❶: *Which groups are equal? How do you know?*
- Question ❶ a): *Which picture shows four equal groups with two in each group?*
- Question ❶ b): *Which picture shows three equal groups with five in each group?*

IN FOCUS In question ❶, children should discuss in pairs what is represented in each picture. They should focus on the number of groups and the number of items that are in each group. They should be able to identify any groups that are not equal and explain why not.

PRACTICAL TIPS Provide children with counters to represent each picture.

ANSWERS

Question ❶ a): 2 + 2 + 2 + 2 = 8

$\qquad$ 4 × 2 = 8

Question ❶ b): 3 × 5 = 15

❶ a) Can you see 4 groups of 2?

Write this as a repeated addition and multiplication.

b) Can you see 3 groups of 5?

Write this as a multiplication.

164

PUPIL TEXTBOOK 3A PAGE 164

Share

WAYS OF WORKING Whole class teacher led

ASK

- Question ❶ a): *What are the groups? What is in each group? What does the addition and multiplication work out?*
- Question ❶ b): *What are the groups? What is in each group? What does the multiplication work out?*

IN FOCUS Children should recognise the number of groups and the number of items in each group. For example, in question ❶ a), the boats are the groups and the children are the 'items' in each group. Encourage correct language, such as 'There are 4 boats with 2 children in each boat, so there are 4 groups with 2 in each group.'

Share

a) There are 4 boats of children.

There is a group of 2 children in each boat.

There are 4 groups of 2.

2 + 2 + 2 + 2 = 8

4 × 2 = 8

I can see the total number of children is 8.

b) There are 3 groups (rows) of 5 muffins on the tray.

I can see 3 rows of 5 muffins. So the total is 15.

3 × 5 = 15

165

PUPIL TEXTBOOK 3A PAGE 165

Think together

Think together

WAYS OF WORKING Whole class teacher led (I do, We do, You do)

ASK

- Question ①: *Do the groups look equal? Why/why not? How can you check?*
- Question ①: *How many groups are there? How many counters are in each group? How can you write this as a repeated addition? How can you write this as a multiplication sentence?*
- Question ②: *How many groups are there? How many jam tarts are in each group?*

IN FOCUS These questions provide children with the opportunity to consolidate their understanding of equal groups and associated calculations. Question ① draws children's attention to the common misconception that equal groups have to look the same. Children should recognise that, since there are the same number of counters in each group, the groups are equal.

STRENGTHEN Provide children with cubes or counters to represent the items in the pictures and use these to support their understanding of the associated calculations.

DEEPEN In question ②, children could be stretched to consider what would happen if there is another plate of jam tarts. Ask: *What happens if there is another jam tart on each plate? Which number changes in each scenario?* In question ③, children need to think about how to rearrange the cubes to show equal groups

ASSESSMENT CHECKPOINT Use questions ① and ② to assess whether children can use multiplication to represent equal groups and find the total.

ANSWERS

Question ①: $3 \times 5 = 15$

Question ②: $3 \times 4 = 12$

Question ③ a): There is a different number of cubes in each group.

Question ③ b): The cubes can be rearranged to show 4 equal groups of 4 cubes.

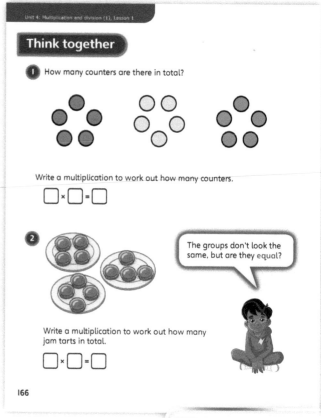

1. How many counters are there in total?

Write a multiplication to work out how many counters.

☐ × ☐ = ☐

2.

The groups don't look the same, but are they equal?

Write a multiplication to work out how many jam tarts in total.

☐ × ☐ = ☐

166

PUPIL TEXTBOOK 3A PAGE 166

3.

a) Explain why this does not show equal groups.

b) How can you make the groups equal?

I wonder if I can use the same number of cubes to make a different number of equal groups.

I am going to rebuild the towers and try to make them equal.

167

→ Practice book 3A p121

PUPIL TEXTBOOK 3A PAGE 167

Practice

WAYS OF WORKING Independent thinking

IN FOCUS These questions provide opportunity for children to consolidate their understanding of equal groups. In question ❶, the stem sentences help children to understand the groups and what is in each group. Questions ❷ and ❸ focus on writing multiplications from pictures, linking the pictorial representations to the abstract calculations. Question ❹ focuses on the common misconception that groups have to look the same in order to be equal.

STRENGTHEN Provide children with cubes or counters to represent the pictures and support their understanding.

DEEPEN In question ❻, children make equal groups and explore what equal groups can be made using exactly twenty cubes. Ask: *What if there were eighteen cubes rather than twenty? Can you make more or fewer different equal groups? What do you predict? Use cubes to check.*

ASSESSMENT CHECKPOINT Use questions ❶ to ❸ to assess whether children can write multiplication sentences to represent equal groups and find the total. Use question ❹ to assess whether children can identify and explain why groups do not have to look the same to be equal. Use question ❺ to ensure children understand the connection between multiplication and repeated addition.

ANSWERS Answers for the **Practice** part of the lesson can be found in the *Power Maths* online subscription.

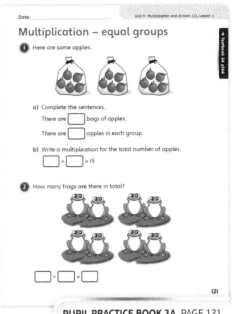

PUPIL PRACTICE BOOK 3A PAGE 121

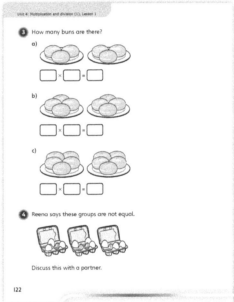

PUPIL PRACTICE BOOK 3A PAGE 122

Reflect

WAYS OF WORKING Independent thinking

IN FOCUS This activity provides opportunity for children to explain their learning on equal groups in their own words.

ASSESSMENT CHECKPOINT Children should be able to explain what it means for groups to be equal. They could draw pictures to support their answers.

ANSWERS Answers for the **Reflect** part of the lesson can be found in the *Power Maths* online subscription.

After the lesson ⏸

- Can children recognise equal groups?
- Can children use repeated addition to find the total?
- Can children write a multiplication sentence to work out the total?

205

Use arrays

Learning focus

In this lesson, children explore using arrays and finding corresponding multiplication sentences.

Before you teach

- Do children know the difference between rows and columns?
- Can children write multiplication sentences from pictures?
- Do children understand equal groups?

NATIONAL CURRICULUM LINKS

Year 3 Number – multiplication and division

Write and calculate mathematical statements for multiplication and division using the multiplication tables that they know, including for two-digit numbers times one-digit numbers, using mental and progressing to formal written methods.

Recall and use multiplication and division facts for the 3, 4 and 8 multiplication tables.

ASSESSING MASTERY

Children can make and use arrays to represent multiplication sentences. They recognise the commutative property of multiplication and can find two multiplication sentences for each array.

COMMON MISCONCEPTIONS

Children may not 'complete' an array. For example, they may show four rows of seven counters and one row of three counters and then think that $5 \times 7 = 31$. Ask:

- *Are there an equal number of counters in each row?*
- *How many counters should be in each row?*

If children do not line their counters up correctly when making arrays, they may find incorrect answers. Ask:

- *How should the counters be lined up so that you have equal groups? Are they lined up correctly?*

STRENGTHENING UNDERSTANDING

Provide children with counters so they can make the arrays themselves. Encourage children to count aloud when building their rows and columns to strengthen their understanding of the numbers in the multiplication sentences.

GOING DEEPER

Prompt children to explore what different arrays they can make for the same number. Ask: *Does a greater number always have a greater number of different arrays?*

KEY LANGUAGE

In lesson: arrays, columns, rows, total, multiply, equal, commutative

STRUCTURES AND REPRESENTATIONS

Arrays

RESOURCES

Mandatory: counters

 In the eTextbook of this lesson, you will find interactive links to a selection of teaching tools.

Quick recap

Give children some counters. Ask them to show 3 equal groups of 4 counters. Ask them to show 5 equal groups of 2 counters. Then ask them to show other equal groups and write the associated multiplication sentence for each.

Discover

Unit 4: Multiplication and division (1), Lesson 2

Use arrays

Discover

WAYS OF WORKING Pair work

ASK

- Question ① a): *How do you know the array is not complete?*
- Question ① a): *How many more counters are needed in the third row?*
- Question ① a): *How many columns and rows are there?*
- Question ① a): *Can you see equal groups in the array?*

IN FOCUS In question ① b), children explore making arrays using counters. They should make links to their previous learning on equal groups, with either the columns or the rows being the equal groups, and then use this to write the associated multiplication sentence.

PRACTICAL TIPS Provide children with twenty counters to recreate the **Discover** scenario and to explore what other arrays can be made in question ① b).

ANSWERS

Question ① a): The array shows the multiplication
$5 \times 4 = 20$ or $4 \times 5 = 20$.

Question ① b): The counters can be rearranged to show
$2 \times 10 = 20$, $10 \times 2 = 20$, $1 \times 20 = 20$ or
$20 \times 1 = 20$.

① a) Complete the array for Max.

   Write down a multiplication for the array.

   b) Arrange the same number of counters into a different array.

168

PUPIL TEXTBOOK 3A PAGE 168

Share

WAYS OF WORKING Whole class teacher led

ASK

- Question ① a): *How many rows are there? How many counters are in each row? How do you write this as a multiplication sentence?*
- Question ① a): *How many columns are there? How many counters are in each column? How do you write this as a multiplication sentence?*
- Question ① b): *Are there any other arrays? How do you know? Why can't you make an array with three columns?*

IN FOCUS Children explore arrays. They link this to their knowledge of equal groups from the previous lesson, to find the associated multiplication sentence, and to find the total. In question ① a), they should recognise that either the columns or the rows could be the equal groups, so they can write the multiplication in two different ways. They can see that the answer does not change because multiplication is commutative.

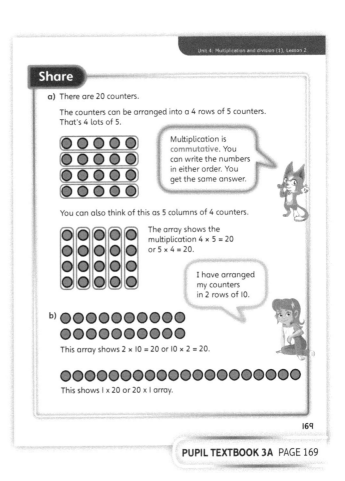

Share

a) There are 20 counters.

The counters can be arranged into a 4 rows of 5 counters. That's 4 lots of 5.

Multiplication is commutative. You can write the numbers in either order. You get the same answer.

You can also think of this as 5 columns of 4 counters.

The array shows the multiplication $4 \times 5 = 20$ or $5 \times 4 = 20$.

I have arranged my counters in 2 rows of 10.

b)

This array shows $2 \times 10 = 20$ or $10 \times 2 = 20$.

This shows 1×20 or 20×1 array.

169

PUPIL TEXTBOOK 3A PAGE 169

Think together

Whole class teacher led (I do, We do, You do)

ASK

- Question **1**: *How many rows are there? How many strawberries are in each row? How many columns are there? How many strawberries are in each column?*
- Question **2**: *How many rows are there? How many counters are in each row? How many columns are there? How many counters are in each column?*

IN FOCUS These questions provide an opportunity for children to consolidate their understanding of arrays and writing the associated multiplication sentences. In question **2**, they should recognise that they can write the multiplication either way around, because they could treat either the rows or the columns as the equal groups.

STRENGTHEN Provide children with counters to recreate the arrays and support their understanding.

DEEPEN In question **3**, children should be able to explain how each child is seeing the array differently. They could be stretched to explore different-sized arrays and find similar ways to partition them.

ASSESSMENT CHECKPOINT Use questions **1** and **2** to assess whether children can use arrays to write multiplication sentences.

ANSWERS

Question **1**: $2 \times 6 = 12$
$6 \times 2 = 12$

Question **2**: $5 \times 10 = 50$ or $10 \times 5 = 50$

Question **3**: There are 30 stars in total. The groups can be reversed: 6 groups of 5, 10 groups of 3 and 2 groups of 15 are possible. This could also be expressed as 30×1 or 1×30.

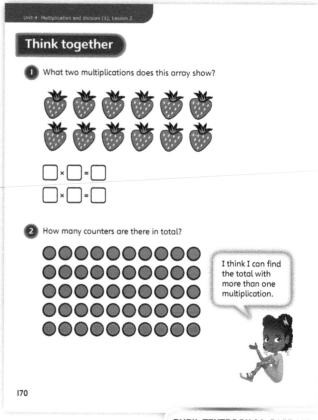

PUPIL TEXTBOOK 3A PAGE 170

PUPIL TEXTBOOK 3A PAGE 171

Practice

WAYS OF WORKING Independent thinking

IN FOCUS These questions provide an opportunity for children to practise using arrays to find multiplication sentences. Question ❶ reinforces that multiplication is commutative by asking children to write two multiplication sentences for each array. Questions ❷ and ❹ ask children to draw/complete an array, linking the abstract calculations to pictorial representations. They should be encouraged to explain each array to understand what each number in the multiplications represent. For example, there are five rows with two apples in each row. There are ten apples in total. $2 \times 5 = 10$.

STRENGTHEN Provide children with counters so they can make arrays to represent each question and support them in finding the associated multiplication sentences.

DEEPEN In question ❺, children see an array represented differently, as a grid. They could be encouraged to draw a similar style array for other numbers.

THINK DIFFERENTLY Question ❸ more formally links the idea of equal groups with arrays. Ask: *Can you see any other equal groups in the image?*

ASSESSMENT CHECKPOINT Use questions ❶ and ❷ to assess whether children can write multiplication sentences for arrays. Use question ❸ to assess whether children understand the link between arrays and equal groups.

ANSWERS Answers for the **Practice** part of the lesson can be found in the *Power Maths* online subscription.

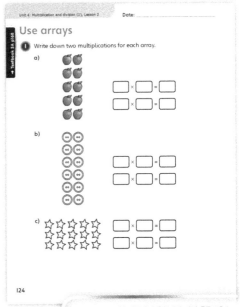

PUPIL PRACTICE BOOK 3A PAGE 124

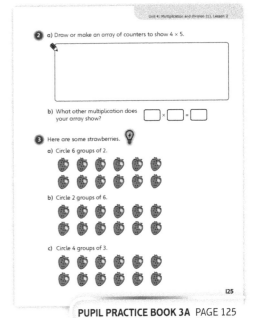

PUPIL PRACTICE BOOK 3A PAGE 125

Reflect

WAYS OF WORKING Independent thinking

IN FOCUS Children explore making arrays and writing multiplication sentences to match their partner's arrays.

ASSESSMENT CHECKPOINT Children should be able to make arrays and write two multiplication sentences for their partner's arrays. They should be able to repeat this with different numbers.

ANSWERS Answers for the **Reflect** part of the lesson can be found in the *Power Maths* online subscript

---

After the lesson ⏸

- Can children make arrays?
- Can children write a multiplication sentence to match their array?
- Can children write two multiplication sentences to match their array?

PUPIL PRACTICE BOOK 3A PAGE 126

Multiples of 2

Learning focus

In this lesson, children explore multiples of 2 and recognise that any number in the 2 times-table is a multiple of 2. They link this to their understanding of even numbers to see that all multiples of 2 are even.

Before you teach

- Do children know their 2 times-table?
- Do children understand the difference between odd and even numbers?
- Can children count in 2s beyond 24?

NATIONAL CURRICULUM LINKS

Year 3 Number – multiplication and division

Write and calculate mathematical statements for multiplication and division using the multiplication tables that they know, including for two-digit numbers times one-digit numbers, using mental and progressing to formal written methods.

Recall and use multiplication and division facts for the 3, 4 and 8 multiplication tables.

ASSESSING MASTERY

Children can find multiples of 2. They can use their understanding of multiples of 2 to decide whether a given number is or is not a multiple of 2.

COMMON MISCONCEPTIONS

Children may think that a number like 211 is a multiple of 2 because there is a 2 in the number. Ask:
- Is this number odd or even? How do you know?
- Is this number a part of the 2 times-table?

STRENGTHENING UNDERSTANDING

Provide children with a printed 100 square for reference throughout the lesson. Encourage them to explain what they notice about multiples of 2 and what digit they end in.

GOING DEEPER

Encourage children to explain anything they notice. Ask: A 2-digit number has six 1s. Is the number odd or even? A 2-digit number has six 10s. Can you tell whether it is odd or even?

KEY LANGUAGE

In lesson: two (2), multiple, times-table, even

Other language to be used by the teacher: ones (1s), digit, even, odd

STRUCTURES AND REPRESENTATIONS

100 square, number tracks

RESOURCES

Mandatory: 100 square

Optional: digit cards

 In the eTextbook of this lesson, you will find interactive links to a selection of teaching tools.

Quick recap

Play 'Odd or even'. Tell children one half of the room is even and the other is odd. Say a number and they have to point or go to the side of the room depending on whether the number is even or odd.

Discover

Unit 4: Multiplication and division (1), Lesson 3

Multiples of 2

Discover

Multiples of 2 are numbers that are in the 2 times-table.

WAYS OF WORKING Pair work

ASK

- Question ① a): *What is the 2 times-table?*
- Question ① b): *Colour in the 2 times-table on your 100 square. What do you notice?*
- Question ① b): *Can you continue the pattern?*

IN FOCUS In question ① a), children explore multiples of 2 on a 100 square. In question ① b), they should recognise that they need to colour every other column on the 100 square fully, and be encouraged to discuss with a partner why this happens.

PRACTICAL TIPS Provide children with a printed 100 square that they can colour in. Encourage them to say the numbers aloud as they colour them in.

ANSWERS

Question ① a): The first ten numbers in the 2 times-table are 2, 4, 6, 8, 10, 12, 14, 16, 18 and 20.

Question ① b): Children will colour in every other column on the 100 square. They should notice that the multiples of 2 end with 0, 2, 4, 6 or 8.

① a) Point to or colour in the first ten numbers in the 2 times-table.

b) Colour in all the multiples of 2 on a 100 square. Use a different colour.

What do you notice?

172

PUPIL TEXTBOOK 3A PAGE 172

Share

WAYS OF WORKING Whole class teacher led

ASK

- Question ① a): *What was Dexter's next calculation? What number did he colour in next?*
- Question ① b): *What do you notice about the numbers you have coloured in? Why do you have to colour every other number?*

IN FOCUS Children should explain what they notice and why they think this happens. In question ① b), they should recognise that, because they are counting in 2s, they miss a square then colour one each time. You may want to discuss with children at this stage that numbers that end in 0, 2, 4, 6 or 8 are also known as even numbers. Do they remember this from last year?

This could prompt a discussion about odd numbers too. Explain that odd numbers end in 1, 3, 5, 7 or 9.

Share

a) I went through my 2 times-table, so 1 × 2 = 2 ...

b) I can see that all the coloured numbers are in the same columns.

All the coloured numbers end with 0, 2, 4, 6 or 8.

173

PUPIL TEXTBOOK 3A PAGE 173

Think together

WAYS OF WORKING **WAYS OF WORKING** Whole class teacher led (I do, We do, You do)

ASK

• Question ❶: *What are the number tracks counting up in? How do you know? What is the same about the number tracks? What is different?*

• Question ❷: *What do all multiples of 2 end in? How can you use this to help you answer the question? Does it matter that you do not know how many 10s one of the numbers has?*

IN FOCUS These questions provide an opportunity for children to consolidate their learning on multiples of 2. In question ❶, they should recognise that the 10s in each place of the number tracks are the same and it is the 1s that are different. When they find the missing numbers they will see that the 10s digit changes when they bridge the 10s. In question ❷, they use their knowledge of multiples of 2 ending in either 0, 2, 4, 6 or 8 to identify multiples of 2.

STRENGTHEN Ensure children have access to their 100 squares from the **Discover** activity to support them in this section. Encourage them to focus on the pattern in the 1s digits.

DEEPEN In question ❸, children explore different numbers that can be made from the digit cards that are multiples of 2. They make a connection between multiples of 2 and even numbers to recognise that all multiples of 2 are even. They could be stretched to consider how their answers change if a fourth digit card is added and whether it matters if that digit is even or odd.

ASSESSMENT CHECKPOINT Use question ❶ to assess whether children can count in 2s beyond multiplication facts that they know from the 2 times-table. Use question ❷ to assess whether children can identify whether or not a number is a multiple of 2.

ANSWERS

Question ❶: ..., 18, 20, 22, 24, 26, 28
..., 78, 80, 82, 84, 86, 88

Question ❷: We only need to see the 1s digit.
87 is not a multiple of 2.
30, 56, 134 and ✳8 are multiples of 2
because they end in either 0, 2, 4, 6 or 8.

Question ❸ a): 48, 74, 78, 84

Question ❸ b): 478, 748, 784, 874

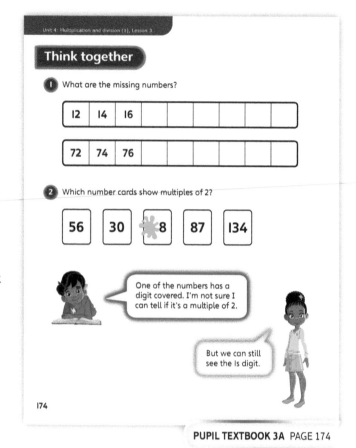

PUPIL TEXTBOOK 3A PAGE 174

PUPIL TEXTBOOK 3A PAGE 175

Practice

WAYS OF WORKING Independent thinking

IN FOCUS These questions provide an opportunity for children to consolidate their learning on multiples of 2. Question ❶ focuses on counting in 2s beyond the multiplication facts that they know from the 2 times-table. Questions ❷ and ❸ focus on identifying multiples of 2.

STRENGTHEN Ensure children have access to their 100 squares from the Discover activity to support them in this section. Encourage them to focus on the pattern in the 1s digits.

DEEPEN In question ❹, ask children to think about whether their answers would change if Ambika makes a 3-digit even number. Ask: *What can you still say about the last digit of Ambika's number?*

ASSESSMENT CHECKPOINT Use question ❶ to assess whether children can count in 2s beyond the multiplication facts in the 2 times-table. Use questions ❷ and ❸ to assess whether children can identify multiples of 2.

ANSWERS Answers for the **Practice** part of the lesson can be found in the *Power Maths* online subscription.

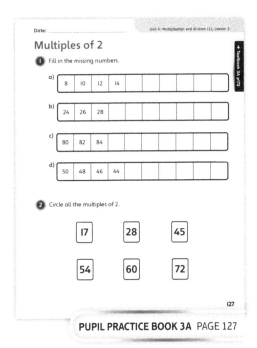

PUPIL PRACTICE BOOK 3A PAGE 127

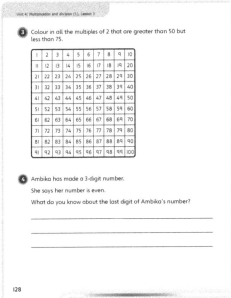

PUPIL PRACTICE BOOK 3A PAGE 128

Reflect

WAYS OF WORKING Pair work

IN FOCUS This activity asks children to explain their learning and recognise that all multiples of 2 are even.

ASSESSMENT CHECKPOINT Children should be able to explain in their own words why all multiples of 2 are even.

ANSWERS Answers for the **Reflect** part of the lesson can be found in the *Power Maths* online subscription.

After the lesson ⏸

- Can children count in 2s?
- Can children identify multiples of 2?
- Do children know that all multiples of 2 are even?

PUPIL PRACTICE BOOK 3A PAGE 129

Multiples of 5 and 10

Learning focus

In this lesson, children build on their learning of multiples from the previous lesson and explore multiples of 5 and 10. They recognise that any number in the 5 times-table is a multiple of 5 and any number in the 10 times-table is a multiple of 5 and of 10.

Before you teach

- Do children know their 5 times-table?
- Do children know their 10 times-table?
- Can children count in 5s and 10s beyond the multiplication facts in the 5 and 10 times-tables?

NATIONAL CURRICULUM LINKS

Year 3 Number – multiplication and division

Write and calculate mathematical statements for multiplication and division using the multiplication tables that they know, including for two-digit numbers times one-digit numbers, using mental and progressing to formal written methods.

Recall and use multiplication and division facts for the 3, 4 and 8 multiplication tables.

ASSESSING MASTERY

Children can find multiples of 5 and multiples of 10. They can use their understanding of multiples of 5 and 10 to decide whether or not a given number is a multiple of 5 or 10.

COMMON MISCONCEPTIONS

Children may think that all multiples of 5 are multiples of 10 because it works the other wayround. Ask:
- *What digit do all multiples of 10 end in?*
- *Which two digits can a multiple of 5 end in?*

Children may think that a number like 507 is a multiple of 5 because there is a 5 in the number. Ask:
- *What digit does this number end in? Can it be a multiple of 5?*

STRENGTHENING UNDERSTANDING

Provide children with a printed 100 square for reference throughout the lesson. Encourage them to explain any patterns that they notice in the 1s digits of the numbers. Children could be given base 10 equipment to support their understanding of multiples of 10.

GOING DEEPER

Encourage children to explain anything they notice, particularly with multiples of 10 also being multiples of 5. Ask: *A 2-digit number has five 1s. Is it a multiple of 5 or 10? A 2-digit number has five 10s. It is a multiple of 5 and 10. How many 1s does the number have?*

KEY LANGUAGE

In lesson: five (5), ten (10), multiple, times-table

Other language to be used by the teacher: ones (1s), digit

STRUCTURES AND REPRESENTATIONS

100 square, number tracks

RESOURCES

Mandatory: 100 square

Optional: base 10 equipment, digit cards

 In the eTextbook of this lesson, you will find interactive links to a selection of teaching tools.

Quick recap

Play times-table bingo. List on the board the multiples of 5 and 10 up to 100 and ask children to choose six of the numbers. Read out 5 or 10 times-table calculations and children cross out the answer if they have it. The first person to cross out all of their numbers wins.

Discover

Multiples of 5 and 10

Discover

WAYS OF WORKING Pair work

ASK

- Question ① a): *Where are the numbers in the 10 times-table? What do you notice?*
- Question ① b): *What numbers has the teacher covered? How do you know?*
- Question ① b): *What times-table are all of these numbers in?*

IN FOCUS This activity provides an opportunity for children to explore multiples of 5 and 10 on a 100 square. In question ① a), they should use multiplication facts that they know to support them in identifying all the multiples of 10 on the 100 square and explain any patterns they notice in the position of the numbers.

PRACTICAL TIPS Provide children with a printed 100 square and ask them to count aloud when identifying numbers. They could complete the teacher's pattern and count how many places they move along each time to find the times-table they are in.

ANSWERS

Question ① a): The multiples of 10 are:
10, 20, 30, 40, 50, 60, 70, 80, 90, 100.
They are all in the end column and they all end in 0.

Question ① b): The numbers covered with stars all come from the 5 times-table.

① **a)** Where are all the multiples of 10?

What do you notice about the multiples of 10?

b) The numbers covered with stars all come from the same times-table. Which one?

176

PUPIL TEXTBOOK 3A PAGE 176

Share

WAYS OF WORKING Whole class teacher led

ASK

- Question ① a): *What is 1 × 10? Where is this on the 100 square? What about 2 × 10? What about 3 × 10? What do you notice?*
- Question ① b): *How do you know which numbers are covered? How many do you count along to get to the next number covered? What times-table is it? What are all the numbers a multiple of?*

IN FOCUS This section provides an opportunity for children to explore multiples of 5 and 10 on a 100 square. In question ① b), they should spot patterns in the position of the multiples on the 100 square as well as in the 1s digits of the numbers.

PUPIL TEXTBOOK 3A PAGE 177

Think together

WAYS OF WORKING Whole class teacher led (I do, We do, You do)

ASK

- Question **1**: *What is each number track counting up in? How do you know?*
- Question **2** a): *What do all multiples of 5 end in? How many 1s could the number have? How many 10s could the number have?*
- Question **2** b): *What do all multiples of 10 end in? How many 1s must the number have? How many 10s could it have?*

IN FOCUS These questions provide an opportunity for children to consolidate their learning on multiples of 5 and 10. Question **1** asks children to count on and back in 5s and 10s beyond the multiplication facts that they know from the 5 and 10 times-tables. Question **2** uses what children have noticed about the number of 1s in multiples of 5 and 10 to find other multiples of 5 and 10. In question **2** c), children should realise that, because any multiple of 10 must have a 0 in the 1s place and they only have one card with a 0, they can make the same number of 2-digit multiples of 10 as 3-digit multiples of 10.

STRENGTHEN Ensure children have access to their 100 square from earlier in the lesson to support them. Children could use base 10 equipment to support them in finding multiples of 10.

DEEPEN In question **3**, ask children to explain what they notice about the numbers in each section. Is it possible for a number to go in more than one part of the table? Can they add another number in each section?

ASSESSMENT CHECKPOINT Use question **1** to assess whether children can count in 5s and 10s. Use question **2** to assess children's understanding of multiples of 5 and 10.

ANSWERS

Question **1** a): ..., 25, 30, 35, 40, 45, 50

Question **1** b): ..., 160, 170, 180, 190, 200, 210

Question **1** c): ..., 80, 75, 70, 65, 60, 55

Question **2** a): 50, 80, 85

Question **2** b): 580, 850

Question **2** c): Any multiple of 10 must have a 0 in the 1s place.
There is only one card with a 0.
The 2-digit multiples of 10 are 50 and 80.
The 3-digit multiples of 10 are 580 and 850.
So Lee is incorrect: Emma can make the same number of 2-digit and 3-digit multiples of 10.

Question **3**:

	Multiple of 5	Not multiple of 5
Even numbers	70, 120, 300	158
Odd numbers	95	63

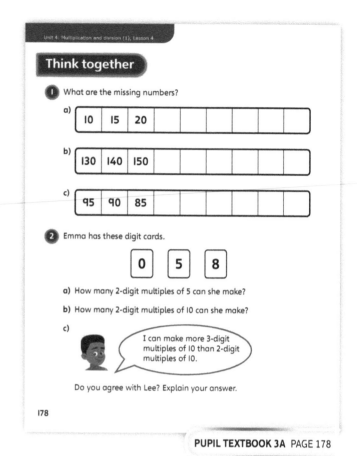

Unit 4: Multiplication and division (1), Lesson 4

Think together

1 What are the missing numbers?

a)

10	15	20					

b)

130	140	150					

c)

95	90	85					

2 Emma has these digit cards.

0   5   8

a) How many 2-digit multiples of 5 can she make?

b) How many 2-digit multiples of 10 can she make?

c)

I can make more 3-digit multiples of 10 than 2-digit multiples of 10.

Do you agree with Lee? Explain your answer.

178

PUPIL TEXTBOOK 3A PAGE 178

Unit 4: Multiplication and division (1), Lesson 4

CHALLENGE

3 Here are some number cards.

63   70   95   120   158   300

Put the numbers into the correct part of the table.

	Multiple of 5	Not multiple of 5
Even numbers		
Odd numbers		

I'm going to try to write my own cards to put in the table.

I wonder if there are any multiples of 5 that are not multiples of 10.

179

→ Practice book 3A p130

PUPIL TEXTBOOK 3A PAGE 179

Practice

Independent thinking

IN FOCUS These questions provide an opportunity for children to consolidate their learning on multiples of 5 and 10. Question ❶ asks children to count in 5s and 10s beyond the multiplication facts that they know from the 5 and 10 times-tables. Questions ❷ and ❸ focus on their understanding of multiples of 5. Question ❹ encourages children to think about possible digits if a number is a multiple of 5.

STRENGTHEN Encourage children to count aloud to support their understanding of multiples. Ensure they have access to their 100 square from the Discover section to support their understanding.

DEEPEN In question ❻, encourage children to explain how they found their answers. In question ❻ a), ask: *If your numbers are multiples of 5 and are also even, what else must these numbers be multiples of?* Give children a fourth digit card and ask: *Does this change your answers?*

THINK DIFFERENTLY Question ❹ encourages children to think about possible missing digits in numbers if they know that each number is a multiple of 5.

ASSESSMENT CHECKPOINT Use question ❶ to assess whether children can count in 5s and 10s beyond the multiplication facts from the 5 and 10 times-tables. Use questions ❷ to ❺ to assess children's understanding of multiples of 5 and 10.

ANSWERS Answers for the **Practice** part of the lesson can be found in the *Power Maths* online subscription.

Reflect

WAYS OF WORKING Independent thinking

IN FOCUS This question gives children the chance to explain their learning about multiples of 5 in their own words.

ASSESSMENT CHECKPOINT Children should be able to explain that all multiples of 5 end in either 0 or 5.

ANSWERS Answers for the **Reflect** part of the lesson can be found in the *Power Maths* online subscription.

After the lesson ⏸

- Can children count in 5s and 10s?
- Can children identify multiples of 5?
- Can children identify multiples of 10?

Unit 4: Multiplication and division (1), Lesson 4   Date:

Multiples of 5 and 10

❶ Fill in the missing numbers.

a) | 15 | 20 | 25 | | | | |

b) | 75 | 80 | | | 95 | 100 |

c) | 225 | 230 | 235 | | | | |

d) | 50 | 45 | 40 | | | | |

❷ Circle all the multiples of 5.

15   145   58

208   60   320

130

PUPIL PRACTICE BOOK 3A PAGE 130

❸ Is Kate correct?

390 is not a multiple of 5 as it does not end in a 5.

❹ Each of these numbers is a multiple of 5.
Complete the missing digits.
Write two answers for each one.

a) 52☐   b) 3☐5
   52☐   3☐5

❺ Write all the multiples of 5 that come between the two numbers shown on the number line.

273 ——————————— 322

131

PUPIL PRACTICE BOOK 3A PAGE 131

❻   0   4   5   CHALLENGE

Arrange all the digit cards to make:

a) Two even multiples of 5. ☐☐

b) Two odd multiples of 5. ☐☐

c) Two numbers that are multiples of both 5 and 10. ☐☐

Reflect

Sam says you can tell if a number is a multiple of 5 by looking at one of the digits.
Is Sam correct?

132

PUPIL PRACTICE BOOK 3A PAGE 132

Share and group

Learning focus

In this lesson, children answer sharing and grouping division questions. They should recognise the difference in context between the problems and understand whether this makes them a sharing or grouping problem.

Before you teach ⏸

- Can children share things equally into groups?
- Can children divide by 2, 5 and 10?
- Do children understand the link between multiplication and division?

NATIONAL CURRICULUM LINKS

Year 3 Number – multiplication and division

Write and calculate mathematical statements for multiplication and division using the multiplication tables that they know, including for two-digit numbers times one-digit numbers, using mental and progressing to formal written methods.

Recall and use multiplication and division facts for the 3, 4 and 8 multiplication tables.

ASSESSING MASTERY

Children can answer sharing and grouping division questions and explain whether the question is a sharing or a grouping one and why.

COMMON MISCONCEPTIONS

Children may confuse sharing and grouping questions. Ask:
- *What is the question asking you to do? Do you need to share the items out or put them into groups?*

Children may not interpret their answer correctly in the context of the question. Ask:
- *What is the question asking you to do? Is it a sharing or a grouping question?*

STRENGTHENING UNDERSTANDING

Provide children with counters to represent the questions and help them answer them. Ask them if they are 'sharing the counters out' or 'putting the counters into groups' to understand the difference between the two structures.

GOING DEEPER

Ask children to come up with different sharing and grouping problems for different calculations.

KEY LANGUAGE

In lesson: share, group, divide, equal

Other language to be used by the teacher: division

STRUCTURES AND REPRESENTATIONS

Bar models

RESOURCES

Mandatory: counters

Optional: cups, straws, cubes

 In the eTextbook of this lesson, you will find interactive links to a selection of teaching tools.

Quick recap

As a class, complete some calculations that require division by 2, 5 or 10. Ask children to explain how they arrived at their answers.

Discover

WAYS OF WORKING Pair work

ASK

- Question ① a): *How many flowers are there? How many vases are there?*
- Question ① a): *If Amal puts one flower in each vase, will he use all the flowers?*
- Question ① a): *If Amal puts two flowers in each vase, will he use all the flowers?*
- Question ① a): *Is Amal sharing or grouping the flowers?*
- Question ① b): *What do the numbers in the division represent?*

IN FOCUS This question provides an opportunity for children to explore division as sharing in a practical way. In question ① b), they should link each number in their division calculation to the context.

PRACTICAL TIPS Provide children with objects to represent the flowers and vases, such as straws and cups, to support them in finding the answer.

ANSWERS

Question ① a): There are 4 flowers in each vase.

Question ① b): $20 \div 5 = 4$

Share and group

Discover

① a) Amal has 20 flowers.
   He shares the flowers equally between the 5 vases.
   How many flowers are in each vase?

b) Write this as a division.

180

PUPIL TEXTBOOK 3A PAGE 180

Share

WAYS OF WORKING Whole class teacher led

ASK

- Question ① a): *What do Dexter's vases represent? What do his flowers represent?*
- Question ① a): *How does Sparks know that this is a sharing question?*
- Question ① a): *How does the bar model show the problem?*
- Question ① b): *What does each number in the division represent?*

IN FOCUS Question ① a) exposes children to different methods of solving a sharing question. They should explore the similarities and differences between the methods and explain why each one works.

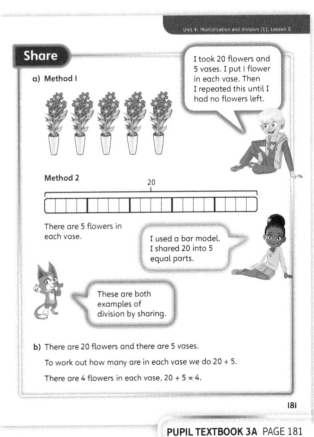

Share

a) Method 1

I took 20 flowers and 5 vases. I put 1 flower in each vase. Then I repeated this until I had no flowers left.

Method 2   20

There are 5 flowers in each vase.

I used a bar model. I shared 20 into 5 equal parts.

These are both examples of division by sharing.

b) There are 20 flowers and there are 5 vases.
   To work out how many are in each vase we do $20 \div 5$.
   There are 4 flowers in each vase, $20 \div 5 = 4$.

181

PUPIL TEXTBOOK 3A PAGE 181

Think together

WAYS OF WORKING Whole class teacher led (I do, We do, You do)

ASK

- Question **1**: *Is Danny sharing or grouping the apples? How do you know?*
- Question **2**: *Is Danny sharing or grouping the apples? How do you know?*
- Question **3**: *What is the same and what is different about questions* **1** *and* **2***?*

IN FOCUS These questions provide an opportunity for children to consolidate their understanding of sharing and grouping to solve division questions. Questions **1** and **2** both lead to the same abstract calculation, but children should recognise that the context changes the question and the answer of 5 represents something different each time.

STRENGTHEN Provide children with counters to represent the questions to support them with their understanding. Ask: *Are you sharing the counters or are you putting them into groups?*

DEEPEN In question **3**, children compare word problems that provide the same abstract calculation, except for the fact that one focuses on the idea of sharing and the other on grouping. They should explain what the answer represents in each case and could be stretched to make up their own sharing and grouping word problems for a different division.

ASSESSMENT CHECKPOINT Use questions **1** and **2** to assess whether children can answer sharing and grouping division questions and identify which one is which. They should be able to explain the differences between the questions.

ANSWERS

Question **1**: They each have 5 apples.

Question **2**: $10 \div 2 = 5$; Danny needs 5 bags.

Question **3**: $15 \div 3 = 5$

15 is the number of sweets.
3 is the number of children sharing the sweets.
5 is the number of sweets each child gets.

15		
5	5	5

15 is the number of sweets.
3 is the number of sweets Paul puts into each group.
5 is the number of groups of sweets.

15				
3	3	3	3	3

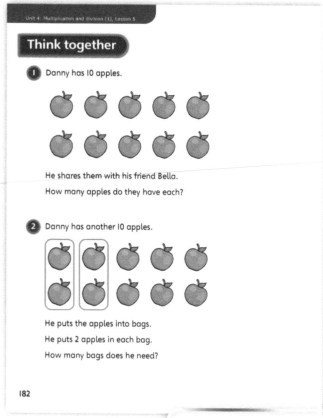

PUPIL TEXTBOOK 3A PAGE 182

PUPIL TEXTBOOK 3A PAGE 183

Practice

WAYS OF WORKING Independent thinking

IN FOCUS These questions provide an opportunity for children to consolidate their understanding of sharing and grouping. Questions **1** and **2** use the same numbers in different contexts to highlight the difference in structure of the questions. Question **3** uses a similar approach with children physically sharing or grouping counters themselves.

STRENGTHEN Provide children with counters to represent the questions. They could draw a bar model to represent the questions and physically share or group their counters into the bar model to support their understanding.

DEEPEN In question **4**, encourage children to predict whether each person will get more or less money in part b) than in part a), before they work out the answer. In question **5**, ask children to make predictions first, based on their earlier learning on multiples, and then test their answers. Give them a different number of cubes and ask them to repeat the activity.

ASSESSMENT CHECKPOINT Use questions **1** to **3** to assess whether children can solve sharing and grouping division questions and identify which is which. They should be able to explain their answers in the context of the question and explain how they know whether it is sharing or grouping.

ANSWERS Answers for the **Practice** part of the lesson can be found in the *Power Maths* online subscription.

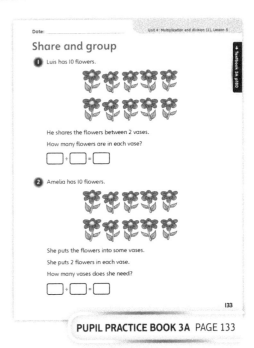

PUPIL PRACTICE BOOK 3A PAGE 133

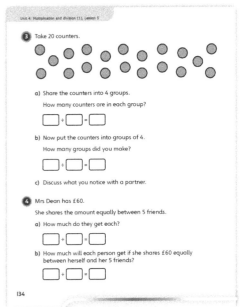

PUPIL PRACTICE BOOK 3A PAGE 134

Reflect

WAYS OF WORKING Independent thinking

IN FOCUS Children explain their learning about sharing and grouping in their own words. They can link this to the questions from earlier in the lesson to support them in explaining the difference between sharing or grouping.

ASSESSMENT CHECKPOINT Children should be able to explain the difference between a sharing and grouping problem, and could provide an example of each to support their explanation.

ANSWERS Answers for the **Reflect** part of the lesson can be found in the *Power Maths* online subscription.

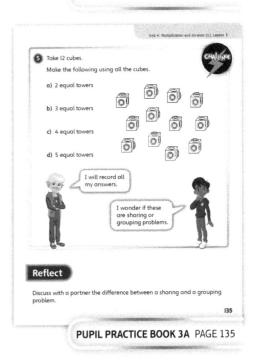

PUPIL PRACTICE BOOK 3A PAGE 135

After the lesson ⏸

- Can children answer division questions?
- Can children identify whether a question is about sharing or grouping?
- Can children interpret their answers in the context of the question?

End of unit check

Don't forget the unit assessment grid in your *Power Maths* online subscription.

WAYS OF WORKING Group work adult led

IN FOCUS These questions cover the multiplication and division methods from this unit. They are designed to draw out particular misconceptions and misunderstandings.

- Question **1** ensures children know what equal groups look like.
- Question **2** checks children's understanding of arrays.
- Questions **3** and **4** focus on different multiples and checks that children know how to count in 2s, 5s and 10s.
- Question **7** checks that children recognise when numbers are multiples of 2, 5 and 10.

All other questions check understanding of simple multiplication and division and provide a couple of typical, simple examples. Questions **6** and **7** are SATs style questions.

ANSWERS AND COMMENTARY Children who have mastered this unit will be able to count in multiples of 2, 5 and 10 and recognise multiples of these numbers (including 3-digit multiples). They should also have a basic understanding of how to answer simple multiplication and division problems.

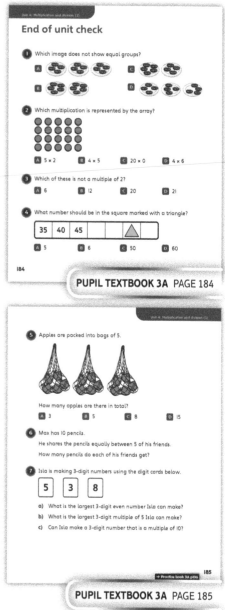

PUPIL TEXTBOOK 3A PAGE 184

PUPIL TEXTBOOK 3A PAGE 185

Q	A	WRONG ANSWERS AND MISCONCEPTIONS	STRENGTHENING UNDERSTANDING
1	C	D suggests that children think that equal groups must look the same.	If children need support, ask them to use concrete objects to demonstrate grouping and sharing, making equal groups and arrays.
2	B	C indicates that children have counted the number of counters, but the question asks for a multiplication.	
3	D	This question checks whether children know that numbers that end in an odd number are not multiples of 2. They may think 21 is a multiple of 2 because it contains a 2.	For example, for questions 5 and 6, use counters to demonstrate the situations.
4	D	C suggests that children may think they need to find the next multiple from the numbers that are shown. A indicates that children have identified what the numbers are going up in.	To strengthen understanding of multiples, keep practising counting in 2s, 5s and 10s and reinforce what makes numbers a multiple of 2, 5 and 10.
5	D	Check children know they need to multiply and work out 3 × 5. Any of the other answers show that children do not understand multiplication.	
6	2	Check that children understand they have to share 10 pencils between 5 children. Some children may try and put the pencils into five groups of 2.	
7	a) 538 b) 835 c) No	For part b), children need to realise that their number needs to end in 5. For part c), children should realise they cannot make a multiple of 10 as there is no 0.	

My journal

ANSWERS AND COMMENTARY

In this activity, children have to sort numbers into multiples of 2, 5 and 10. Remind them that some numbers may appear in more than one column. They should discuss with each other what makes a number a multiple of 2, for example, that they end in 2, 4, 6, 8 or 0, and what makes numbers multiples of 5 and 10. If numbers do not satisfy these criteria, then they are not multiples of any of these numbers. By the end of this activity, children should have a greater understanding of what makes a number a multiple of 2, 5 or 10.

Power check

WAYS OF WORKING Independent thinking

ASK

- *Can you draw some equal groups?*
- *Do you feel confident making a 2 × 5 array?*
- *Can you show how to share 10 objects between 5 people? Can you show how to put 10 objects into groups of 5? How many groups would there be?*
- *Do you know how to work out if a number is a multiple 2, 5 or 10?*

Power play

WAYS OF WORKING Pair or individual work

IN FOCUS Children should use different coloured counters to identify multiples of 2, 5 and 10. They may notice that some numbers are multiples of more than one of these numbers. They should start to see patterns, for example, numbers that are multiples of 2 are all in five particular columns.

ANSWERS AND COMMENTARY

a) There are 50 multiples of 2 on the 100 square. These are in the five columns of even numbers.

There are 20 multiples of 5 on the 100 square. These are in the columns where all the numbers end in 5 or 0.

Finally, there are 10 multiples of 10 on the 100 square. These are all in the last column.

b) There are 10 even multiples of 5. These are all in the last column. They are all also multiples of 10.

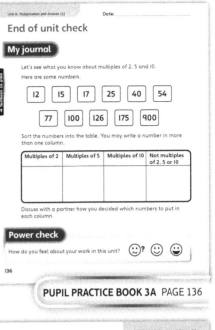

End of unit check

My journal

Let's see what you know about multiples of 2, 5 and 10. Here are some numbers.

| 12 | 15 | 17 | 25 | 40 | 54 |

| 77 | 100 | 126 | 175 | 900 |

Sort the numbers into the table. You may write a number in more than one column.

Multiples of 2	Multiples of 5	Multiples of 10	Not multiples of 2, 5 or 10

Discuss with a partner how you decided which numbers to put in each column.

Power check

How do you feel about your work in this unit?

136

PUPIL PRACTICE BOOK 3A PAGE 136

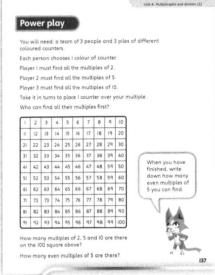

Power play

You will need: a team of 3 people and 3 piles of different coloured counters.

Each person chooses 1 colour of counter.

Player 1 must find all the multiples of 2.

Player 2 must find all the multiples of 5.

Player 3 must find all the multiples of 10.

Take it in turns to place 1 counter over your multiple.

Who can find all their multiples first?

When you have finished, write down how many even multiples of 5 you can find.

How many multiples of 2, 5 and 10 are there on the 100 square above?

How many even multiples of 5 are there?

137

PUPIL PRACTICE BOOK 3A PAGE 137

After the unit ⏸

- Do children know when a number is a multiple of 2, 5 or 10?
- Can they show you how to share and group?
- Can they solve simple multiplication problems using concrete objects to support them?

Strengthen and **Deepen** activities for this unit can be found in the *Power Maths* online subscription.

Unit 5
Multiplication and division ②

Mastery Expert tip! 'Rapid recall and knowledge of times-tables is vital to free up valuable working memory. I make time for children to regularly practise these, so that they can improve their speed of recall. We use arrays to help children visualise 2 × 3 as an array of 6 counters, and then they are more likely to remember the fact.'

Don't forget to watch the Unit 5 video!

WHY THIS UNIT IS IMPORTANT

This unit builds on recognising equal groups. Three lessons are spent exploring in depth each of the times-tables that children need to know in Year 3, encouraging rapid recall. Children are reminded of the difference between equal sharing and equal grouping and then move on to look at when division problems may have a remainder of sorts. Although a full understanding of remainder is not essential in Year 3, children do need to have a basic understanding of it. There are two lessons that focus on problem solving, and using the bar model to represent simple one-step multiplication and division problems. This reinforces multiplication as repeated addition. Children then move on to solve simple two-step problems that involve all of the four operations.

WHERE THIS UNIT FITS

→ Unit 4: Multiplication and division (1)
→ **Unit 5: Multiplication and division (2)**
→ Unit 6: Multiplication and division (3)

This unit builds on the previous unit, where multiples of 2, 5 and 10 are introduced and equal and unequal groups are explored. It also builds on equal sharing and equal grouping. This unit provides essential preparation for beginning to multiply and divide 2-digit numbers by 1-digit numbers in the spring term, and also for working with fractions. Knowledge of multiplication facts is also essential.

Before they start this unit, it is expected that children:
- know what it means when groups are equal and not equal
- know that multiplication can be seen as repeated addition and division as repeated subtraction
- know that an array shows two multiplications, such as $5 × 4 = 4 × 5$.

ASSESSING MASTERY

Children who have mastered this unit will start to know their 3, 4, and 8 times-tables off by heart. They will know when to multiply and will understand the difference between equal grouping and sharing. They will know that some divisions do not always give a whole answer and can have a remainder. They will be able to represent multiplication and division problems using a bar model.

COMMON MISCONCEPTIONS	STRENGTHENING UNDERSTANDING	GOING DEEPER
Children may think that $0 × a = a$. This is a common mistake that is made with multiplication facts.	Reinforce the importance of knowing times-tables. 0 groups of anything is 0 and any number of groups of 0 is 0.	Children use their knowledge of times-tables to work out problems such as 13 × 4, or 24 × 4.
Children may not always recognise if they need to do a multiplication or division to find the solution.	Use counters to represent the objects in groups and then use a bar model to represent the situation. You can use counters initially to represent the bars.	Ask children to write simple one-step and multi-step multiplication and division problems and to explain when a problem requires more than one step.

Unit 5: Multiplication and division ②

Introduce the unit using teacher-led discussion. Ask children to think back to their work in Year 2 on equal and unequal groups. Question children to see which of Flo's words they recognise.

STRUCTURES AND REPRESENTATIONS

Arrays: This model shows the total of a multiplication and reinforces commutativity. It can also be used to demonstrate sharing and grouping. Question children to see which of Flo's words they recognise.

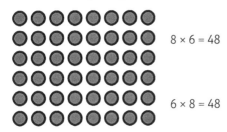

$8 \times 6 = 48$

$6 \times 8 = 48$

Bar model: This model represents the situation in multiplication and division word problems and shows the link between multiplication and repeated addition.

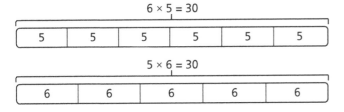

$6 \times 5 = 30$

| 5 | 5 | 5 | 5 | 5 | 5 |

$5 \times 6 = 30$

| 6 | 6 | 6 | 6 | 6 |

Number lines: This model helps understand multiplication and division. $3 + 3 + 3 + 3 + 3 + 3 = 18$; $3 \times 6 = 18$.

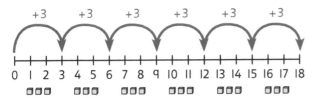

KEY LANGUAGE

There is some key language that children will need to know as part of the learning in this unit.

→ equal, unequal
→ multiply (×), divide (÷)
→ multiple, times-table
→ grouping, sharing
→ remainder
→ array, bar model
→ repeated addition
→ multiplication sentence
→ division sentence
→ multiplication fact
→ division fact

PUPIL TEXTBOOK 3A PAGE 186

PUPIL TEXTBOOK 3A PAGE 187

225

Multiply by 3

Learning focus

In this lesson, children will start to understand what it means to multiply by 3. Children will see the link between repeated addition, counting up in 3s and multiplying by 3.

Before you teach ▮▮

- Can children count up in 3s?
- Do children know how to form a multiplication statement from repeated addition?

NATIONAL CURRICULUM LINKS

Year 3 Number – multiplication and division

Recall and use multiplication and division facts for the 3, 4 and 8 multiplication tables.

Write and calculate mathematical statements for multiplication and division using the multiplication tables that they know, including for two-digit numbers times one-digit numbers, using mental and progressing to formal written methods.

ASSESSING MASTERY

Children can form multiplication sentences involving multiplying by 3. Children can work out the answer to multiplication sentences by knowing the link with repeated addition, can use a number line to count up in 3s and can start to remember some multiplication facts.

COMMON MISCONCEPTIONS

When children are presented with 3 groups of objects as opposed to groups of 3 objects, they may not see this as multiplying by 3. For example, they may not see that 12 groups of 3 is the same as 3 groups of 12, and that they both mean 12 × 3. Show a 12 by 3 array and a 3 by 12 array and ask:
- *What is the same? What is different?*

Children often start the count from 0 when working out, for example, 11 × 3, even if they already know 10 × 3. Ask:
- *How many 3s do you have? How many more 3s do you need?*

STRENGTHENING UNDERSTANDING

To help children multiply by 3, ensure that you show the direct link between the objects, a repeated addition, counting up in 3s on a number line and the multiplication sentence. As children count up in 3s, put counters on the number line each time, to show the 3 objects.

GOING DEEPER

Ask children to use the multiplication facts they know to work out other multiplication facts. Children can also start to explore how to work out 4 × 3 + 5 × 3 as one multiplication, showing by adding arrays together that this is the same as 9 × 3.

KEY LANGUAGE

In lesson: multiply (×), ~~multiplication sentence~~, repeated addition, total, count up, number line, array

Other language to be used by the teacher: multiplication statement, times-table, equal groups, *x* groups of *y*

STRUCTURES AND REPRESENTATIONS

Number lines, arrays

RESOURCES

Mandatory: cubes, counters

Optional: balls, plastic cups

 In the eTextbook of this lesson, you will find interactive links to a selection of teaching tools.

Quick recap 🔁

Children count up in 3s to 30. They should be encouraged to start at 0. They can use a number track to help them, if needed. Can they count on and back?

Discover

Unit 5: Multiplication and division (2), Lesson 1

Multiply by 3

Discover

1 a) There are 3 balls under each cup.

How many balls are there in total?

Write down a multiplication sentence to work out the answer.

b) Work out 8 × 3.

188

PUPIL TEXTBOOK 3A PAGE 188

WAYS OF WORKING Pair work

ASK

- Question **1** a): *How many groups are there? How many are there in each group? How can we write this as a multiplication sentence? How can we work out the total? Should we count up in 1s or 3s? What can we use to help us?*
- Question **1** b): *How many groups are there this time? How many are there in each group? What has changed? How can you work out the answer? Do you need to start from 0 again? Can you start from somewhere else? How do you know?*

IN FOCUS Question **1** a) is about children realising that they have 7 equal groups and there are 3 balls in each group. Using their knowledge from the previous unit, they should start to see this as a multiplication. In order to work out the total, children should be encouraged to explore different ways of working it out and compare methods. Is the best method to count up in 1s, for example? Children should start to realise that an effective way of working out the total is to count up in 3s. They have done this in Year 2, using a number line to support them. In question **1** b), look for children who start the count at 0 again and those who start the count from where they left off previously.

PRACTICAL TIPS You could use balls, counters or cubes under plastic cups to recreate the picture and help children get a feel for this activity.

ANSWERS

Question **1** a): There are 7 cups.
There are 3 balls under each cup.
7 × 3 = 21
There are 21 balls.

Question **1** b): 8 × 3 = 24

Share

PUPIL TEXTBOOK 3A PAGE 189

WAYS OF WORKING Whole class teacher led

ASK

- Question **1** a): *How many groups can you see? How many are there in each group? How can you work out the total? What did you use to help you keep track of the count? Why is counting in 1s not the best method? What multiplication can you see?*
- Question **1** b): *Do we have to start at 0 again? Why not? What would happen if we did not have the answer to the first part, would we still start at 21? What if I had 9 cups?*

IN FOCUS In question **1** a), discuss the method of counting up in 1s and count aloud with children. Reiterate that this is not the most effective way of counting as it takes time. Encourage children to realise there must be a quicker way. Many children are likely to count in 3s as Flo suggests. Show the clear link between each cup and the count. In question **1** b), discuss the idea of starting from 0 to work out 8 × 3 and compare this with starting at 21 and going up another 3. Ask children to decide which is quicker.

Think together

Think together

WAYS OF WORKING Whole class teacher led (I do, We do, You do)

ASK

- Question **1**: *How many groups are there? How many are in each group? What has changed? If we count from 0, where do we get to? How many 3s are we adding together? What multiplication is this? What is the answer? Could we have used an answer from earlier? How?*
- Question **2**: *Do we have equal groups? How many equal groups are there? How many are in each group? What multiplication sentence is this? Do you know the answer straight away? If not, how can you work it out?*
- Question **3**: *What multiplications can you see? What is the same about each of these? What is different? How do the objects show that 3 × 5 is the same as 5 × 3?*

IN FOCUS Question **1** builds on from the **Discover** section by adding an extra cup. Discuss counting from 0 first to allow children to see the link between repeated addition and multiplication. Question **2** sets out a different context. Some children may know that the answer is 18 straight away because of the previous example. Question **3** looks at different ways of representing 5 × 3 through asking what is the same and what is different. Discuss how the total is the same each time, but the objects may be presented in many different ways. Use the arrays to show the link between 3 × 5 and 5 × 3.

STRENGTHEN At each stage, use concrete objects alongside a number line to help children see the link between repeated addition, counting up in 3s and multiplication. For question **1**, encourage children to put 3 counters at each point on the number line to show that this represents one cup. Children need to be secure in knowing what counting in 3s looks like (0, 3, 6, 9, and so on).

DEEPEN Encourage further thinking in question **2** by extending the calculation. Ask: *How many different ways can you work out 6 × 3? How can you work out 8 × 3 from your answer to this question?*

ASSESSMENT CHECKPOINT Can children understand what it means to multiply by 3 and use their knowledge of counting in 3s to work out the answers to the multiplication sentences?

ANSWERS

Question **1**: 9 × 3 = 27
There are 27 balls.

Question **2**: 6 × 3 = 18
There are 18 hats.

Question **3**: The total is the same each time; the arrangements are different (some are in arrays, others are not and the arrays are different ways around).

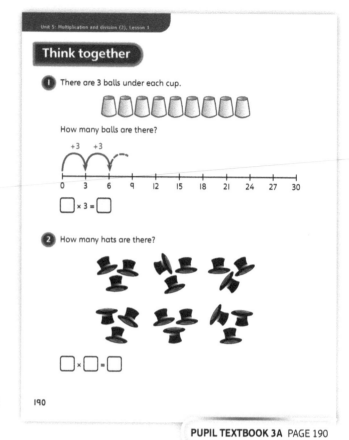

PUPIL TEXTBOOK 3A PAGE 190

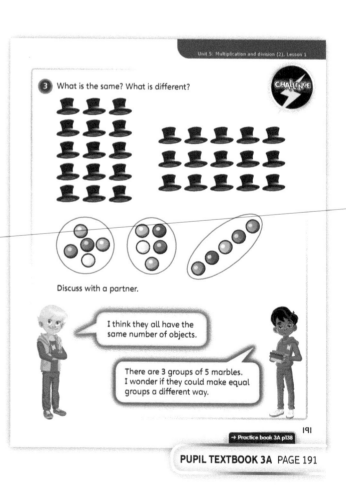

PUPIL TEXTBOOK 3A PAGE 191

Practice

WAYS OF WORKING Independent thinking

IN FOCUS Questions ❶ and ❷ provide practice for children to form the correct multiplication sentences for situations. Encourage children to use a number line and counting up in 3s to help them work out the answers to the calculations. Some children may already start to know the 3 times-table off by heart and so will be able to just write the answer. Encourage checking of answers. Question ❹ links multiplying by 3 with addition. Ash asks children to realise that it is just one multiplication. Question ❺ reinforces the idea that you do not always have to start the count from 0.

STRENGTHEN To support multiplying by 3, use concrete objects alongside a number line. Count aloud with children in 3s from 0, each time showing the jump on the number line. For question ❹, show that this can be written as one multiplication by asking children to join the two arrays together to make one big array. Children could use counters on their desk.

DEEPEN In question ❸ a), children need to realise that 3 × 12 is the same as 12 × 3. For question ❹, ask: *How can this be a single multiplication? How can you show it using counters?* Question ❻ links to children's prior understanding of odd and even numbers and asks them if an odd number multiplied by 3 always gives an odd number. Can children find a way of showing this is true for any odd number? Do even numbers multiplied by 3 ever give odd numbers?

THINK DIFFERENTLY In question ❸, children now have 3 groups as opposed to 3 in each group. This requires children to realise that this is still multiplying 12 by 3. Draw out the fact that 3 groups of 12 is the same as 12 groups of 3.

ASSESSMENT CHECKPOINT Can children represent a question as a multiplication sentence with ×3 and can they use counting in 3s to work out the correct answer to the multiplication?

ANSWERS Answers for the **Practice** part of the lesson can be found in the *Power Maths* online subscription.

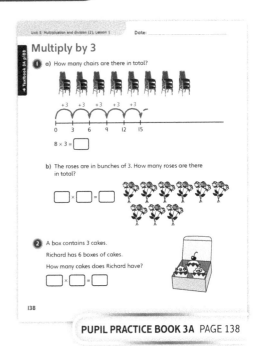

PUPIL PRACTICE BOOK 3A PAGE 138

PUPIL PRACTICE BOOK 3A PAGE 139

Reflect

WAYS OF WORKING Independent thinking

IN FOCUS Children are presented with 9 × 3 = 27 and are asked to think of a story or situation that matches this. Encourage them to write or draw a story that they have not previously met. Share these as a class. Discuss what is the same and what is different. For example, some children may have 3 groups, others may have 3 in a group. Discuss if this is still multiplying by 3.

ASSESSMENT CHECKPOINT Check if children know that, for multiplying by 3, they can either have 3 groups or 3 in each group.

ANSWERS Answers for the **Reflect** part of the lesson can be found in the *Power Maths* online subscription.

After the lesson

- Can children form multiplication sentences involving the 3 times-table from word problems?
- Can children use 10 × 3 = 30 to work out 11 × 3 and 12 × 3?
- Do children know how to use repeated addition and counting up in 3s to multiply by 3?

PUPIL PRACTICE BOOK 3A PAGE 140

Divide by 3

Learning focus

In this lesson, children will start to understand what it means to divide by 3. Children will see that a division sentence can be used to represent either equal grouping or sharing.

Before you teach

- Can children count back in 3s from 30 to 0?
- Can children share a set of objects between a given number of groups?
- Can children put a set of objects into groups of a given size?

NATIONAL CURRICULUM LINKS

Year 3 Number – multiplication and division

Recall and use multiplication and division facts for the 3, 4 and 8 multiplication tables.

Write and calculate mathematical statements for multiplication and division using the multiplication tables that they know, including for two-digit numbers times one-digit numbers, using mental and progressing to formal written methods.

ASSESSING MASTERY

Children can form division sentences from a grouping or sharing situation, and they know the difference between grouping and sharing. Children can use repeated subtraction and counting back in 3s to work out the result of a division, as well as seeing how an array can help them.

COMMON MISCONCEPTIONS

Children may think that dividing always means sharing. Present children with examples to show that it can also mean grouping. Ask:

- *Can you tell me a story where you share and another where you group? What is the difference between your two stories?*

STRENGTHENING UNDERSTANDING

Work with concrete objects to show the difference between grouping and sharing. For grouping, explicitly link each group with the steps in repeated subtraction. Use a number line to show how grouping can be seen as repeated subtraction by counting back in 3s (or subtracting 3 each time). To help understand sharing, first ask children to share the objects one-by-one and then ask children to take three at a time and allocate one to each group. Also ensure that children know what each number in each division sentence represents and allow them to see that division can be either grouping or sharing.

GOING DEEPER

Ask children to show two different word problems for $12 \div 3 = 4$. Ask them to describe what is the same about their stories and what is different. Can they show that 3 is the number of groups, but that it could also be the amount in each group? Can children use $36 \div 3$ to work out $42 \div 3$?

KEY LANGUAGE

In lesson: division sentence, equally, group, method, array, multiplication fact

Other language to be used by the teacher: grouping, sharing, multiply, divide

STRUCTURES AND REPRESENTATIONS

Number lines, arrays

RESOURCES

Mandatory: cubes, counters

 In the eTextbook of this lesson, you will find interactive links to a selection of teaching tools.

Quick recap

Ask children to count on in 3s from 0 to 60. Then ask them to count back in 3s from 60 to 0.

Discover

Unit 5: Multiplication and division (2), Lesson 2

Divide by 3

Discover

WAYS OF WORKING Pair work

ASK

- Question ① a): *How many cupcakes are there? How many cupcakes go into each box? Is this an example of grouping or sharing? What is the difference? How can you work out how many boxes are needed?*
- Question ① b): *How many cakes do we have now? How many people are we sharing them between? How can you share them? Can they be shared equally? Can you share them three at time? How many does each person get?*

IN FOCUS Question ① a) is about showing division as grouping. The division sign is given to help children remember what they need to do. Children may approach this in different ways. They are likely to put the cupcakes into groups of 3 and count the groups. Try to encourage them to see that grouping can be seen as repeated subtraction. Question ① b) provides a contrasting example, this time looking at sharing. Children may use different methods: they might share the cakes out one at a time or three at a time. They should start to understand the two reasons for division (grouping and sharing).

PRACTICAL TIPS Children could use counters to represent the cakes.

ANSWERS

Question ① a): $18 ÷ 3 = 6$
6 boxes are needed.

Question ① b): $27 ÷ 3 = 9$
Each person gets 9 cakes.

① a) Each box holds 3 cupcakes.
How many boxes are needed for all the cupcakes?
Work this out by writing a division sentence.

b) David buys 27 cakes. He shares them equally between 3 people.
How many cakes do they get each?

192

PUPIL TEXTBOOK 3A PAGE 192

Share

WAYS OF WORKING Whole class teacher led

ASK

- Question ① a): *Why has Dexter done a repeated subtraction? What is happening at each stage? Can you see why division can be seen as repeated subtraction? What does each number represent in 18 ÷ 3 = 6?*
- Question ① b): *Is this the same as the previous question? Is this sharing or grouping? How do you know? Do you have to share just one at a time? What does each number represent in 27 ÷ 3 = 9? What does the division sign represent this time?*

IN FOCUS In question ① a), children are likely to have grouped objects into 3s. Dexter talks children through this method, showing it as a repeated subtraction. At each stage, model what children are doing. For example: *I put 3 cakes into a box. I have one box now. I have 18 – 3 = 15 cakes left.* A number line is shown to model each stage of the repeated subtraction. It is important that children see the maths that is going on when they group and not just see that there are 6 groups. Look at the division statement 18 ÷ 3 = 6 and ensure that children know what each number represents. Question ① b) shows sharing the cakes one a time and also three at a time. Discuss with children if these methods give the same answer and why. Look explicitly at the difference between grouping and sharing. Children need to understand that the first method is grouping and the second is sharing.

Share

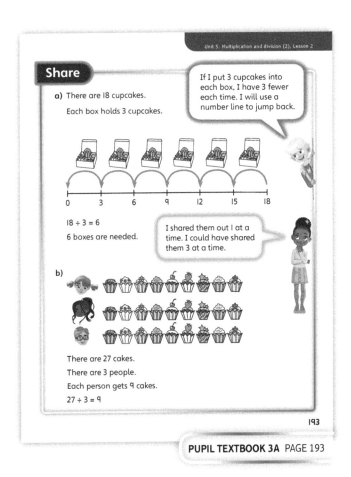

If I put 3 cupcakes into each box, I have 3 fewer each time. I will use a number line to jump back.

a) There are 18 cupcakes.
Each box holds 3 cupcakes.

$18 ÷ 3 = 6$
6 boxes are needed.

I shared them out 1 at a time. I could have shared them 3 at a time.

b) There are 27 cakes.
There are 3 people.
Each person gets 9 cakes.
$27 ÷ 3 = 9$

193

PUPIL TEXTBOOK 3A PAGE 193

Think together

WAYS OF WORKING Whole class teacher led (I do, We do, You do)

ASK

- Question **1**: *How many bread rolls are there? How many are in each pack? Is this grouping or sharing? Can you explain each step of your method?*
- Question **2**: *How many doughnuts are there? How many plates are there? Is this grouping or sharing? What division do you need to do? How does the array help you get the answer quicker?*
- Question **3**: *Could this be grouping or sharing? What has each person done? How can an array help? What would the array look like? How can you use multiplication to help you? Do you know a different way of getting the answer?*

IN FOCUS Question **1** is an example of grouping. Although the full structure for repeated subtraction is not shown, children may need to go through it. Encourage children to group the objects into 3s. Children may explore whether they can draw an array to help them. Question **2** is an example of sharing. Children may share one at a time or three at time. The doughnuts are presented in a 7 × 3 array. Ask children if the array can help them get the answer to the division. Explain that there are 21 doughnuts and 3 plates (number of rows), and there are 7 doughnuts on each plate (the number in each row). Link this with the division statement. Question **3** shows that a division can be seen as both equal sharing and grouping, and shows different methods that children may use. The last method links to their work on multiplication. Children can see that the answer to the division is the same as ☐ × 3 = 33.

STRENGTHEN For questions **1** and **2**, carry out the grouping and sharing with objects. Explicitly link each action to the steps in repeated subtraction or sharing. Also ensure that children know what each number in each division statement represents and allow children to see that division can be either grouping or sharing.

DEEPEN Ask children to use their division knowledge to answer a missing number multiplication problem. Ask: *How could you use division to work out ☐ × 3 = 36? Can you explain why? Can children make up their own division problems?*

ASSESSMENT CHECKPOINT Can children understand that division can either be equal grouping or sharing as well as form a division sentence to answer a question? Can children work out the answer to division sentences, practically, using an array or through knowledge of multiplication facts?

ANSWERS

Question **1**: There are 12 bread rolls.
$12 ÷ 3 = 4$
There are 4 packs of bread rolls.

Question **2**: $21 ÷ 3 = 7$
7 doughnuts will go on each plate.

Question **3**: Zac used grouping.
Olivia drew a 3 × 11 array or an 11 × 3 array.
Lee used the fact 11 × 3 = 33 to help him.
You could also use sharing or repeated subtraction.

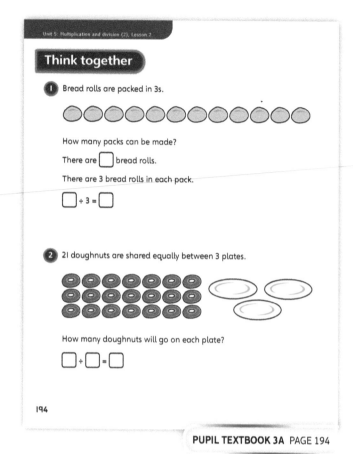

PUPIL TEXTBOOK 3A PAGE 194

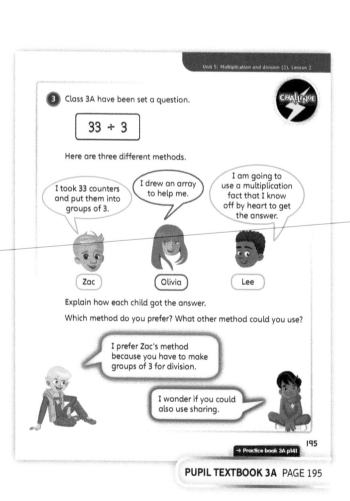

PUPIL TEXTBOOK 3A PAGE 195

Practice

WAYS OF WORKING Independent work

IN FOCUS Question ② starts to look at how arrays can be used to help solve divisions. Children should start to see that if a group of objects can be shown as either 3 rows or 3 columns then they can work out the answer.

STRENGTHEN To support dividing by 3, first ensure children understand the difference between equal grouping and equal sharing. Provide simple contexts and ask them to use counters to create the problems. For example, in question ①, use 12 counters to represent the grapes and put them into 3 piles. Link any division sentences clearly with the equipment and the context. For example, in 12 ÷ 3, 12 is the number of grapes; 3 is the number of children. Division in this case means sharing. For grouping, ask children to take three counters at a time and show how many groups can be made. Link this to counting back on a number line and repeated subtraction. Again, link the concrete and abstract division sentence.

DEEPEN In question ③ b), Ash asks how many more are needed to share the cubes equally. Ask children to use this to work out which numbers divide equally by 3. Do they notice anything about these numbers?

Question ⑤ is a two-step question. Children first need to count in 2s and determine that there are 18 cubes. They then choose a method of dividing 18 by 3.

THINK DIFFERENTLY Question ④ provides children with a division sentence and asks them to work out another. Ask: *Can we start with the previous answer? How many more do you have now? If you are grouping, how many more groups have you got?*

ASSESSMENT CHECKPOINT Can children represent a question as a division sentence involving ÷ 3 and use their knowledge of counting back in 3s to work out the answer? Can children understand the difference between sharing and grouping, and know that the division sign can be used to represent both?

ANSWERS Answers for the **Practice** part of the lesson can be found in the *Power Maths* online subscription.

Reflect

WAYS OF WORKING Pair work

IN FOCUS After independently working out their explanation, children should explain their thinking to a partner. Encourage the use of concrete objects to help them. Check whether they use a grouping or a sharing strategy. Ask them if they can also demonstrate the other method.

ASSESSMENT CHECKPOINT Check if children know that, for dividing by 3, they can either group or share, and that they know the differences between the methods. They should also know that the answer to the division will be the same whichever method they use.

ANSWERS Answers for the **Reflect** part of the lesson can be found in the *Power Maths* online subscription.

After the lesson ⏸

- Can children form division sentences involving grouping or sharing?
- Do children know the difference between division as grouping and division as sharing?
- Do children know how to use repeated subtraction and counting back in 3s to divide by 3? Do they know how an array can help?

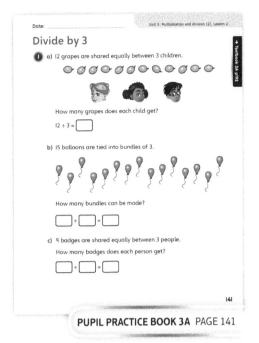

PUPIL PRACTICE BOOK 3A PAGE 141

PUPIL PRACTICE BOOK 3A PAGE 142

PUPIL PRACTICE BOOK 3A PAGE 143

The 3 times-table

Learning focus

In this lesson, children will focus on the 3 times-table. They will realise that the 3 times-table contains multiplication facts and they should be able to recall associated division facts from multiplication facts.

Before you teach ⏸

- Can children count on and back in 3s?
- Do they understand the link between multiplication and division?

NATIONAL CURRICULUM LINKS

Year 3 Number – multiplication and division

Recall and use multiplication and division facts for the 3, 4 and 8 multiplication tables.

Write and calculate mathematical statements for multiplication and division using the multiplication tables that they know, including for two-digit numbers times one-digit numbers, using mental and progressing to formal written methods.

ASSESSING MASTERY

Children can recognise a multiplication fact from a given image and begin to develop a rapid recall of multiplication facts and associated division facts for the 3 times-table.

COMMON MISCONCEPTIONS

Children think that $0 \times 3 = 3$. This is a common mistake and it is mainly because children have not thought about the answer and instead try to recall it too quickly. Remind children that $0 \times$ anything always gives the answer 0. Ask:

- *Apples are in groups of 3. There is one group of apples, how many apples are there? There are no groups of apples, how many apples are there now?*

To work out if 9×3 is greater than or equal to 7×3, children may think they have to work out each multiplication fact, but this is not the case. They could draw the groups in a straight line to compare them. Ask:

- *Can you show me why 9 groups of 3 must be greater than 7 groups of 3, without working them out?*

STRENGTHENING UNDERSTANDING

To strengthen understanding of multiplication facts, show the facts as visual or concrete representations. For example, 1×3 could be represented by 1 row of 3 counters, 2×3 as 2 rows of 3 counters, etc. Link this to counting in 3s on a number line.

At this stage, it is fine if some children still need to use cubes or counters to represent multiplications and divisions. What is important is that children can see what the multiplication facts are. Reinforce that they need to eventually develop rapid recall of the multiplication and division facts, but also need to understand what they are.

GOING DEEPER

Can children reason why $8 \times 3 > 7 \times 3$ without working out the two multiplications? Can they work out missing numbers to make statements correct, for example, $12 \div 3 > \boxed{} \div 3$. What patterns can they see on a 100 square if they shade the numbers from the 3 times-table? What do they notice if they add up these digits?

KEY LANGUAGE

In lesson: times-table, multiply (×), divide (÷), group, multiplication fact, division, total, method

Other language to be used by the teacher: recall, number line, pattern

STRUCTURES AND REPRESENTATIONS

Number lines, arrays

RESOURCES

Mandatory: cubes, counters

Optional: times-tables written on stairs

 In the eTextbook of this lesson, you will find interactive links to a selection of teaching tools.

Quick recap 🔄

Play times-table bingo with the 2, 5 and 10 times-tables. Children draw a 2×3 grid and fill the grid by writing the answer to six multiplication facts for the 2, 5 and 10 times-tables, up to ×12. Read out a multiplication fact. If they have the answer, they cross it out. The first to cross out all six answers wins. You can play this as separate times-tables or as mixed times-tables.

Discover

The 3 times-table

WAYS OF WORKING Pair work

Discover

ASK

- Question ❶ a): *What times-table is shown here? How do you know? What do you notice about the answers to the multiplications? How do they link to the divisions? Can you convince me that 5 × 3 = 15?*
- Question ❶ b): *What different multiplication facts for the 3 times-table can you see here? What can the facts help you work out? Can you draw or write a statement to show 10 × 3? How can you work out how many bananas there are in total?*

IN FOCUS Question ❶ a) asks children to work out missing multiplication and division facts. Children may use their knowledge from the previous two lessons where they were multiplying and dividing by 3. Question ❶ b) asks children to identify the relevant multiplication fact shown by the image. This is important in helping children see that multiplication facts are not just abstract representations.

PRACTICAL TIPS If you have the multiplication facts on a poster in your classroom, cover them up. Times-tables written on stairs are becoming increasingly common and could be something that your school considers.

ANSWERS

Question ❶ a): 18 ÷ 3 = 6   11 × 3 = 33
             9 ÷ 3 = 3   4 × 3 = 12

Question ❶ b): 5 × 3 = 15

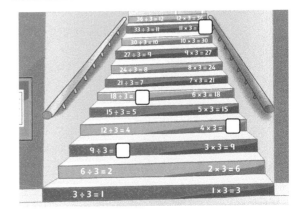

❶ a) What are the missing answers?
How did you work them out?

b) Which multiplication fact can help you work out the total number of bananas shown?

196

PUPIL TEXTBOOK 3A PAGE 196

Share

WAYS OF WORKING Whole class teacher led

ASK

- Question ❶ a): *How can you show 4 × 3? What does 4 × 3 look like as an array? What method did you use to work out the answer? How did you work out the divisions?*
- Question ❶ b): *What multiplication fact is shown? What does it help you work out? Could you write this another way?*

IN FOCUS In question ❶ a), arrays are shown as a visual representation of each question. To work out the answer, children may need to count up in 3s. For the divisions, this is shown as grouping. Most children understand division as grouping when it is read out, as teachers often say 'how many 3s go into 15?' Dexter shows how many groups of 3 are in each number and how this leads to the answer. For question ❶ b), children interpret the visual representation as a multiplication fact. Discuss with children that they can use the 3 times-table to work out totals if you know there are 3 in each group. Children should be encouraged to start to remember multiplication facts.

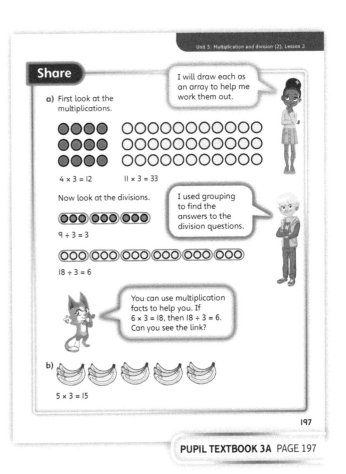

PUPIL TEXTBOOK 3A PAGE 197

Think together

Unit 5: Multiplication and division (2), Lesson 3

Think together

WAYS OF WORKING Whole class teacher led (I do, We do, You do)

ASK

- Question **1**: *What multiplication facts can you see? How can you use these facts to work out the total?*
- Question **2**: *What answers did you know straight away? How many did you complete in one minute? Which ones did you need help with? What can you do to help you learn them?*
- Question **3**: *How can you use the answer to 6 × 3 to work out 12 × 3? What else can you do to work out 12 × 3?*

IN FOCUS Question **1** presents images where the total can be worked out using multiplication facts for 3. Children should be able to explain which multiplication fact they can use. Encourage them to talk about the number of groups and the number of objects in each group. Question **2** aims to develop fluency with multiplication facts by asking how fast children can find the answers. Explain that it is important for children to develop speed of recall with multiplication facts and that they should also learn the division facts. Show children how the multiplication facts can be used to derive the division facts. In question **3**, children use multiplication facts to derive other multiplication facts such as using the fact that 12 × 3 is 10 × 3 plus 2 × 3, or that 12 × 3 is double 6 × 3. This might help children with strategies for working out multiplication facts that they do not know.

STRENGTHEN To strengthen understanding of multiplication facts, show the facts as visual or concrete representations such as 1 × 3 represented by a tower of 3 cubes, or 2 × 3 by 2 towers of 3 cubes. It is important for children to see what multiplication facts are. Reinforce that they need to develop rapid recall of the multiplication and division facts, but also need to understand what they are.

DEEPEN For question **3**, ask: *If you know that 8 × 3 = 24, how many division facts can you show? What other multiplication facts can you work out from this?*

ASSESSMENT CHECKPOINT Have children begun to know multiplication and division facts using rapid recall? Can they work out which fact will help them work out the total, or the answer to a division question?

ANSWERS

Question **1** a): 4 × 3 = 12

Question **1** b): 11 × 3 = 33

Question **2**:

33	27	0	4
21	24	12	6
10	6	8	12

Question **3** a): Double 6 × 3
10 × 3 + 2 × 3

Question **3** b): 3 × 3 × 3 = 27 (use 9 × 3)
13 × 3 = 39 (add 3 to the answer to 12 × 3)
3 × 20 = 60 (double the answer to 10 × 3)

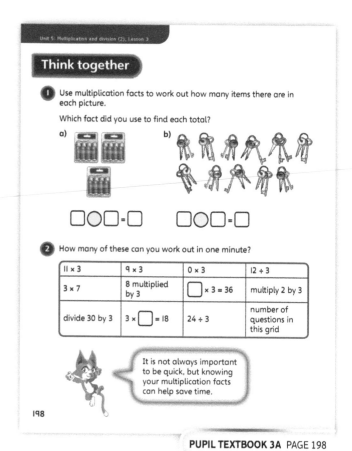

1 Use multiplication facts to work out how many items there are in each picture.

Which fact did you use to find each total?

a)   b)

□○□-□   □○□-□

2 How many of these can you work out in one minute?

11 × 3	9 × 3	0 × 3	12 ÷ 3
3 × 7	8 multiplied by 3	□ × 3 = 36	multiply 2 by 3
divide 30 by 3	3 × □ = 18	24 ÷ 3	number of questions in this grid

It is not always important to be quick, but knowing your multiplication facts can help save time.

198

PUPIL TEXTBOOK 3A PAGE 198

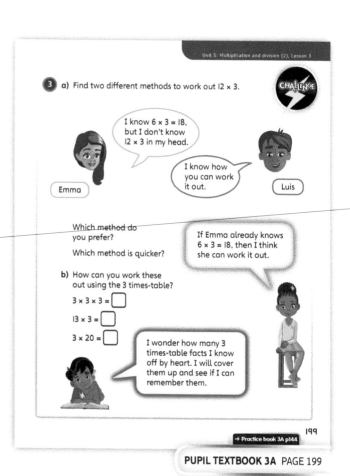

3 a) Find two different methods to work out 12 × 3.

CHALLENGE

I know 6 × 3 = 18, but I don't know 12 × 3 in my head.

I know how you can work it out.

Emma   Luis

Which method do you prefer?

Which method is quicker?

If Emma already knows 6 × 3 = 18, then I think she can work it out.

b) How can you work these out using the 3 times-table?

3 × 3 × 3 = □

13 × 3 = □

3 × 20 = □

I wonder how many 3 times-table facts I know off by heart. I will cover them up and see if I can remember them.

→ Practice book 3A p144

199

PUPIL TEXTBOOK 3A PAGE 199

Practice

WAYS OF WORKING Independent thinking

IN FOCUS Question **2** checks if children know their multiplication facts for the 3 times-table. Children should try to do these without referring back and should rely on mental recall. For those who are unable to do this, ask if they can work any out using other multiplication facts. If they are still unable to work them out, then ask them to go back to counting up in 3s. In question **3**, children might need to work out the answers to compare them to 21, or they may develop a reasoning strategy. For example, if they know 8 × 3 is greater than 21, then 9 × 3 must also be greater than 21. Question **5** asks children to put in inequality or equals signs to make statements correct. Encourage children to use reasoning instead of working out all the answers. Question **6** looks at the pattern that the 3 times-table makes on a 100 square.

STRENGTHEN To strengthen understanding of multiplication facts, show the facts as visual or concrete representations. Question **1** will help to strengthen this. As children move through the rest of the questions, they become more abstract so they may still need to use cubes or counters to represent multiplications and divisions. Reinforce that they need to develop rapid recall of multiplication facts.

DEEPEN Ask children if they had to work out each answer in question **3**. If so, ask: *Is there another way you can do it? Which multiplication fact do you need to know to start?* For question **5**, ask children to explain how and why they chose the sign for each statement. After shading the 100 square in question **6**, can children come up with a rule for numbers that are in the 3 times-table? What do they notice if they add up all the digits?

ASSESSMENT CHECKPOINT Can children recognise multiplication facts for the 3 times-table from images? Children should start to develop a secure recall of the 3 times-table (both multiplication and associated division facts). Encourage children to practise at home and build in daily practice in school, if necessary.

ANSWERS Answers for the **Practice** part of the lesson can be found in the *Power Maths* online subscription.

Reflect

WAYS OF WORKING Pair work

IN FOCUS Children should try to work out the answers mentally and as quickly as possible to encourage rapid recall. Ask children to check each other's answers. Give support to any who are struggling. Ask children to make a note of any facts that they are consistently answering incorrectly and focus on them.

ASSESSMENT CHECKPOINT Check if children have instant recall of multiplication facts for the 3 times-table up to 12 × 3. Look to see if there are any that they are consistently answering incorrectly.

ANSWERS Answers for the **Reflect** part of the lesson can be found in the *Power Maths* online subscription.

After the lesson

- Can children recall division and multiplication facts for the 3 times-table?
- Which facts do children find more difficult?
- Do children know which multiplication or division fact they need to know in order to work out a total?

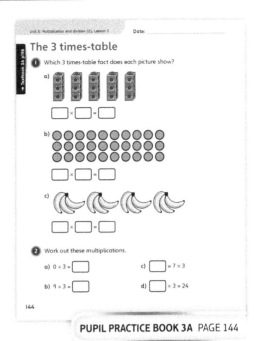

PUPIL PRACTICE BOOK 3A PAGE 144

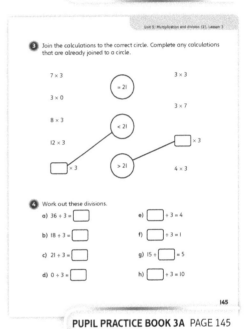

PUPIL PRACTICE BOOK 3A PAGE 145

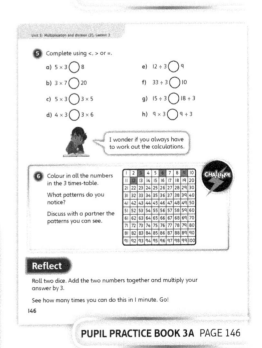

PUPIL PRACTICE BOOK 3A PAGE 146

Multiply by 4

Learning focus

In this lesson, children will start to understand what it means to multiply by 4. Children will see the link between repeated addition, counting up in 4s and multiplying by 4.

Before you teach

- Can children count in 2s?
- Do children know their 2 times-table?
- Do children know the link between repeated addition and the answer to a multiplication?

NATIONAL CURRICULUM LINKS

Year 3 Number – multiplication and division

Recall and use multiplication and division facts for the 3, 4 and 8 multiplication tables.

Write and calculate mathematical statements for multiplication and division using the multiplication tables that they know, including for two-digit numbers times one-digit numbers, using mental and progressing to formal written methods.

ASSESSING MASTERY

Children can form a multiplication sentence involving multiplying by 4. Children can work out the answer to a multiplication sentence by knowing its link with repeated addition and using a number line to count up in 4s.

COMMON MISCONCEPTIONS

When children are presented with 4 groups of objects as opposed to 4 objects in groups, they may not see this as multiplying by 4. For example, the language used so far is that 12×4 is linked to 12 groups of 4. When children are presented with 4 groups of 8, they may only see this as 4×8 and so may not see the connection with 8×4. Show an array with 4 rows of 8. Work out the total and then rotate the array. Ask:

- *What is the total now? What has changed? What is still the same?*

Some children start the count from 0 when working out 11×4, for example, even if they know 10×4. Ask:

- *What multiplication fact for 4 do you already know? How could that help you here?*

STRENGTHENING UNDERSTANDING

Ensure that direct links are really clear between the objects in a picture, a repeated addition, counting up in 4s on a number line and the multiplication sentence. As children count up in 4s, put counters on the number line to show each group.

GOING DEEPER

Can children use multiplication facts that they know in order to work out other multiplication facts? For example, can they use 12×4 to work out 13×4, or 24×4? They could start to explore how to work out $2 \times 4 + 5 \times 4$ as one multiplication, showing by adding arrays together that this is the same as 7×4.

KEY LANGUAGE

In lesson: multiply (×), ~~multiplication~~ sentence, total, double, count up

Other language to be used by the teacher: multiplication statement, equal groups, x groups of y, times-table

STRUCTURES AND REPRESENTATIONS

Number lines, arrays

RESOURCES

Mandatory: cubes, counters

Optional: plastic animals

 In the eTextbook of this lesson, you will find interactive links to a selection of teaching tools.

Quick recap

On the board, draw a 10×4 array of apples or counters. Explain to children that you are going to count them. What methods do they know of counting them? They may count the rows of 10s or in 2s. As a class, count in 4s. If any of the children are unsure, give them strategies to get the next number.

Discover

Multiply by 4

Discover

WAYS OF WORKING Pair work

ASK

- Question ① a): *How many groups (donkeys) do you have? How many legs are there in each group? How can we write this as a multiplication sentence? How can we work out the total? How do you get from 6 × 2 to 6 × 4? What is the connection?*
- Question ① b): *How much does each person pay? How much do they pay in total? What is the multiplication sentence you have worked out?*

IN FOCUS Question ① a) asks children to find the number of legs on 6 donkeys. Children should be able to write this as a multiplication calculation and should start to understand this as 6 groups of 4, which can be worked out using 6 × 4. Children may use different methods for working this out; try to encourage efficient methods as much as possible. Some children may work out 6 × 4 by counting in 2s and will see the connection between multiplying by 2 and by 4. Children should start to realise that an effective way for working out the total is to count up in 4s or to double the number and double again. Question ① b) asks a similar question, but this time an answer is given and children have to reason whether it is correct. They could use counters to represent the coins.

PRACTICAL TIPS For this activity, you can use plastic animals to represent the donkeys. Ensure there are counters and cubes available for each child.

① a) There are 6 donkeys.

How many donkey legs are there in total?

Write a multiplication sentence to work out the answer.

b) A family of 5 people are going donkey trekking.

Mr Peters pays 20 £1 coins in total for him and his family.

Is this the correct amount?

200

PUPIL TEXTBOOK 3A PAGE 200

ANSWERS

Question ① a): 6 × 4 = 24
There are 24 donkey legs.

Question ① b): 5 × 4 = 20
This is the correct amount.

Share

WAYS OF WORKING Whole class teacher led

ASK

- Question ① a): *Why do we count in 4s and not in 1s? Can we count in 2s to get the answer? Why can this be seen as a multiplication?*
- Question ① b): *How does the multiplication relate to the array?*

IN FOCUS For question ① a), encourage children to share their methods with the rest of the class. Astrid encourages children to find a more efficient way of finding the total rather than counting in 1s. Counting in 4s is used to show how children might have arrived at the answer. Explain that the situation can be seen as 6 groups of 4 and therefore can be written as a multiplication calculation. The method of multiplying by 2 and by 2 again is not covered until later, but you may find that you want to discuss it here. Question ① b) shows a multiplication in the form of a 5 × 4 array and shows how the total is 20. Children should understand how an array links to a multiplication.

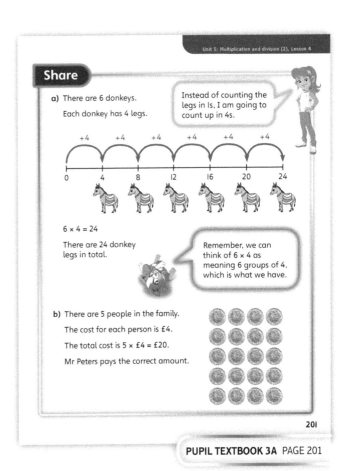

PUPIL TEXTBOOK 3A PAGE 201

Think together

Whole class teacher led (I do, We do, You do)

ASK

- Question ❶: *How many groups are there? How many are in each group? If we count from 0, where do we get to? How many 4s are we adding together? What multiplication is this? What is the answer? Could we have used an answer from earlier? How?*
- Question ❷: *How can we work out the number of donkeys? Do we count in 4s again? Where do we stop? Why is it 8 donkeys? Could we have got the answer quicker by using the answer to question ❶?*
- Question ❸: *Is Ebo right? Can you use Ebo's method to work out the other number multiplied by 4? Why does Ebo's method work?*

IN FOCUS Question ❶ builds on the **Discover** section by adding an extra donkey. This reinforces the link between counting up in 4s (repeated addition) and multiplying by 4. The question starts by counting in 4s from 0 again. Discuss with children how they can use their answer from **Discover** to get the answer by simply counting on another 4. In question ❷, children need to see this as counting up in 4s to 32, as opposed to a division. They should point to the number line and count up in 4s as they go. Encourage them to show why it is 8 donkeys. Question ❸ draws out the method of multiplying by 2 and then multiplying by 2 again to multiply by 4. Children should follow Ebo's method and then apply it to other multiplications. Children are asked to use diagrams or equipment to explain why multiplying by 2 and 2 again is the same as multiplying by 4.

STRENGTHEN At each stage, use concrete objects alongside a number line to help children see the link between repeated addition, counting up in 4s and the eventual multiplication. In question ❶, you might want to encourage children to put 4 counters at each point on the number line to show that this represents a donkey. You may find it easier to support multiplying by 4 initially by multiplying by 2 and then multiplying by 2 again.

DEEPEN In question ❸, can children explain why multiplying by 2 and then by 2 again is equal to multiplying by 4? Ask: *How can you use 10 × 4 to work out 20 × 4, or 14 × 4?*

ASSESSMENT CHECKPOINT Can children understand what it means to multiply by 4 and use their knowledge of counting in 4s to work out the answers to the repeated addition and multiplication sentences that they form?

ANSWERS

Question ❶ a): 7 × 4 = 28
There are 28 donkey legs.

Question ❶ b): 4 × 4 = 16
It costs £16 in total to go donkey trekking.

Question ❷: There are 8 donkeys because 8 × 4 = 32.

Question ❸: 9 × 4 = 36
Yes, Ebo's method always works.

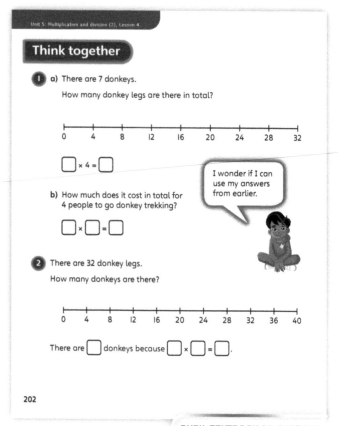

PUPIL TEXTBOOK 3A PAGE 202

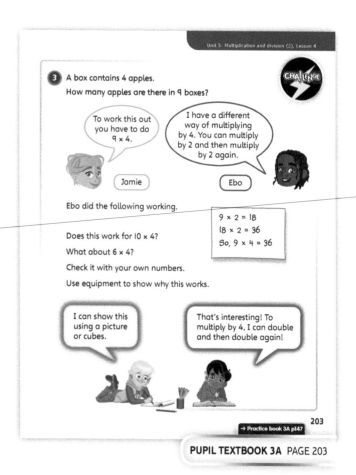

PUPIL TEXTBOOK 3A PAGE 203

Practice

WAYS OF WORKING Independent thinking

IN FOCUS Question ❶ provides practice for children to form the correct multiplication sentences from given situations and pictorial representations. Encourage children to use a number line and to count up in 4s to help them work out the answers. Some children may use the answer to part a) to help them with part b). Questions ❸ and ❹ provide different contexts involving measure and money. Children should write down the correct multiplication sentence from the context given. Children may already know some of the multiplication facts for the 4 times-table off by heart and will simply write down the answer, so encourage checking.

STRENGTHEN To support multiplying by 4, use concrete objects alongside a number line. Count aloud with children in 4s from 0, each time showing the jump on the number line. If children are confident with multiplying by 2, or can double, then you might want to encourage children to double and double again in order to multiply by 4.

DEEPEN In question ❹ c), how many different ways can children write their answer? For example, do they write it as an addition or can they write it as a single multiplication? Question ❻ is similar, but this time the problem is not separated into parts. Encourage children to attempt the question in their own way. If children have worked out two separate multiplications and added them, ask if there is a quicker way.

THINK DIFFERENTLY Question ❺ asks children to use the method of doubling and doubling again to multiply by 4. Children have already looked at this method in the **Think together** section. Children may not be able to work this out mentally and so you may want to ask them to use another method, such as partitioning 21 groups of 4 into: 10 groups of 4 plus 10 groups of 4 plus 1 group of 4.

ASSESSMENT CHECKPOINT Can children represent a question as a multiplication sentence involving × 4 and can they use counting in 4s to work out the correct answer to the multiplication? Do children know that multiplying by 2 and multiplying by 2 again is the same as multiplying by 4?

ANSWERS Answers for the **Practice** part of the lesson can be found in the *Power Maths* online subscription.

Reflect

WAYS OF WORKING Independent thinking

IN FOCUS Children may approach this question by explaining a specific example or by explaining a generic method. They may discuss how they could use a number line to count up in 4s or they may use the method of doubling and doubling again. After explaining a method, ask them to apply it.

ASSESSMENT CHECKPOINT Check if children have an appropriate method for multiplying by 4.

ANSWERS Answers for the **Reflect** part of the lesson can be found in the *Power Maths* online subscription.

After the lesson ⏸

- Can children form multiplication sentences involving × 4 from a word problem?
- Do children know how to work out an answer to a multiplication by counting in 4s?
- Do children know the method of doubling and doubling again?

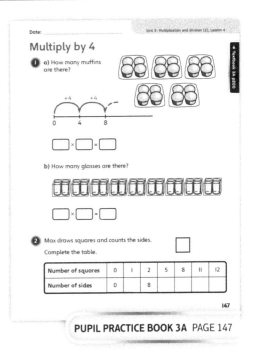

PUPIL PRACTICE BOOK 3A PAGE 147

PUPIL PRACTICE BOOK 3A PAGE 148

PUPIL PRACTICE BOOK 3A PAGE 149

Divide by 4

Learning focus

In this lesson, children will start to understand what it means to divide by 4. Children will understand that division sentences can be used to represent either equal grouping or equal sharing.

NATIONAL CURRICULUM LINKS

Year 3 Number – multiplication and division

Recall and use multiplication and division facts for the 3, 4 and 8 multiplication tables.

Write and calculate mathematical statements for multiplication and division using the multiplication tables that they know, including for two-digit numbers times one-digit numbers, using mental and progressing to formal written methods.

ASSESSING MASTERY

Children can form a division sentence from either a grouping or a sharing situation as they know the difference between grouping and sharing. Children are able to use repeated subtraction and counting back in 4s to work out the result of a division; and they know that a method for dividing by 4 is to halve and halve again.

COMMON MISCONCEPTIONS

Children may think that ÷ always means sharing. Present children with examples to show that it can also mean grouping. Ask:
- *Can you tell me a story where you share and another where you group? What is the same in your stories? What is different?*

When dividing by 4, children may lose track of the count on or back and, therefore, arrive at the wrong answer. Explain that children need to commit the multiplication facts and associated division facts to memory. Ask:
- *Have you checked your answer? Do you know where your mistake came from? Is there a quicker way to find the answer?*

STRENGTHENING UNDERSTANDING

Arrays may help children find the answers to divisions. If the problem asks to divide by 4 and they are sharing, children can put objects into 4 columns; if they are grouping, they should put 4 objects in a column. The number of rows and columns formed will give the answer to the division.

GOING DEEPER

Ask children to show two different word problems for 12 ÷ 4 = 3. What is the same about their stories? What is different? Can children use 48 ÷ 4 to work out 52 ÷ 4? Ask children to solve missing number problems, such as ☐ × 4 = 44.

KEY LANGUAGE

In lesson: divide (÷), division sentence, group, share, shared equally, array, left over

Other language to be used by the teacher: equal, count back

STRUCTURES AND REPRESENTATIONS

Number lines, arrays

RESOURCES

Mandatory: cubes, counters

Optional: 48 playing cards

 In the eTextbook of this lesson, you will find interactive links to a selection of teaching tools.

Quick recap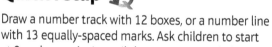

Draw a number track with 12 boxes, or a number line with 13 equally-spaced marks. Ask children to start at 0 and count in 4s until they get to 48. Ask them to compare their answers with a partner. Did they get the same answers?

Discover

Divide by 4

WAYS OF WORKING Pair work

Discover

ASK

- Question ① a): *How many cards are being shared? How many people are we sharing them between? How many does each person receive? Is this equal grouping or sharing? How can you write it as a division?*
- Question ① b): *How many cards are left over after the 20 have been shared? How did you work this out? How many cards are going into each pile? Is this equal sharing or grouping? How many piles do you get? Can you write this as a division?*

IN FOCUS Question ① a) is about showing division as equal sharing. Children may share cards, cubes or counters out equally to imitate sharing the 20 cards between 4 people. Ask them how they could record their working. For example, they may draw 20 cards and 4 people and draw arrows from each card to a person. They may share one at a time or they may take four cards at a time and give one to each person. Question ① b) provides a contrasting example and this time looks at grouping. There is an additional step in this question as children first have to work out how many leftover cards there are. They may approach this by counting the ones remaining or by doing a subtraction, which is the method that should be encouraged. This activity should reinforce the different instances where the division sign is used (grouping and sharing).

PRACTICAL TIPS Put children into groups of four and play a game using 48 cards to help them understand the context and the problem.

ANSWERS

Question ① a): 20 ÷ 4 = 5
Each player gets 5 cards.

Question ① b): 28 ÷ 4 = 7
There are 7 piles.

① a) 20 cards are dealt equally between 4 players.
How many cards does each player get?
Write this as a division sentence.

b) The 28 left over cards are put into piles of 4.
How many piles are there?

204

PUPIL TEXTBOOK 3A PAGE 204

Share

WAYS OF WORKING Whole class teacher led

ASK

- Question ① a): *How will Dexter share out the cards? Is there another way? Will each person get the same number of cards? How do you know?*
- Question ① b): *How can you work out how many cards are left over? What is the difference between this question and the one before? Why do you still use the division sign? What does each number in the division mean?*

IN FOCUS For question ① a), ensure that the method of sharing is demonstrated and explain each number in the division sentence. It is important to link each number in the calculation to what it means in the image. The most efficient method for sharing would be to take four cards from the box each time and give one to each child. Explain that children could also have shared the cards one by one, like Dexter. Question ① b) shows a method for grouping. Children first need to find how many cards are left over. Children need to understand the methods for grouping and sharing.

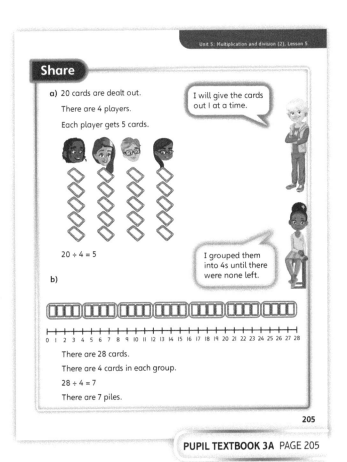

Share

a) 20 cards are dealt out.
There are 4 players.
Each player gets 5 cards.

I will give the cards out 1 at a time.

20 ÷ 4 = 5

I grouped them into 4s until there were none left.

b)

0 1 2 3 4 5 6 7 8 9 10 11 12 13 14 15 16 17 18 19 20 21 22 23 24 25 26 27 28

There are 28 cards.
There are 4 cards in each group.
28 ÷ 4 = 7
There are 7 piles.

205

PUPIL TEXTBOOK 3A PAGE 205

Think together

Whole class teacher led (I do, We do, You do)

ASK

- Question ❶: *How many coins are there? How many money boxes are there? Is this grouping or sharing? Can you explain each step of your method? What does the division look like? What does each number in the division sentence mean?*
- Question ❷: *How many flowers are there? How many flowers are put into a bunch? Is this grouping or sharing? What division do you do? Can you explain what each number in the division means?*
- Question ❸: *What is Jamie's method? Can you explain why Jamie's method works? Can you share 44 marbles equally? Why can you not share 22 marbles equally between 4 boxes? How many have you got left over? How many more would you need to share equally?*

IN FOCUS Question ❶ is an example of equal sharing. Children may share out one coin at a time or take a group of four coins. Children should be confident with sharing and may go straight to the division. Question ❷ is an example of equal grouping. Children may circle the flowers with their finger to work out how many groups of 4 there are. As a class, discuss that the flowers may be grouped in different ways, but the same number of groups remains. In both questions, link the numbers in the division to the context. Question ❸ explores the method of halving and halving again as an alternative for dividing by 4. Children may find this easier.

STRENGTHEN For questions ❶ and ❷, carry out the grouping and sharing with counters to represent the objects. Explicitly link each action with the steps in repeated subtraction or sharing. Ensure that children know what each number in each division sentence represents. Children may find making an array with 4 counters in each row an effective way. In question ❸, some children may need to use counters to help them do the division.

DEEPEN Ask children how they could use division to work out the missing number problem ☐ × 4 = 36. Ask: *Can you explain why you use division to work out the missing number? Can you make up your own problems similar to this?*

ASSESSMENT CHECKPOINT Children understand that division can be either equal grouping or equal sharing and are able to form a division sentence to answer a question. They can also work out the answer to division sentences practically, using an array or through knowledge of multiplication facts.

ANSWERS

Question ❶: 12 ÷ 4 = 3
There are 3 coins in each money box.

Question ❷: 32 ÷ 4 = 8
There are 8 bunches.

Question ❸ a): 44 ÷ 4 = 11
There are 11 marbles in each box.

Question ❸ b): You cannot share 22 marbles equally between 4 boxes. There are 2 marbles too many. You would need 2 more marbles to share equally.

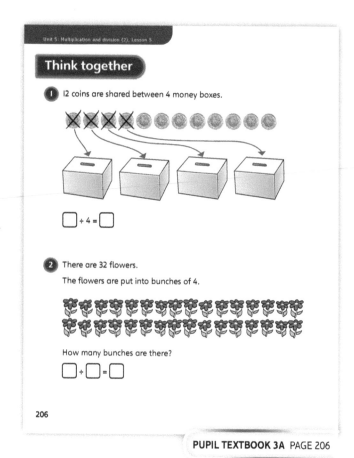

Think together

❶ 12 coins are shared between 4 money boxes.

☐ ÷ 4 = ☐

❷ There are 32 flowers.
The flowers are put into bunches of 4.

How many bunches are there?

☐ ÷ ☐ = ☐

206

PUPIL TEXTBOOK 3A PAGE 206

❸ a) 44 marbles are shared equally between the boxes.

CHALLENGE

I have 44 marbles. I know I need to divide by 4. I can divide by 2 and by 2 again. This will tell me how many I should put in each box.

Jamie

How many marbles are there in each box?

b) Is it possible to divide 22 marbles equally between 4 boxes? Discuss this with a partner.

I think Jamie's method works every time.

I wonder which numbers can be divided by 4 and which cannot. Is there any way of telling?

→ Practice book 3A p150

207

PUPIL TEXTBOOK 3A PAGE 207

Practice

WAYS OF WORKING Independent thinking

IN FOCUS Questions **1** and **2** provide a mixture of grouping and sharing problems. Encourage children to link each number in the division sentence with the context. Questions **3** and **4** focus on using an array. Children should start to see that if a group of objects can be shown as either 4 rows or 4 columns then they can work out the answer to the division.

STRENGTHEN To support dividing by 4, ensure children understand the difference between equal grouping and equal sharing. Provide simple contexts and ask children to show with counters. For example, in question **1**, they could take 12 counters to represent the jelly beans. They should make 4 piles to show the division or put them into an array with 4 columns (or rows). The number of rows or columns gives the answers to the division: make sure they see the link. In question **5**, children can replace the £20 note with twenty £1 coins, which may help them with grouping. If children struggle with division, they may find halving and halving again easier.

DEEPEN Question **6** provides an opportunity to lead into questions such as 12 ÷ 4 ◯ 12 ÷ 3. Ask children to reason which sign goes between the expressions. They may say that 12 ÷ 4 is smaller as you are sharing the 12 between more 'people'.

THINK DIFFERENTLY Question **6** asks children to consider which is bigger, 24 divided by 4 or 24 divided by 3. Some children may work out the answers and make a decision based on the answers. However, children should be encouraged to reason that the answer to 24 ÷ 4 will be smaller as they are sharing amongst more groups.

ASSESSMENT CHECKPOINT Can children represent a question as a division sentence involving ÷ 4 and can they use their knowledge of counting back in 4s to work out the answer? Do children understand the difference between sharing and grouping, and the division sign can represent both? Do children understand that one method to divide by 4 is to halve and halve again?

ANSWERS Answers for the **Practice** part of the lesson can be found in the *Power Maths* online subscription.

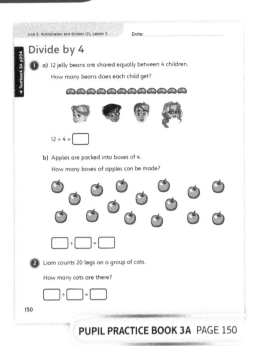

PUPIL PRACTICE BOOK 3A PAGE 150

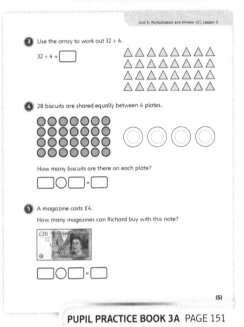

PUPIL PRACTICE BOOK 3A PAGE 151

Reflect

WAYS OF WORKING Pair work

IN FOCUS Children may draw a shape and show that if they divide the shape into 2 equal parts and then divide each half into half again, they end up with 4 equal parts.

ASSESSMENT CHECKPOINT Check whether children understand that a method for dividing by 4 is to halve and halve again.

ANSWERS Answers for the **Reflect** part of the lesson can be found in the *Power Maths* online subscription.

After the lesson

- Can children form division sentences involving 4 from a word problem?
- Do children know how to use repeated subtraction and counting back in 4s to divide by 4?
- Do children know that a method for dividing by 4 is to halve and halve again?

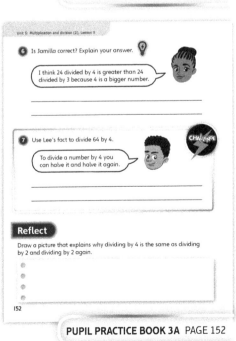

PUPIL PRACTICE BOOK 3A PAGE 152

The 4 times-table

Learning focus

In this lesson, children will focus on learning the 4 times-table and should be able to recall associated division facts from multiplication facts. They will know how the 4 times-table can be derived from the 2 times-table.

Before you teach ⏸

- Can children multiply by 4?
- Can children divide by 4?
- Do children know the 2 times-table?

NATIONAL CURRICULUM LINKS

Year 3 Number – multiplication and division

Recall and use multiplication and division facts for the 3, 4 and 8 multiplication tables.

Write and calculate mathematical statements for multiplication and division using the multiplication tables that they know, including for two-digit numbers times one-digit numbers, using mental and progressing to formal written methods.

ASSESSING MASTERY

Children can recognise a multiplication fact from the 4 times-table in a given image. Children should be developing a rapid recall of multiplication facts and associated division facts from the 4 times-table.

COMMON MISCONCEPTIONS

Children may think that 0 × 4 = 4. This is a common mistake, mainly because children have answered too quickly and not thought about their response. Remind children that 0 × anything always gives the answer 0. Ask:
- *What is 0 × 3? Can that help you work out 0 × 4? How many objects or groups are there when you multiply by 0?*

To work out if 9 × 4 is greater than or equal to 7 × 4, children may think they have to work out each multiplication fact, but this is not the case. Children can draw the groups in a straight line to show this. Ask:
- *Do you need to work out both calculations? Why not? How else could you work it out?*

STRENGTHENING UNDERSTANDING

To strengthen understanding of multiplication facts, show the facts as visual or concrete representations, for example, towers or blocks of 4 cubes. Link this to children's work on counting in 4s on a number line. It is fine at this stage if some children still need to use cubes or counters to represent multiplications and divisions. What is important is that children can see what multiplication facts are. Reinforce that they need to develop rapid recall of multiplication and division facts, but also need to understand what they are.

GOING DEEPER

Ask children to reason why 6 × 4 = 8 × 3 without working out the two multiplications. Can they also work out missing numbers to complete statements such as 12 ÷ 4 < ☐ ÷ 3?

KEY LANGUAGE

In lesson: times-table, multiplication fact, array, pattern

Other language to be used by the teacher: divide, division, grouping, multiply, recall

STRUCTURES AND REPRESENTATIONS

Number lines, arrays

RESOURCES

Mandatory: cubes, counters, 100 square

Optional: 4 times-table flashcards

 In the eTextbook of this lesson, you will find interactive links to a selection of teaching tools.

Quick recap ↺

Ask children to divide each of these numbers by 4: 12, 28, 40, 48. Discuss what methods they used.
Ask: *Does 18 divide by 4 equally?* Children may halve and halve again, or just know the answer.

Discover

WAYS OF WORKING Pair work

ASK

• Question ① a): *What times-table is shown here? How do you know? What do you notice about the answers to the multiplications? What numbers are being covered by the children? How did you work out the answers?*

• Question ① b): *What multiplication can you use to work out 4 × 7? Why? How can you show this using an array? What about a division? Do you have to share or group? Is there a quicker way of doing this?*

IN FOCUS Question ① a) asks children to work out the two multiplication facts that children in the image are standing on. Children may use their knowledge of multiplying and dividing by 4 from previous lessons to answer the questions. Some children may already know their multiplication facts; you could ask them to convince you of their answers. Others may also link this to their knowledge of the 2 times-table, but some may just count on from previous answers. Question ① b) asks children to work out which facts they can use to work out other multiplication facts. Encourage children to use an array to show that 4 × 7 is the same as 7 × 4. For 48 ÷ 4, they may need to use equal grouping or sharing. Try to encourage these children to use one of the multiplication facts to help them.

PRACTICAL TIPS Ensure any multiplication facts on a poster in your classroom are covered up. You could consider using flashcards with multiplication facts for 4 on them.

ANSWERS

Question ① a): 6 × 4 = 24, 8 × 4 = 32

Question ① b): Use 7 × 4 to work out 4 × 7 = 28.
Use 12 × 4 to work out 48 ÷ 4 = 12.

Share

WAYS OF WORKING Whole class teacher led

ASK

• Question ① a): *What does each array show? Could you use something other than an array?*

• Question ① b): *Which multiplication fact did you use? Why do you think it is important to know your 4 times-table off by heart? Can you work out the 4 times-table from the 2 times-table? How?*

IN FOCUS In question ① a), an array is used to give a visual representation of the two multiplication facts. There are multiple ways that children may get the correct answer. For example, they may just count on 4 or double and double again. All are valuable methods, but children need to be moving towards instant recall of these multiplication facts. In question ① b), an array is used to show that 4 × 7 and 7 × 4 are the same. Discuss that using the fact 12 × 4 = 48 makes it easier to work out 48 ÷ 4, rather than going through the process of grouping or sharing.

The 4 times-table

Discover

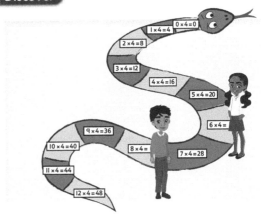

① a) What numbers are the children covering?
How did you work them out?

b) Which multiplication facts will help you work out these calculations?

4 × 7 = ☐   48 ÷ 4 = ☐

Work out the answers.

208

PUPIL TEXTBOOK 3A PAGE 208

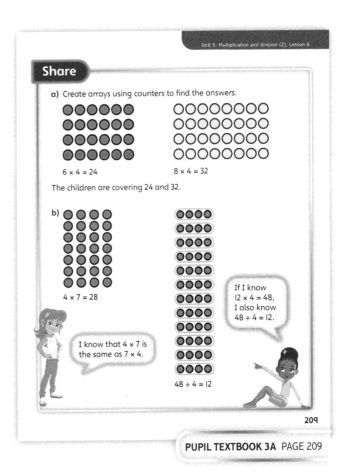

PUPIL TEXTBOOK 3A PAGE 209

Think together

Whole class teacher led (I do, We do, You do)

ASK

- Question **1**: *What multiplication facts can you see? How can you use these facts to work out the total? How can you work out 4 × 6 if you do not know the 6 times-table yet?*
- Question **2**: *How many answers has Mary got right? Which one can you tell is wrong straight away? How do you know? Which ones did Mary get wrong? What mistakes has she made?*
- Question **3**: *Which numbers are in the 4 times-table? Can you see any patterns? What do you notice about the numbers? Are they all even?*

IN FOCUS Question **1** presents images where the total can be worked out using the 4 times-table. Children should be able to explain which multiplication fact they can use. Encourage them to talk about the number of groups and the number of objects in each group. Children may notice that 4 × 6 can be worked out using 6 × 4. Again, children may use methods they have explored in previous lessons if they do not have rapid recall of the facts. Question **2** aims to develop fluency with multiplication facts. Can children work out how many answers Mary has got right? Can they explain the mistakes that have been made? Explain that they should also learn the division facts and show them how multiplication facts can be used to derive division facts. In question **3**, children start to explore times-table patterns on a 100 square. They should see that the numbers are all even and the pattern repeats every two rows.

STRENGTHEN To strengthen understanding of multiplication facts, show the facts as visual or concrete representations, for example, with towers of cubes. It is important that children can visualise and understand multiplication facts. Reinforce that eventually they will need to develop rapid recall.

DEEPEN In question **3**, ask children if they can work out the pattern of the ones digits. Can they use this pattern to work out if 192 can be divided by 4?

ASSESSMENT CHECKPOINT Children start to know and remember their times-table multiplication and division facts using rapid recall. They can work out which fact will help them to work out the total or the answer to a division question.

ANSWERS

Question **1** a): 5 × 4 = 20; there are 20 cubes.

Question **1** b): 11 × 4 = 44; there are 44 boxed pens.

Question **1** c): 4 × 6 = 24; there are 24 bread rolls.

Question **2**: Mary has got five answers right and five answers wrong. The correct answers are:

7 × 4 = 28	12 ÷ 4 = 3
4 × 9 = 36	4 ÷ 4 = 1
4 × 1 = 4	8 ÷ 4 = 2
0 × 4 = 0	24 ÷ 4 = 6
10 × 4 = 40	44 ÷ 4 = 11

Question **3**: 4, 8, 12, 16, 20, 24, 28, 32, 36, 40, 44, 48, 52, 56, 60, 64, 68, 72, 76, 80, 84, 88, 92, 96, 100
They are all even numbers and the pattern repeats every two rows.

PUPIL TEXTBOOK 3A PAGE 210

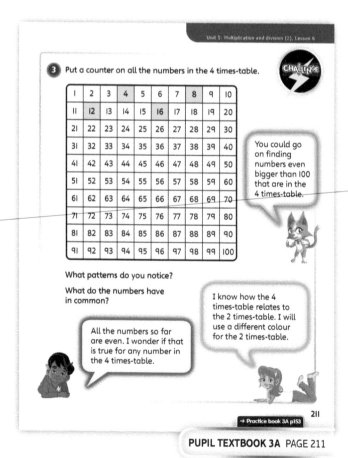

PUPIL TEXTBOOK 3A PAGE 211

Practice

WAYS OF WORKING Independent thinking

IN FOCUS Question ① reinforces visual images of several multiplication facts from the 4 times-table. Questions ② and ④ check whether children know the multiplication and division facts. Children should try to do these without referring back, and should be relying on mental recall. For any that they are still unable to work out, ask them to think about multiplication facts from the 2 times-table. In question ③, children identify numbers that are in the 4 times-table. They use reasoning to try to explain the two children's statements. Question ⑤ asks children to put in inequality or equals signs to make statements correct. Encourage children to use reasoning instead of working out all the answers and then comparing.

STRENGTHEN To strengthen understanding of multiplication facts, show the facts as visual or concrete representations; question ① will help with the strengthening of this. As children move through the rest of the questions, they become more abstract. Children may still need to use cubes or counters to see what multiplication facts are, although they will also need to develop instant recall.

DEEPEN In question ⑤, ask children to explain how they can work out which sign completes the sentence without having to work out the answers. Which ones do they need to work out in order to compare? Can children find the pattern in question ⑥? Can they work out how the missing numbers are formed?

ASSESSMENT CHECKPOINT Can children recognise multiplication facts for 4 from images? They should be starting to develop a secure recall of their 4 times-table (both multiplication and associated division facts). Help children to develop quick recall by setting daily practice at home and at school.

ANSWERS Answers for the **Practice** part of the lesson can be found in the *Power Maths* online subscription.

Reflect

WAYS OF WORKING Pair work

IN FOCUS This brings together work on the 3 times-table and the 4 times-table. Children work out which numbers appear in both the 3 and the 4 times-tables. Encourage children to take a strategic approach rather than just guessing. Ask children to list the numbers in the times-tables and to circle those numbers that appear in both. They should start to notice that the numbers go up in 12s.

ASSESSMENT CHECKPOINT Check whether children have instant recall of the multiplication facts up to 12 × 3 and 12 × 4. Check if there are any facts that they are consistently getting wrong.

ANSWERS Answers for the **Reflect** part of the lesson can be found in the *Power Maths* online subscription.

After the lesson

- Can children recall division and multiplication facts from the 4 times-table?
- Which facts are children struggling with? How can these be reinforced further at school?

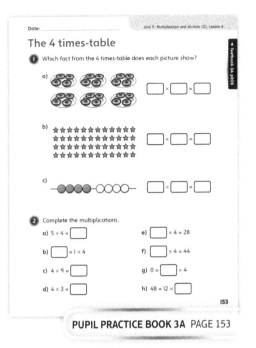

PUPIL PRACTICE BOOK 3A PAGE 153

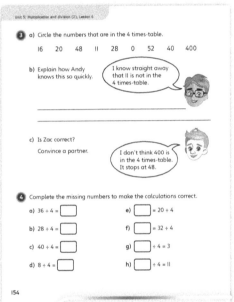

PUPIL PRACTICE BOOK 3A PAGE 154

PUPIL PRACTICE BOOK 3A PAGE 155

Multiply by 8

Learning focus

In this lesson, children will start to understand what it means to multiply by 8. Children will use counting up in 8s to work out the answers to multiplications.

Before you teach

- Can children count in 2s and 4s?
- Do children know the 2 and 4 times-tables?
- Do children know the link between repeated addition and the answer to a multiplication?

NATIONAL CURRICULUM LINKS

Year 3 Number – multiplication and division

Recall and use multiplication and division facts for the 3, 4 and 8 multiplication tables.

Write and calculate mathematical statements for multiplication and division using the multiplication tables that they know, including for two-digit numbers times one-digit numbers, using mental and progressing to formal written methods.

ASSESSING MASTERY

Children can form a multiplication sentence involving multiplying by 8 and can work out the answer to a multiplication sentence by knowing its link with repeated addition and using a number line to count up in 8s. They will start to remember some of the multiplication facts for multiplying by 8 and they will know the link between multiplying by 2, by 4 and by 8.

COMMON MISCONCEPTIONS

Children often start counting in 8s from 0 when working out, for example, 11 × 8, even if they already know 10 × 8. Encourage children to start counting on from known facts. Ask:

- *How will you work this one out? Is there a quicker way?*

Children may lose count, for example, when working out 6 × 8 in their head; they may count up too many or too few 8s. This should reinforce the need for them to know their multiplication facts. Ask:

- *How can you be more efficient? Could a number line help you to keep count? Is there a quicker way to find the answer?*

STRENGTHENING UNDERSTANDING

Children often find it easier to multiply by 8 by doubling the number, then doubling again and doubling again. You can show children this visually by using simple arrays: the array gets twice as big each time. This is a useful method, although children will eventually need to know their multiplication facts for the 8 times-table.

GOING DEEPER

Can children use multiplication facts they know in order to work out other multiplication facts? For example, can they use 12 × 8 to work out 13 × 8, or 24 × 8? Ask children to explore how to work out 4 × 8 + 6 × 8 as one multiplication. They could show, by joining arrays together, that this is the same as 10 × 8.

KEY LANGUAGE

In lesson: multiply (×), total, multiplication sentence, group, double, method

Other language to be used by the teacher: multiplication statement, equal groups, count in 8s, times-table, *x* groups of *y*, altogether

STRUCTURES AND REPRESENTATIONS

Number lines, arrays

RESOURCES

Mandatory: cubes, counters

Optional: circles divided into 2, 4 and 8 equal pieces

 In the eTextbook of this lesson, you will find interactive links to a selection of teaching tools.

Quick recap

Put children into groups of four. Ask each child to make 8 on a ten frame with counters. How many counters do they have? How many counters do they and a partner have? What about three of them? How many counters are there now? Finally, how many counters in total between all four of them? Look for methods that children use to work this out.

Discover

Multiply by 8

Discover

WAYS OF WORKING Pair work

ASK

- Question ① a): *How many groups (or pies) do you have? How many slices are there in each group (or pie)? How can we write this as a multiplication sentence? How can we work out the total? Do you know how to count in 8s?*
- Question ① b): *What is the connection between multiplying by 2, 4 and 8? Can you come up with a rule to help other children multiply by 8? Can you show why the rule works?*

IN FOCUS Question ① a) asks children to find the total number of slices. Each pie is cut into 8 slices. Children should be able to write this as a multiplication sentence. They should start to understand this as 4 groups of 8, which you can work out using 4 × 8. Children may use different methods for working this out. For example, some children may use a number line to count up in 8s, or they may link multiplying by 8 with multiplying by 2 and/or 4. However, try to encourage the use of efficient methods as much as possible. Question ① b) asks children to see the connection between multiplying by 2, 4 and 8. Children should notice that the answer doubles each time. Discuss how this can be used to help form a rule to multiply by 8.

PRACTICAL TIPS For this activity, you could provide circles that are divided into 2, 4 and 8 equal pieces to represent the pies.

ANSWERS

Question ① a): 4 × 8 = 32

There are 32 slices in total.

Question ① b): 5 × 2 = 10, 5 × 4 = 20, 5 × 8 = 40

The answer doubles each time.

① a) Each pie has been cut into 8 slices.

How many slices are there in total?

Write down a multiplication sentence to work out the answer.

b) 5 × 2 = ☐   5 × 4 = ☐   5 × 8 = ☐

What do you notice?

212

PUPIL TEXTBOOK 3A PAGE 212

Share

WAYS OF WORKING Whole class teacher led

ASK

- Question ① a): *Why do we count in 8s and not in 1s? Can we count in 2s or 4s to get the answer? Do you know the answer to 4 × 8 from your knowledge of the 4 times-table?*
- Question ① b): *What is the connection between the answers? How can you use the answer to 5 × 4 to find the answer to 5 × 8?*

IN FOCUS In question ① a), counting in 8s on a number line is used to show how children might have arrived at the answer. Use what Sparks says to explain that the situation can be seen as 4 groups of 8 and can, therefore, be written as a multiplication sentence: 4 × 8 = 32. You can also link this to the 4 times-table: 8 × 4 = 32. In question ① b), draw out the connection between the three multiplications. Children should see from the arrays that the answer doubles each time. Discuss the implications of using this method to help multiply by 8. For example, you can multiply by 4 and double the answer, or you could double, double and double again.

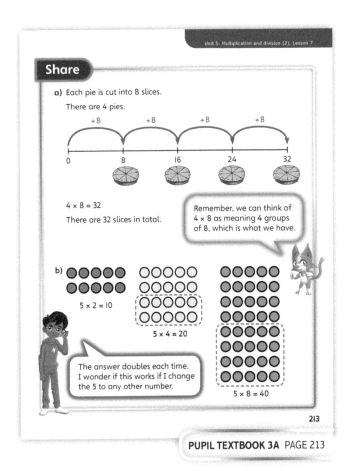

Share

a) Each pie is cut into 8 slices. There are 4 pies.

4 × 8 = 32

There are 32 slices in total.

Remember, we can think of 4 × 8 as meaning 4 groups of 8, which is what we have.

b) 5 × 2 = 10

5 × 4 = 20

5 × 8 = 40

The answer doubles each time. I wonder if this works if I change the 5 to any other number.

213

PUPIL TEXTBOOK 3A PAGE 213

Think together

WAYS OF WORKING Whole class teacher led (I do, We do, You do)

ASK

- Question **1**: *How many legs does 1 spider have? How many spiders are there? How many legs are there in total? What multiplication can you do to work this out?*
- Question **2**: *How many people are in the queue? How much is a ticket? How can you work out the total cost?*
- Question **3**: *Can you explain to a friend why Isla's method works?*

IN FOCUS Question **1** asks children to work out how many legs are on six spiders. Children use a number line to find the total, with the first few jumps given. In question **2**, children have to find 11 × 8. This time there is a unit of money in the answer. Question **3** aims to get children to think about different methods for multiplying by 4 and 8. Children use the different methods to multiply numbers by 4 and 8. Discuss if there is a particular method that children prefer for multiplying by 8. For example, children who struggle to remember the count in 8s may prefer a calculation method such as doubling the number three times. Emphasise the importance of committing these multiplication facts to memory.

STRENGTHEN At each stage, use concrete objects alongside a number line to help children see the link between repeated addition, counting up in 8s and multiplication. In question **1**, you can encourage children to put 8 counters at each point on the number line to show that this represents the legs on the spider. In question **2**, you can use counters to represent the coins and put them into an array. Children who struggle to remember the count sequence for 8s may find it easier to double, double and double again.

DEEPEN Ask: *Why is multiplying by 2 and then by 2 and then by 2 again equal to multiplying by 8? How can you use 12 × 8 = 96 to work out 13 × 8, 14 × 8 or 15 × 8?*

ASSESSMENT CHECKPOINT Children understand what it means to multiply by 8 and use their knowledge of counting in 8s to work out the answers to the repeated addition and multiplication sentences that they form. Children know that to multiply by 8 they can multiply by 4 and double, or multiply by 2, then by 2 and then by 2 again.

ANSWERS

Question **1**: 6 × 8 = 48
There are 48 legs altogether.

Question **2**: 11 × 8 = £88
The total cost is £88.

Question **3** a): 9 × 2 = 18, 18 × 2 = 36

Question **3** b): 9 × 2 = 18
18 × 2 = 36
36 × 2 = 72

Question **3** c): 15 × 4 = 60, 15 × 8 = 120

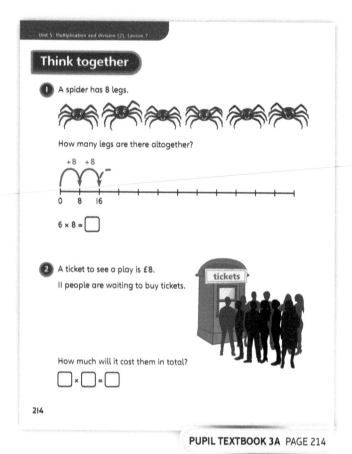

PUPIL TEXTBOOK 3A PAGE 214

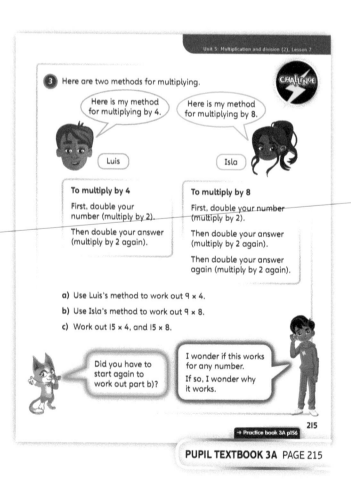

PUPIL TEXTBOOK 3A PAGE 215

Practice

WAYS OF WORKING Independent thinking

IN FOCUS Questions ❶ to ❹ provide practice for children to form the correct multiplication sentences from given situations. Encourage children to use a number line and count up in 8s to help them work out the answers to the calculations; or they may use the rule of doubling, doubling and doubling again. Children may also use their knowledge of other times-tables to work out some of the answers: for example, they may use 8 × 4 to work out 4 × 8. They should write down the correct multiplication sentence from the context given. Children may already know some of the multiplication facts for the 8 times-table off by heart and will simply write down the answer, so encourage checking.

STRENGTHEN To support multiplying by 8, use concrete objects alongside a number line. Count aloud with children in 8s from 0, each time showing the jump on the number line. If children are confident with multiplying by 2, or can double, encourage them to double, double and double again.

DEEPEN Can children make up their own questions similar to the ones in question ❺? Ask: *If I know that 2 × a number is 18, what is 4 × the number and 8 × the same number?* Can children explain what they did and why it works? Question ❻ requires children to use a rule to multiply by 8. Encourage them to come up with other rules for multiplying by 8. For example, they may multiply by 10 and then subtract 8 from the number twice.

THINK DIFFERENTLY Question ❺ gives children a multiplication fact from the 4 times-table and requires them to work out what the number is when multiplied by 8. They may think that they need to do a division to work out the missing number, but instead they should be encouraged to think of a way of doing it using their knowledge of the connection between multiplying by 4 and multiplying by 8.

ASSESSMENT CHECKPOINT Can children represent a question as a multiplication sentence involving × 8 and use counting in 8s to work out the correct answer? Do children know that multiplying by 2, then multiplying by 2 and multiplying by 2 again, is the same as multiplying by 8?

ANSWERS Answers for the **Practice** part of the lesson can be found in the *Power Maths* online subscription.

Reflect

WAYS OF WORKING Independent thinking

IN FOCUS Children should be aware of the connection between multiplying by 2, 4 and 8. Children need to use this knowledge to work out 6 × 8 from 6 × 4, which they can do by doubling. Children may use other valid methods, such as starting from 0 and counting up in 8s, or counting up in 6s from 24.

ASSESSMENT CHECKPOINT Check whether children can use a multiplication fact of 4 to multiply the same numbers by 8.

ANSWERS Answers for the **Reflect** part of the lesson can be found in the *Power Maths* online subscription.

After the lesson ⏸

- Can children form multiplication sentences involving × 8 from a word problem?
- Can they work out multiplications by counting in 8s?
- Do children know the connection between multiplying by 2, 4 and 8?

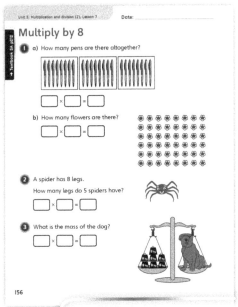

PUPIL PRACTICE BOOK 3A PAGE 156

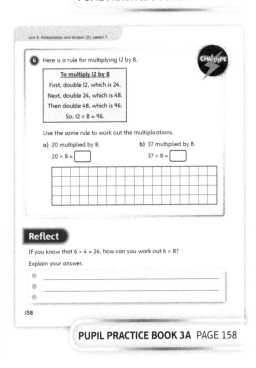

PUPIL PRACTICE BOOK 3A PAGE 157

PUPIL PRACTICE BOOK 3A PAGE 158

Divide by 8

Learning focus

In this lesson, children will understand how they can divide a number by 8.

Before you teach

- Can children count back in 8s from 80 to 0?
- Can children share a set of objects between a given number of groups?
- Can children put a set of objects into groups of a given size?

NATIONAL CURRICULUM LINKS

Year 3 Number – multiplication and division

Recall and use multiplication and division facts for the 3, 4 and 8 multiplication tables.

Write and calculate mathematical statements for multiplication and division using the multiplication tables that they know, including for two-digit numbers times one-digit numbers, using mental and progressing to formal written methods.

ASSESSING MASTERY

Children can form a division sentence from either a grouping or a sharing situation and know the difference between grouping and sharing. Children can use repeated subtraction and counting on or back in 8s to work out the result of a division, as well as know that a method for dividing by 8 is to halve the number, halve the answer and then halve again.

COMMON MISCONCEPTIONS

When dividing by 8, children may lose track of the count on or back. Explain that the sooner they can commit the multiplication facts and associated division facts to memory, the better. Ask:
- *Is there a quicker way to work this out? What multiplication fact could have helped you with this?*

When using the method where they have to halve, halve again and halve again, often children may only halve twice (and, therefore, only divide by 4). Ask:
- *What is the link between the 2, 4 and 8 times-tables? Can you use the 4 times-table to check your answer?*

STRENGTHENING UNDERSTANDING

Work with concrete objects to show the difference between grouping and sharing. Explicitly link each action in grouping with the steps in repeated subtraction. Use a number line to show how grouping can be seen as repeated subtraction by counting back in 8s (or subtracting 8 each time).

To understand sharing, first ask children to share the objects one by one and then ask children to take 8 at a time and allocate to each group. Ensure children know what each number in the division sentence represents.

GOING DEEPER

Ask children to show two different word problems for $32 \div 8 = 4$. What is the same about their stories? What is different? Can children use $96 \div 8$ to work out $104 \div 8$?

KEY LANGUAGE

In lesson: divide ($\div$), division, division sentence, share equally

Other language to be used by the teacher: equal, grouping, array, halve, count in 8s, method, multiply, multiplication fact

STRUCTURES AND REPRESENTATIONS

Number lines, arrays

RESOURCES

Mandatory: cubes, counters

Optional: ice lolly moulds, fruit juice, paper circles

 In the eTextbook of this lesson, you will find interactive links to a selection of teaching tools.

Quick recap

Ask children to divide 20 by 4. What method did they use? Ask children to find one-quarter of 20. What do they notice? Finding one-quarter is the same as dividing by 4. Ask children to divide each of these numbers by 4: 80, 100, 200, 800, 1,000.

Discover

Divide by 8

Discover

WAYS OF WORKING Pair work

ASK

- Question ① a): *How many lollies are in each mould? How many complete moulds can you make? What division calculation should you do? What does each number and sign represent?*
- Question ① b): *What is different about this question? Does this question work? Do you have enough lollipop sticks to fill the last mould? How can you write the division? What do you think your answer could look like?*

IN FOCUS Question ① a) focuses on equal grouping. Children must work out how many 8s are in 24. Children may count on in 8s until they get to 24, or use repeated subtraction of 8. Those children who group counters in 8s should be encouraged to think about the mathematical process that is going on by saying, for example: *Each time I group 8, I reduce the original amount by 8.* Encourage children to show this on a number line. Question ① b) is still about grouping, however, this time children have some lollipop sticks left over. Some children may think that the question is wrong, because they do not have enough lollipop sticks to complete the last container. At this stage, we are not writing formal remainders; we are just interested in how many complete moulds can be filled. For both questions, encourage children to write down the correct division sentence and ensure that children know what each number and sign represents.

PRACTICAL TIPS You could make ice lollies out of fruit juice in groups of 8, as in the image.

ANSWERS

Question ① a): $24 \div 8 = 3$; 3 moulds can be filled.

Question ① b): She can fill 4 moulds.

Share

WAYS OF WORKING Whole class teacher led

ASK

- Question ① a): *Can you explain each of Flo's steps? Why does the number line help? What other methods could you have used? Could an array help you?*
- Question ① b): *What is the same about this question and the way you can answer it? What is different about what you do? How could you record your answer?*

IN FOCUS For question ① a), ensure children demonstrate the method of grouping. Make each step of the subtraction clear alongside a demonstration of grouping. You can use counters to represent the lollipop sticks. It is important to link any numbers in the calculation to what it means in the context of the image. Children should notice that there are three groups of 8.

Question ① b) is similar, but this time children may notice that there are some left over. At this point, we are only interested in the complete number of containers that can be filled. Flo points out that there are not enough lollipop sticks to fill another container. Discuss this with children.

① a) Each ice lolly mould uses 8 lollipop sticks.
   Mr Jones has 24 sticks.
   How many moulds can he fill?

   b) Miss Hall has 38 sticks.
   How many moulds can she fill?

216

PUPIL TEXTBOOK 3A PAGE 216

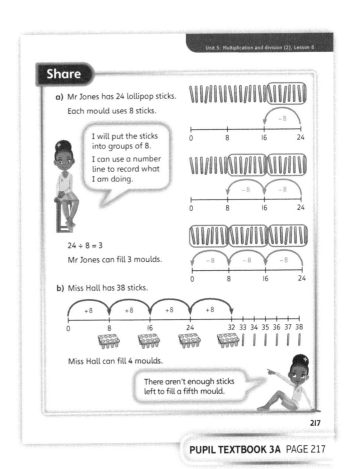

Share

a) Mr Jones has 24 lollipop sticks.
Each mould uses 8 sticks.

I will put the sticks into groups of 8. I can use a number line to record what I am doing.

$24 \div 8 = 3$
Mr Jones can fill 3 moulds.

b) Miss Hall has 38 sticks.

Miss Hall can fill 4 moulds.

There aren't enough sticks left to fill a fifth mould.

217

PUPIL TEXTBOOK 3A PAGE 217

Think together

Think together

WAYS OF WORKING Whole class teacher led (I do, We do, You do)

ASK

- Question ❶: *How many cupcakes are there? How many chocolate chips are there? Is this grouping or sharing? Can you explain each step of your method? What does the division look like? What does each number in the division sentence mean? Could you find the answer more quickly?*
- Question ❷: *What array is shown? How can you use the array to work out the answer to the division?*
- Question ❸: *Can you explain Lexi's method? How do the candles help you work out the answer to the problem? What divisions can Lexi work out from her method?*

IN FOCUS Question ❸ explores the method of halving, halving again and halving again as an alternative to dividing by 8. Children should come up with this method themselves as they see a birthday cake gradually cut into 8 pieces. Draw children's attention to the number of candles on the cake at the start, and the number on each part of the cake as they go through each of the divisions, to help them see the connection.

STRENGTHEN In question ❶, children can share practically using counters to represent the chocolate chips. Some children may group these into 8s in order to share them more quickly. Ask children what each number in the division represents and how they know they are doing a division. What clues tell them it is a division?

Children may find making an array with 8 counters in each row an effective way of finding the answers to the divisions. They may find it easier to understand dividing by 8 as dividing by 2, dividing by 2 again and then dividing by 2 again. Question ❸ provides a demonstration of why this works. Also, talk through the steps and consider modelling it with a paper circle and counters.

DEEPEN To deepen understanding, discuss with children how they could use division to work out the missing number problem ☐ × 8 = 72? Ask: *Can you explain how you can do this? Can you make up your own problems similar to this?* Follow on from question ❸ by asking children how they can use this method to work out 288 ÷ 8.

ASSESSMENT CHECKPOINT Children understand that division can either be equal grouping or sharing and they are able to form a division sentence to answer a question. They can use a variety of different methods to work out an answer to a division by 8, including counting on and back, using an array or through knowledge of multiplication facts.

ANSWERS

Question ❶: 40 ÷ 8 = 5.
    She can use 5 chocolate chips on each cupcake.

Question ❷ a): 72 ÷ 8 = 9

Question ❷ b): 48 ÷ 8 = 6

Question ❸ a): 16 ÷ 2 = 8, 16 ÷ 4 = 4, 16 ÷ 8 = 2

Question ❸ b): Halve it, halve it again and halve it again.
    88 ÷ 2 = 44,
    44 ÷ 2 = 22,
    22 ÷ 2 = 11

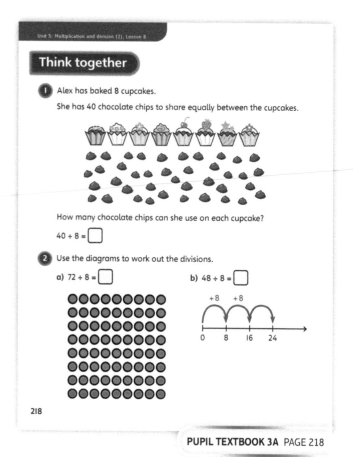

PUPIL TEXTBOOK 3A PAGE 218

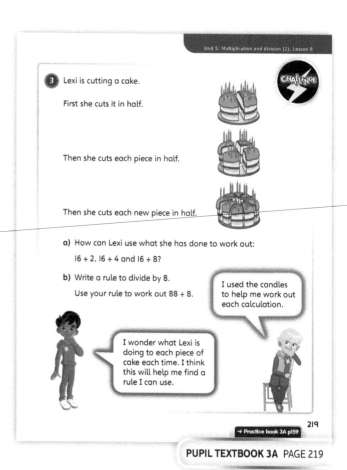

PUPIL TEXTBOOK 3A PAGE 219

Practice

WAYS OF WORKING Independent thinking

IN FOCUS Questions **1**, **2** and **3** provide a mixture of grouping and sharing problems. Encourage children to link each number in the division sentence with the context. Ask them to consider whether the problem is grouping or sharing. Question **4** a) asks children to use arrays with groups circled to find the answers to divisions. In question **4** b), children may circle the groups in different ways. Question **5** gives the answer to a division and asks children to use their related multiplication facts to work out the total. Children need to understand that 8 is the result of the division, rather than dividing by 8.

STRENGTHEN To support dividing by 8, ensure that children understand the difference between equal grouping and equal sharing. Provide simple contexts and ask children to show with counters what they are doing. Use counters or other materials to show the actions of grouping and sharing. For example, in question **1**, 24 counters can represent the pegs.

For grouping problems, ask children to take 8 counters at a time. For sharing problems, you may still want children to take 8 counters, but explain that, by taking 8, you are giving one to each group.

DEEPEN Question **6** provides an opportunity to look at the connection between multiplying and dividing by 2, 4 and 8. Give children a multiplication or division fact about a number, such as: *My number divided by 4 is 12. What is my number divided by 8? What is my number divided by 2?*

Ask questions, such as: *24 ÷ 8 ◯ 24 ÷ 4: which sign goes between the expressions?* Ask children to reason what the answer might be and why.

ASSESSMENT CHECKPOINT Can children represent a question as a division sentence involving ÷ 8, and use their knowledge of counting on or back in 8s to work out the answer? Children should understand the difference between sharing and grouping, but also know that the division sign can be used to represent both. Do children understand that one method to divide by 8 is to halve the number, halve it again and then halve it again?

ANSWERS Answers for the **Practice** part of the lesson can be found in the *Power Maths* online subscription.

Reflect

WAYS OF WORKING Pair work

IN FOCUS Children explore the problem of dividing 16 by 8. Ask children to decide what method they will use to solve this and to share their method. Children may count on, count back, or use equipment or pictures to show their thinking. Try to highlight that some children have grouped and others have shared. Draw out the connection between the numbers and what children may have drawn or written.

ASSESSMENT CHECKPOINT Check whether children understand the methods for dividing by 8.

ANSWERS Answers for the **Reflect** part of the lesson can be found in the *Power Maths* online subscription.

After the lesson ⏸

- Can children form division sentences involving 8 from a word problem?
- Do children know the difference between division as grouping and division as sharing?
- Do children know at least one method to divide by 8?

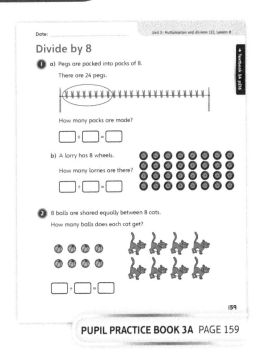

PUPIL PRACTICE BOOK 3A PAGE 159

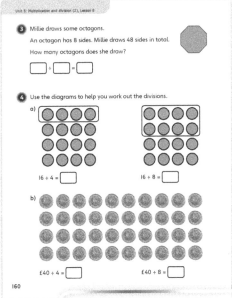

PUPIL PRACTICE BOOK 3A PAGE 160

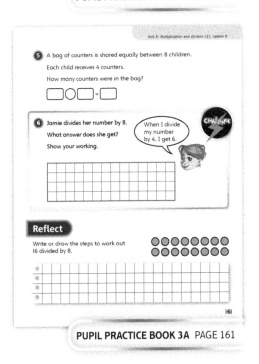

PUPIL PRACTICE BOOK 3A PAGE 161

The 8 times-table

Learning focus

In this lesson, children will focus on learning the 8 times-table and associated division facts from multiplication facts. They will show that they know how the 8 times-table can be derived from the 2 and 4 times-tables.

Before you teach ⏸

- Can children multiply by 8?
- Can children divide by 8?
- Do children know their 2 and 4 times-tables?

NATIONAL CURRICULUM LINKS

Year 3 Number – multiplication and division

Recall and use multiplication and division facts for the 3, 4 and 8 multiplication tables.

Write and calculate mathematical statements for multiplication and division using the multiplication tables that they know, including for two-digit numbers times one-digit numbers, using mental and progressing to formal written methods.

ASSESSING MASTERY

Children can recognise a multiplication fact from the 8 times-table given as an image. Children should be developing a rapid recall of multiplication facts and associated division facts from the 8 times-table.

COMMON MISCONCEPTIONS

Children may think that $0 \times 8 = 8$. This is a common mistake because children may answer too quickly and not think about their response. Ask:

- *Have you tried dividing by 0 before? What answer do you always get when you divide by 0? Why?*

To work out if 9×8 is greater than or equal to 7×8, children may think you have to work out each multiplication fact. Ask:

- *How do you know that 9 groups of 8 must be greater than 7 groups of 8? Why do you not need to work out both calculations to find the answer?*

STRENGTHENING UNDERSTANDING

At this stage, it is fine if some children still need to use cubes or counters to represent multiplications and divisions. The important thing is that children see what multiplication facts are. Reinforce that they need to develop rapid recall of the multiplication and division facts, but also understand what they are.

Remind children that if they are unsure of how to work out a multiplication fact from the 8 times-table, they can use their knowledge from the previous lesson. For example, to work out 3×8, they could double 3, double again and double again.

GOING DEEPER

Ask children to reason why $6 \times 8 < 7 \times 8$ without working out the two multiplications, and to find missing numbers to complete statements such as $40 \div 8 > \square \div 8$. Children can explore the connection between the 2, 4 and 8 times-tables. For example, if they know that $2 \times \square = 15$, can they find $4 \times$ that number and $8 \times$ the same number?

KEY LANGUAGE

In lesson: times-table, multiplication fact, array, double, pattern, total, compare

Other language to be used by the teacher: multiply, divide, grouping, division fact, recall

STRUCTURES AND REPRESENTATIONS

Number lines, arrays

RESOURCES

Mandatory: cubes, counters

 In the eTextbook of this lesson, you will find interactive links to a selection of teaching tools.

Quick recap

Ask children what strategies they know for multiplying a number by 8, if they cannot immediately recall the answer. Explain one method is to double, double and double again. For example, 6 doubles to 12, which doubles to 24, which doubles to 48. So, $6 \times 8 = 48$. Repeat for other numbers.

Discover

The 8 times-table

Discover

WAYS OF WORKING Pair work

ASK

- Question ① a): *What multiplication facts are missing? How did you work them out? How many of the multiplication facts on the stairs do you know off by heart?*
- Question ① b): *What is 5 × 2, 5 × 4 and 5 × 8? What do you notice about this? What is the connection?*

IN FOCUS Question ① a) asks children to work out three missing multiplication facts. Children may use their knowledge from previous lessons of multiplying and dividing by 8. Some children may already know their multiplication facts; you might want to ask these children to convince you of their answers. Some children may also link this to their knowledge of the 2 or 4 times-tables. Others may just count on from the previous values. Question ① b) asks them to start finding the connection between the 2, 4 and 8 times-tables. Children have previously done work on doubles and so they should start to see that, to get from one to the other, you can double.

PRACTICAL TIPS If you have the multiplication facts on a poster in your classroom, ensure that the relevant answers are covered up so that children do not just look at them. Use counters or cubes to create concrete representations of the missing multiplication facts.

ANSWERS

Question ① a): 2 × 8 = 16, 6 × 8 = 48, 9 × 8 = 72

Question ① b): The 4 times-table is double the 2 times-table.
The 8 times-table is double the 4 times-table.

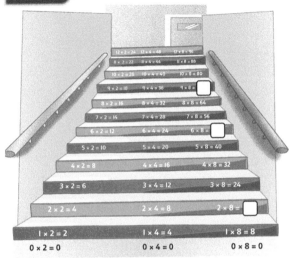

a) What answers are missing?

b) What is the connection between the 2, 4 and 8 times-tables?

220

PUPIL TEXTBOOK 3A PAGE 220

Share

WAYS OF WORKING Whole class teacher led

ASK

- Question ① a): *What does each array show? Could you use something other than an array?*
- Question ① b): *What do you notice about the arrays? Can you explain, using the arrays, why you double each time?*

IN FOCUS In question ① a), an array is used to give a visual representation of the multiplication facts. There are multiple ways that children could find the correct answer. For example, they may have just counted on 8 from the previous multiplication fact; some may have doubled, doubled again and doubled again. All are valid methods, although children do need to be moving towards instant recall of these facts at this stage. Question ① b) shows, using arrays, that to get from the 2 to the 4 times-table you double and then to get to the 8 times-table you double again.

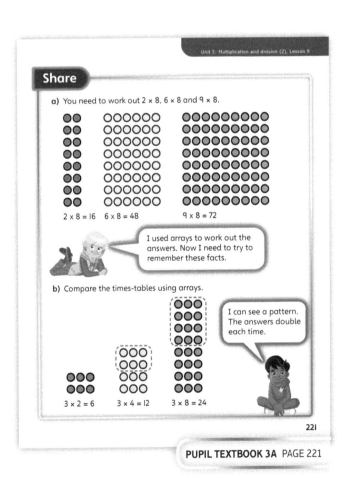

PUPIL TEXTBOOK 3A PAGE 221

Think together

WAYS OF WORKING Whole class teacher led (I do, We do, You do)

ASK

- Question **①**: *What multiplication facts can you see? How can you use these facts to work out the total?*
- Question **②**: *How many answers do you know off by heart? Which ones did you have to work out? How can you work out the divisions from the multiplications?*
- Question **③**: *What is the connection between the multiplications?*

IN FOCUS Question **①** presents images where the total can be worked out using the 8 times-table. Children should be able to explain which multiplication fact they can use. Encourage them to talk about the number of groups and the number of objects in each group. Children may struggle with the problem that shows 1 group of 8. Again, children may use methods from previous lessons if they do not have rapid recall of the multiplication facts. In question **③**, children further explore the connection between the 2, 4 and 8 times-tables by matching facts that give the same answer. They should notice that the answer doubles each time.

STRENGTHEN To strengthen understanding of multiplication facts, show the facts as visual or concrete representations, for example, towers of 8 cubes. It is important that children can visualise and understand what multiplication facts are. Reinforce that, eventually, children will need to develop rapid recall of the multiplication and division facts, but also understand what they are.

DEEPEN Expand question **③** by asking children if they know that a number multiplied by 8 is 100, what is the same number multiplied by 4 and multiplied by 2?

ASSESSMENT CHECKPOINT Children should start to develop rapid recall of multiplication and division facts for 8. They should show an understanding of the connection between the 2, 4 and 8 times-tables and start to explain why this is the case.

ANSWERS

Question **①** a): 5 × 8 = 40
There are 40 bottles of water.

Question **①** b): 11 × 8 = 88
There are 88 eggs.

Question **①** c): 1 × 8 = 8
There are 8 cubes.

Question **②**: a): 24; b): 80; c): 1; d): 96; e): 5; f): 40; g): 16; h) 0

Question **③**:

5 × 8	2 × 4	12 × 2
3 × 8	4 × 20	32 × 2
1 × 8	10 × 4	2 × 40
8 × 11	4 × 22	4 × 2
8 × 10	6 × 4	2 × 44
8 × 8	16 × 4	20 × 2

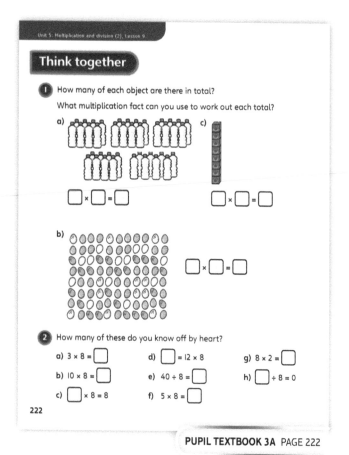

PUPIL TEXTBOOK 3A PAGE 222

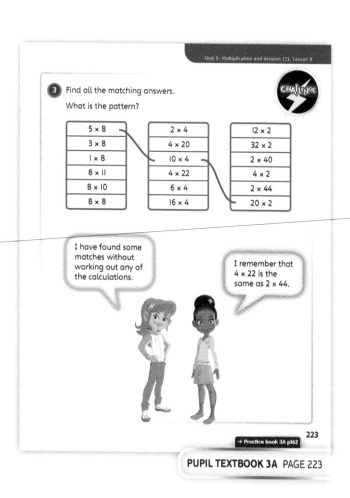

PUPIL TEXTBOOK 3A PAGE 223

Practice

WAYS OF WORKING Independent thinking

IN FOCUS Question ① reinforces visual images of several multiplication facts from the 8 times-table. Questions ② and ④ check whether children know the multiplication and division facts: they should try to do them without looking back in the book. If there are any that they do not know off by heart, they could use methods from the previous lessons. In question ③, children complete number tracks to show the multiples of 8. Question ⑤ asks children to add equals or inequality signs to make statements correct. Encourage children to use reasoning instead of working out all the answers and then comparing. Question ⑥ asks children to use their knowledge of the 3 and 8 times-tables to solve a missing number problem.

STRENGTHEN To strengthen understanding of multiplication facts, show the facts as visual or concrete representations; question ① will help with this. As children move through the rest of the questions, they become more abstract, although some children may still need to use cubes or counters to represent multiplications and divisions. What is important is that they can see what multiplication facts are. Reinforce that eventually they will need to develop rapid recall of the multiplication and division facts.

DEEPEN In question ⑤, ask children to explain how they can work out which sign completes the sentences without having to work out the answers. Which ones do they need to work out to compare? Can children make up similar problems to the ones in question ⑥, involving other times-tables?

ASSESSMENT CHECKPOINT Can children recognise 8 times-table facts from images? Children should start to be developing a secure recall of the 8 times-table multiplication and division facts.

ANSWERS Answers for the **Practice** part of the lesson can be found in the *Power Maths* online subscription.

Reflect

WAYS OF WORKING Pair work

IN FOCUS This brings together work on all the times-tables that children know so far, including those covered in Year 2. Children have to find multiplication facts to write into the relevant boxes. They should reflect on the strategy that they use. For example, do they list out the multiplication facts first or do they just know some answers? Do they notice any patterns in any of the columns?

ASSESSMENT CHECKPOINT Check if children have instant recall of multiplication facts across a range of times-tables they have met in Year 2 and Year 3.

ANSWERS Answers for the **Reflect** part of the lesson can be found in the *Power Maths* online subscription.

After the lesson ⏸

- Can children recall their division and multiplication facts from the 8 times-table?
- Which facts are children struggling with? Why?
- Do children know the connection between the 2, 4 and 8 times-tables?

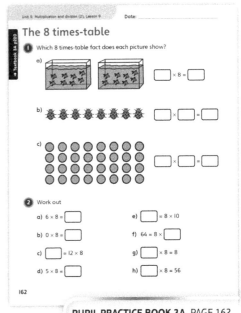

PUPIL PRACTICE BOOK 3A PAGE 162

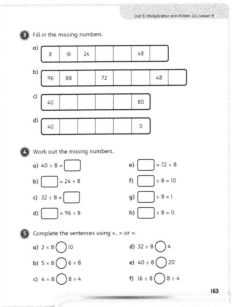

PUPIL PRACTICE BOOK 3A PAGE 163

PUPIL PRACTICE BOOK 3A PAGE 164

Problem solving – multiplication and division ❶

Learning focus

In this lesson, children will solve simple one-step multiplication and division problems. They will recognise when they need to multiply and divide and will draw a simple bar model to represent the problem.

Before you teach

- Do children know multiplication facts for the 2, 3, 4, 5, 8 and 10 times-tables?
- Can children draw a bar model for addition?
- Do children know how to calculate a multiplication or division problem if they do not know the multiplication fact for the relevant times-table?

NATIONAL CURRICULUM LINKS

Year 3 Number – multiplication and division

Solve problems, including missing number problems, involving multiplication and division, including positive integer scaling problems and correspondence problems in which *n* objects are connected to *m* objects.

Write and calculate mathematical statements for multiplication and division using the multiplication tables that they know, including for two-digit numbers times one-digit numbers, using mental and progressing to formal written methods.

ASSESSING MASTERY

Children can solve a range of simple multiplication and division one-step word problems. They are able to use key information in a question to draw a bar model and then work out whether they need to multiply or divide.

COMMON MISCONCEPTIONS

Children may not be sure whether they need to multiply or divide. Drawing a bar model to represent the situation may help them. Encourage children to look at the words in the question for clues. Ask:
- *What will your bar model look like? Do you know how many parts there should be, or the value of each part? What does each number in the question mean?*

STRENGTHENING UNDERSTANDING

Children may not understand the question and what they need to do. To support understanding, read the questions with children line by line. Ask them to identify key information in the questions. If necessary, ask them to represent the objects with counters or cubes. This may help them understand the abstract questions better. Try to encourage children to draw a bar model, which may help them see whether they need to multiply or divide. Some children may not be able to recall multiplication facts quickly, so could use counting on, grouping or sharing strategies.

GOING DEEPER

Ask children to solve problems that involve a single multiplication but are in two parts, such as: *Mary has 4 boxes of 8 apples and 3 more boxes of 8 apples. How many are there in total?* Ask children to find different ways of working out the total. Encourage them to try to find a solution that uses a single multiplication. How did they do it? Can they explain why it works?

KEY LANGUAGE

In lesson: multiplication, division, multiply, divide, group, shared equally, bar model, total

Other language to be used by the teacher: times-table, recall, grouping, repeated addition, equal, sharing, word problem

STRUCTURES AND REPRESENTATIONS

Bar model, number line

RESOURCES

Optional: cubes, counters

 In the eTextbook of this lesson, you will find interactive links to a selection of teaching tools.

Quick recap 🔎

Ask children to work as a class to count in 2s, 3s, 4s, 5s, 8s or 10s, from 0 to 12 times the number. This could be done by going from child to child in order or randomly. All children should be chosen. Time how long it takes and see if they can beat the time in the next lesson.

Discover

WAYS OF WORKING Pair work

ASK

- Question ① a): *How many plants are there? How many plants are in each row? Can you represent this as a bar model? What type of problem are you solving? Is it division or multiplication?*
- Question ① b): *How many bunches of flowers does Amal have? How many are in each bunch? Can you show this on a bar model? How do you work out the total? What calculation do you have to do? Can you solve the answer mentally?*

IN FOCUS Question ① a) requires children to work out how many rows there are, if there are 24 plants. Ensure that children have at least 24 counters which they can group into 4s or put into an array with 4 columns. Children should be encouraged to show their working and write down a division sentence to work out the answer. Children should notice that question ① b) is a multiplication. They need to look at the picture to realise that Amal has 2 bunches of flowers. To work out the answers, children should use mental recall where they can. Encourage children to use a bar model to represent the situations. This will help them determine whether the problem they are trying to solve is a multiplication or a division.

PRACTICAL TIPS Use counters or cubes to represent the plants and flowers.

ANSWERS

Question ① a): 24 ÷ 4 = 6
There will be 6 rows of 4 plants.

Question ① b): 2 × 10 = 20
Amal has 20 flowers in total.

Share

WAYS OF WORKING Whole class teacher led

ASK

- Question ① a): *How many groups of 4 plants are there? How has this been shown using a number line? How does the bar model show this?*
- Question ① b): *What does the bar model look like? Why are there 2 bars, each with 10 in?*

IN FOCUS For question ① a), show children how the grouping bar model is built up, starting with the brace of 24 and then adding bars using repeated addition until they get to 24. For the multiplication in question ① b), children can put the 2 bars of 10 and then use them to work out how many flowers there are in total. Showing multiplication as a bar model reinforces the fact that multiplication is repeated addition.

Problem solving – multiplication and division ①

Discover

Amal

① a) The plants are planted in rows of 4.
There are 24 to plant.
How many rows of plants will there be?

b) Amal has some bunches of flowers.
There are 10 flowers in each bunch.
How many flowers does he have in total?

224

PUPIL TEXTBOOK 3A PAGE 224

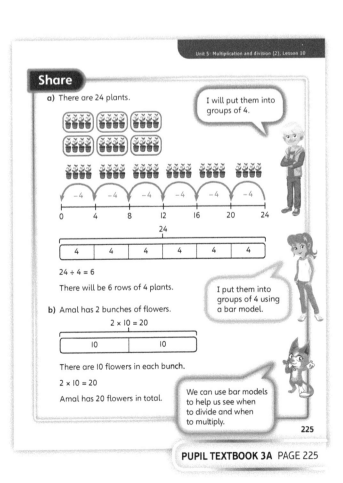

PUPIL TEXTBOOK 3A PAGE 225

Think together

Think together

WAYS OF WORKING Whole class teacher led (I do, We do, You do)

ASK

- Question **1**: *How many equal groups are there? How many in each group? How many plants are there in total? How did you work it out? How does the arrangement help? Can you draw a bar model to show this?*
- Question **2**: *What is the key information in the question that you need to know? How many flowers are there? How many vases? Can you explain the bar model? Do you need to divide or multiply? How do you know?*
- Question **3**: *How much do the small plant pots cost in total? What does this tell you about the cost of the large plant pots? What is the cost of one large pot? Did you do two calculations? Can you represent each calculation on a bar model? Is there a way of working this out more easily?*

IN FOCUS This section takes children through a series of word problems. The key aim is for children to decide whether they need to multiply or divide. They may look at the words, such as 'how many are there in total' or 'shared' to help them decide. Trying to draw a bar model to represent each situation will help them to see whether they have to do a multiplication or a division. Most children are likely to approach question **3** as a two-step problem, first multiplying to find the total and then dividing to find the cost of a large pot. Some children may notice that a large plant pot must be double the price of a small plant pot because 4 large pots cost the same as 8 small pots.

STRENGTHEN Encourage children to work through each problem line by line. Ask them to highlight the key information in the questions. Have they been given the total to start with or do they have to work it out? Support children in drawing a bar model to help them. For question **1**, they will need to draw three bars, each with 8 in it. Ask children what the bars represent and how they can work out the total. If children are struggling with finding the answers, support them by using methods of repeated addition, subtraction, grouping and sharing.

DEEPEN For question **3**, ask children if they have to do both a multiplication and division. Ask: *How could two bars the same length help you work out the cost of the large plant pot?* This might help them see that they can just double the cost of a small plant pot. Challenge children to make up their own word problems. Can they make up a multiplication and then a division word problem with the answer of 12?

ASSESSMENT CHECKPOINT Children can solve simple multiplication and division word problems. They can represent a simple problem as a bar model to help them see whether they need to do a multiplication or a division.

ANSWERS

Question **1**: $8 \times 3 = 24$
There are 24 plants.

Question **2**: $30 \div 5 = 6$
There are 6 flowers in each vase.

Question **3**: $8 \times £3 = £24$; $£24 \div 4 = £6$
A large plant pot costs £6.

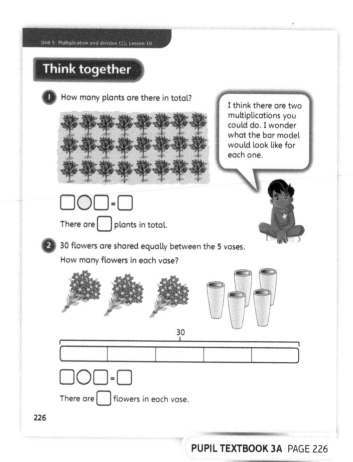

PUPIL TEXTBOOK 3A PAGE 226

PUPIL TEXTBOOK 3A PAGE 227

Practice

WAYS OF WORKING Independent thinking

IN FOCUS These practice questions contain a series of multiplication and division word problems. Gradually, scaffolding is removed so children may find it increasingly difficult to work out whether they need to multiply or divide. For each question, a bar model or part of a model is drawn, except in question **4**, when children are asked to draw their own to represent a situation. Children will start to realise that a bar model will help them determine the calculation that they need to do. Encourage children to highlight any key information in the question, including numbers and key words; for example, 'find a total' implies multiplication and 'each' often implies division. Questions **1** to **3** are fully structured. In question **4**, the structure is removed, which requires children to work out the calculation they need to do. All the questions feature answers from times-tables that they should know, so encourage rapid recall before using other methods to work out an answer.

STRENGTHEN Read the questions line by line and ask children to identify key information. If necessary, ask them to represent the objects with counters or cubes, which may help them to understand the more abstract questions. Encourage children to draw a bar model to help them see whether they need to multiply or divide.

DEEPEN Question **5** is a more complex two-step problem, in the context of weighing. To further deepen understanding, give children multiplication and division bar models for different situations and ask them to write down a word problem that could be solved using this bar model. Can they write another? What is the same about their word problems? What is different?

ASSESSMENT CHECKPOINT Can children represent a word problem as a bar model? Can they use the bar model to work out whether they need to do a multiplication or a division to work out the answer?

ANSWERS Answers for the **Practice** part of the lesson can be found in the *Power Maths* online subscription.

Reflect

WAYS OF WORKING Whole class

IN FOCUS Children need to create a word problem with an answer of 24. Ask them to share their problems with the class and get the other children to check if the answer is 24. What numbers were used? Did children use the same numbers? If children are unsure which multiplications make 24, then ask them to draw a bar model and see how they might be able to divide it up. They can then work out the correct multiplication to make 24.

ASSESSMENT CHECKPOINT Check if children can make up a word problem that requires them to multiply.

ANSWERS Answers for the **Reflect** part of the lesson can be found in the *Power Maths* online subscription.

After the lesson ⏸

- Can children solve a range of multiplication and division word problems?
- Can children identify key information that is important to the problem?
- Can they draw a bar model to represent a multiplication problem and the two types of division problem?

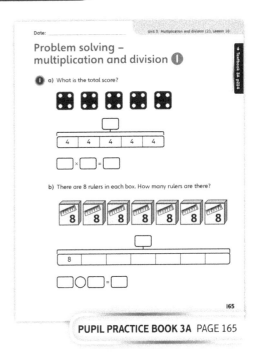

PUPIL PRACTICE BOOK 3A PAGE 165

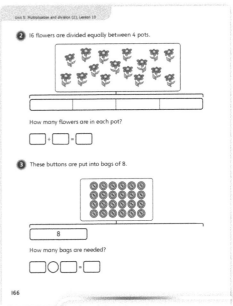

PUPIL PRACTICE BOOK 3A PAGE 166

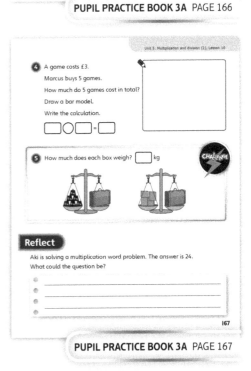

PUPIL PRACTICE BOOK 3A PAGE 167

Problem solving – multiplication and division ❷

Learning focus

In this lesson, children will solve simple one-step multiplication problems. They will also begin to tackle solving simple two- and three-step multiplication and division problems that may involve an addition or a subtraction.

Before you teach

- Do children know the multiplication facts for the 2, 3, 4, 5, 8 and 10 times-tables?
- Can children draw a bar model for addition?
- Can they work out the answer to simple one-step multiplication and division word problems?

NATIONAL CURRICULUM LINKS

Year 3 Number – multiplication and division

Solve problems, including missing number problems, involving multiplication and division, including positive integer scaling problems and correspondence problems in which *n* objects are connected to *m* objects.

Write and calculate mathematical statements for multiplication and division using the multiplication tables that they know, including for two-digit numbers times one-digit numbers, using mental and progressing to formal written methods.

ASSESSING MASTERY

Children can solve a range of multiplication and division problems that may involve more than one step. They are able to use key information from a question to draw a bar model and work out whether they need to multiply or divide first.

COMMON MISCONCEPTIONS

Children may not be able to work out whether they need to multiply or divide. Drawing a bar model to represent the situation may help with this. Ask:

- *Is this multiplication or division? How do you know? What clues are there in the question?*

STRENGTHENING UNDERSTANDING

To support children in understanding what they need to do, read through the questions line by line, asking children to identify the key information. If necessary, they can represent the objects with counters or cubes. Making towers of cubes (or arrays) for multi-step problems will help them see that they have to carry out two separate multiplications as opposed to just one. This may help them to understand better the more abstract questions. Encourage children to draw a bar model to help them see whether they need to multiply or divide.

GOING DEEPER

Encourage children to make up their own one-step and multi-step problems. Working out the answers is not as important as designing the question. If children can recognise when a problem is one-step or two-step, this will help them gain a deeper understanding of problem-solving questions.

KEY LANGUAGE

In lesson: multiplication, multiply (×), division, total, bar model, pattern

Other language to be used by the teacher: method, times-table, divide (÷), multiplication sentence, recall

STRUCTURES AND REPRESENTATIONS

Bar model, number line

RESOURCES

Optional: cubes, counters, wooden blocks

 In the eTextbook of this lesson, you will find interactive links to a selection of teaching tools.

Quick recap

Repeat the activity from yesterday; ask children to work as a class to count in 2s, 3s, 4s, 5s, 8s or 10s from 0 to 12 times the number. This could be done by going from child to child in order or randomly. All children should be chosen. See if they can beat their time from Lesson 10.

Discover

ASK

• Question ① a): *What 3D shape is the block? What is the length of the block? What about the width and the height of the block? What does it mean when the blocks are said to be put 'end to end'? How can you work out the length of the new shape? Do you need to do a multiplication or division? What shape is the new shape?*

• Question ① b): *How does this question differ from the last one? What calculation can you do to work out the answer? Is there more than one possible answer?*

IN FOCUS Ensure that children fully understand the context. Encourage drawing around the shapes if this helps children to work out the total length. Question ① a) asks children to work out the length of 3 blocks that have been put together. Some children may want to add three 8s together; they should realise that this is a multiplication. Children could be encouraged to build on their work on bar models from the previous lesson. Question ① b) gives children the total length of a number of blocks and asks them to work out how many blocks have been used. The structure of the question should help them realise they need to do a division. Some children may count up in 8s until they get to 32. Encourage them to see the link with the division. Ask children to show how they can lay the blocks in a different way to make a different shape with a length of 32 cm. Draw out the fact that the blocks can be laid end to end in different orientations.

PRACTICAL TIPS Using wooden building blocks and blocks of cubes will be useful in this context.

ANSWERS

Question ① a): $3 \times 8 = 24$ cm

Question ① b): $32 \div 8 = 4$ blocks, or $32 \div 4 = 8$ blocks.

Share

ASK

• Question ① a): *How has Astrid answered the question? Why has she used a bar model? What does each part of the bar model represent? How can you use the bar model to work out if you need to do a multiplication or a division?*

• Question ① b): *What is the difference between the two methods? Is there more than one way to lay the blocks end to end? How does the answer change?*

IN FOCUS Discuss as a class the different methods used to get the two answers. In question ① a), Astrid shows how you can use a bar model to represent the scenario. Ensure children know what each part of the bar model represents and how it leads to a multiplication. In question ① b), discuss the difference between the two methods used. Explain to children that eventually they need to try to do the division like Flo, but that Dexter's method explains exactly what is going on. Discuss that there are two solutions to question ① b), as the blocks can be laid in different orientations.

Problem solving – multiplication and division ②

Discover

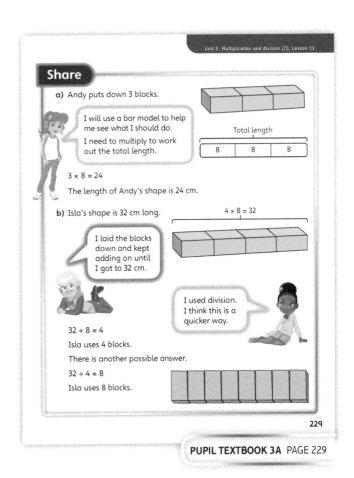

① a) Andy has put 3 blocks end to end to make a new shape. What is the length of Andy's shape?

b) Isla makes a shape that is 32 cm long. How many blocks does she use?

228

PUPIL TEXTBOOK 3A PAGE 228

Share

a) Andy puts down 3 blocks.

I will use a bar model to help me see what I should do.
I need to multiply to work out the total length.

$3 \times 8 = 24$

The length of Andy's shape is 24 cm.

b) Isla's shape is 32 cm long.

I laid the blocks down and kept adding on until I got to 32 cm.

I used division. I think this is a quicker way.

$32 \div 8 = 4$

Isla uses 4 blocks.

There is another possible answer.

$32 \div 4 = 8$

Isla uses 8 blocks.

229

PUPIL TEXTBOOK 3A PAGE 229

Think together

WAYS OF WORKING Whole class teacher led (I do, We do, You do)

ASK

- Question ❶: *How many blocks are there? How long is each block? Why is it a multiplication? How does the bar model help you understand it is a multiplication?*
- Question ❷: *What is the height of each block in each tower? How many blocks are in each tower? How can you work out the total height of each tower? Which tower is taller? How do you know?*
- Question ❸: *Can you make this using your own blocks? Why does Astrid suggest that you need to do two multiplications and add them together?*

IN FOCUS Question ❶ builds on the **Discover** task by giving children seven blocks for which they should find the length. The bar model should help them understand why it is a multiplication. You may also want to use blocks to help children see this. Question ❷ looks at the height of some towers with the blocks stacked in different ways. This is the first time that children will need to realise that the way the blocks are stacked will affect the height of the tower. Question ❸ looks at a different arrangement of blocks. Encourage children to write two multiplications as opposed to a single multiplication sentence. For each multiplication, ask children if they understand what length it finds and why they need to add the two together to find the total length.

STRENGTHEN Encourage children to work through each problem line by line. Ask children to highlight what the key information is in the question. Ask children to draw or use a bar model to explain each of the steps. For question ❶, they will need to draw 7 bars, each with 4 in. Ask children what the bars represent and how they can work out the total length. If children find the shapes in question ❷ hard to visualise, they can use wooden blocks or cubes to help them. Children need to see how stacking the blocks in different ways will give different heights.

DEEPEN For question ❷, probe the question further and ask children what would happen if the tower was 32 cm high. Ask: *How many blocks might you use? Are there two different answers?* In question ❸, can children write their answer as a single multiplication? Why not? Can they find the distance all the way around the shape?

ASSESSMENT CHECKPOINT Children can solve simple multiplication word problems and represent a simple problem as a bar model to help them see whether they need to do a multiplication or division. They recall multiplication facts to work out the answers.

ANSWERS

Question ❶: 7 × 4 = 28
The shape is 28 cm long.

Question ❷: Tower A: 5 × 4 = 20 cm tall.
Tower B: 2 × 8 = 16 cm tall.
Tower A is the taller tower.
It is 4 cm taller.

Question ❸: 3 × 4 = 12 cm, 2 × 8 = 16 cm, 12 + 16 = 28 cm
The pattern is 28 cm long.

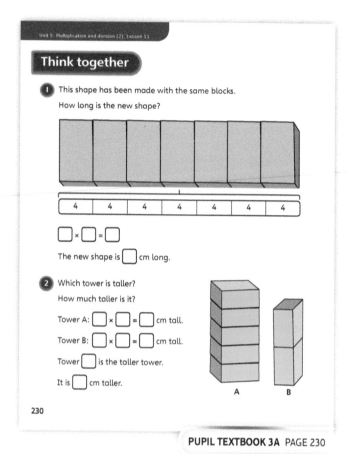

PUPIL TEXTBOOK 3A PAGE 230

PUPIL TEXTBOOK 3A PAGE 231

Practice

WAYS OF WORKING Independent thinking

IN FOCUS In questions ❶ and ❷, children have to do two multiplications and then add them together to find the total. Using a single bar model may help children understand the steps that they need to take. They will see that there are two different-sized bars in the bar model and they need to work out the separate lengths and then add them together. Question ❺ presents children with a common type of repacking problem. They first use multiplication to work out the total and then do the division to work out how many packs there will be if they are repacked into different-sized packs. Due to the numbers involved, children may do this in just one step.

STRENGTHEN To support understanding, use counters and cubes to represent the different groups of objects; this will help children see that, in questions ❶ and ❷, they first need to do two multiplications and then an addition. For question ❸, children may find that drawing a simple array helps them to compare the two groups of objects. For two-step problems, it is helpful to see a visualisation of why they have to work out one part before the other. Encourage children to draw a bar model, which may help them to see whether they need to multiply or divide.

DEEPEN In question ❹, children can make up their own multi-step problems. For example, how much do 5 buckets and spades and 2 beach balls cost? How much change will you get from £50?

ASSESSMENT CHECKPOINT Can children represent a word problem as a bar model? Can they use the bar model to work out whether they need to do a multiplication or division? Can children solve two-step problems across all four operations?

ANSWERS Answers for the **Practice** part of the lesson can be found in the *Power Maths* online subscription.

Reflect

WAYS OF WORKING Independent thinking

IN FOCUS There are two types of problem that children may come up with: either a simple one-step problem working out a total cost or a two-step problem (such as finding a total cost and then calculating change). Some children may want to use the structure of the earlier practice and devise a problem where they have to work out the result of two multiplications and add or subtract them. Encourage children to make up several problems to demonstrate the difference between one-step and multi-step problems. Children can share their problems for a partner to work out.

ASSESSMENT CHECKPOINT Check whether children can make up a word problem that requires either division or multiplication. Do children write a simple one-step problem or problems where multiple steps are involved?

ANSWERS Answers for the **Reflect** part of the lesson can be found in the *Power Maths* online subscription.

After the lesson ⏸

- Can children solve a range of multiplication and division multiple-step word problems?
- Can children identify key information that is important to a problem and realise what type of calculation they have to do?
- Can they draw bar models to represent a multiplication problem and the two types of division problem?

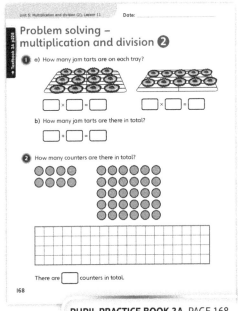

PUPIL PRACTICE BOOK 3A PAGE 168

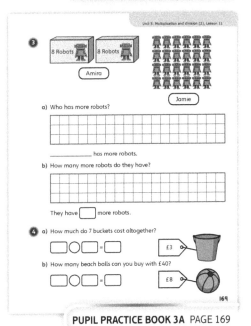

PUPIL PRACTICE BOOK 3A PAGE 169

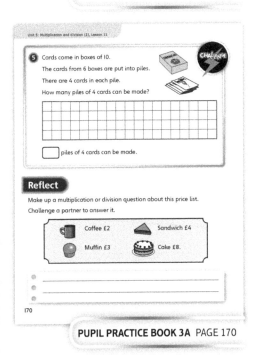

PUPIL PRACTICE BOOK 3A PAGE 170

Understand divisibility

Learning focus

In this lesson, children will realise that some division problems leave a remainder. They will also realise that the greatest possible remainder is 1 less than the number they divide by.

Before you teach

- Do children understand simple division by grouping?
- Do they know the names of basic shapes and the number of sides they have?

NATIONAL CURRICULUM LINKS

Year 3 Number – multiplication and division

Solve problems, including missing number problems, involving multiplication and division, including positive integer scaling problems and correspondence problems in which *n* objects are connected to *m* objects.

Write and calculate mathematical statements for multiplication and division using the multiplication tables that they know, including for two-digit numbers times one-digit numbers, using mental and progressing to formal written methods.

ASSESSING MASTERY

Children can realise that some divisions are not always exact and can leave a remainder. Children start to understand that the greatest possible remainder is 1 less than the number they are dividing by.

COMMON MISCONCEPTIONS

Children may not recognise that what they are doing in this lesson is division. To support this, remind children that division can be seen as either grouping or sharing. Explain that making squares is like taking a group of 4 lollipop sticks from the starting amount. Children should continue to do this until they have some left over. Ask:
- *What does this remind you of? Is it grouping or sharing?*

STRENGTHENING UNDERSTANDING

Children should have access to lollipop sticks or strips of paper to make squares, triangles and other shapes. Work with children to present the results of their divisions in the form of a table. Encourage them to record all their working, including diagrams of what they made. Some children may find it easier to begin again each time by taking the correct number of lollipop sticks and making the squares. Eventually, children may see that they can just add 1 more lollipop stick each time.

GOING DEEPER

Children could do the reverse to work out the number of lollipop sticks that are needed to make 5 squares and a remainder of 3. They should also try to predict amounts of lollipop sticks that have a remainder of 1 when they are making squares. For example, they may say that there is a remainder of 1 for numbers that are 1 more than a multiple of 4. They should solve other problems like this, such as: when does a division have a remainder of 0?

KEY LANGUAGE

In lesson: left over, **remainder**, pattern

Other language to be used by the teacher: divide, grouping, complete, whole

RESOURCES

Mandatory: lollipop sticks or strips of paper to make shapes, counters

Optional: cubes

 In the eTextbook of this lesson, you will find interactive links to a selection of teaching tools.

Quick recap

In pairs or small groups, ask children to take 20 counters and make arrays. What arrays can they make? What multiplication and division facts can they write down from the arrays? Ask children to repeat the activity with 24 counters.

Discover

PUPIL TEXTBOOK 3A PAGE 232

WAYS OF WORKING Pair work

ASK

- Question ① a): *How many lollipop sticks do the children have in total? How many whole squares can they make? How many are left over? Why is this a division problem? Why are you dividing by 4?*
- Question ① b): *What happens to your answers as you increase the number of lollipop sticks? Is this still dividing by 4? Why? What is the smallest number of lollipop sticks that are left over? What is the greatest? Why? Do you notice a pattern in the number of lollipop sticks left over?*

IN FOCUS In question ① a), children have to deal with an amount left over in division and are introduced to the term 'remainder' for the first time. In question ① b), children start to investigate further by looking at how the answer changes if they have 14, 15, 16 or 17 lollipop sticks. They may start to realise that what they are doing is a division (equal grouping). In this case, they are dividing by 4, because they are making squares. The squares represent the whole and the ones left over represent the remainder(s). Children may start to notice a pattern in the number of whole squares and remainders.

PRACTICAL TIPS Provide children with lollipop sticks or strips of paper to make the squares.

ANSWERS

Question ① a): 13 sticks – 3 whole squares and 1 left over

Question ① b): 14 sticks – 3 whole squares and 2 left over
15 sticks – 3 whole squares and 3 left over
16 sticks – 4 whole squares and 0 left over
17 sticks – 4 whole squares and 1 left over

Share

WAYS OF WORKING Whole class teacher led

ASK

- Question ① a): *How many squares have been made? How many lollipop sticks are left over? What name do we give to the amount left over? What calculation is represented here?*
- Question ① b): *How has Flo organised her work? Do you think this is a good way? Why? Did you need to start again each time? What has Flo put into her table? Do you notice any patterns in the table? Why will the remainder never be greater than 3?*

IN FOCUS Question ① a) is an opportunity to look at wholes and remainders. Explain to children that what they have done is the calculation 13 divided by 4. This will be introduced formally in the next lesson, though children do need to understand that they are doing a division. Explain that they had 13 lollipop sticks to start with and were grouping into 4s; making squares helps children see the number of wholes. Sparks explains that the amount left over is called the remainder.

Understand divisibility ①

Discover

① a) Lexi and Zac are using lollipop sticks to make squares.
   How many squares can they make?
   How many lollipop sticks are left over?

b) How would the answer change if they had 14 lollipop sticks?
   What about 15, 16 or 17 lollipop sticks?

232

PUPIL TEXTBOOK 3A PAGE 232

Share

We call the amount left over the remainder.

a) 4 lollipop sticks make 1 square.

13 lollipop sticks make 3 squares with 1 stick left over.

I will try organising my work in a table.

b)

Number of sticks	Working	Number of squares	Number of sticks left over
14		3	2
15		3	3
16		4	0
17		4	1

233

PUPIL TEXTBOOK 3A PAGE 233

Think together

Think together

WAYS OF WORKING Whole class teacher led (I do, We do, You do)

ASK

- Question **1**: *How many lollipop sticks are we going to use this time? What do you need to record in the table? Do you need to start again each time? Can you do this without making the squares?*
- Question **2**: *What patterns can you see? Why do you think that this is the case? What is the greatest remainder possible? Can you explain why?*
- Question **3**: *What shape are we making this time? What are we dividing by this time? Can you explain why? Can you predict the greatest remainder possible? What are the reasons for your prediction?*

IN FOCUS In question **1**, children extend the table from earlier in the lesson. They should not need to start again each time and should now be able to build on the squares they already have, adding one lollipop stick each time. They should now be looking for patterns. Remind children that they are dividing by 4. Can they explain why? This leads into question **2**, where children are asked to spot patterns. Ask children what patterns they notice. For example, why does the number of whole squares stay the same for a while? Children should be able to explain the pattern of remainders going 0, 1, 2, 3. They should also reason that the remainder will never be 4 or above as this would make another square. Question **3** is a similar investigation, but this time looking at making triangles with the lollipop sticks. Children should see this as dividing by 3 this time, because they are making triangles.

STRENGTHEN Ensure that children have lollipop sticks or strips of paper to make the squares and triangles. They should be able to record the number of whole squares (or triangles) and the number of lollipop sticks left over.

DEEPEN Deepen understanding of this being a division task by seeing if children can write some of the answers as divisions. They could predict the remainder if they had 41 lollipop sticks and they were making squares. How do they know that the remainder is 1? (It is 1 more than a multiple of 4.)

ASSESSMENT CHECKPOINT Children recognise that they have been grouping objects. They see that the number of squares or triangles is the whole and the number left over is the remainder.

ANSWERS

Question **1**: 18 sticks – 4 squares and 2 left over.
               19 sticks – 4 squares and 3 left over.
               20 sticks – 5 squares and 0 left over.

Question **2** a): The amount left over goes 0, 1, 2, 3 and then repeats itself.

Question **2** b): The amount left over can never be 4 as otherwise they would be able to make another square.

Question **3**: 11 sticks – 3 triangles and 2 left over.
               12 sticks – 4 triangles and 0 left over.
               13 sticks – 4 triangles and 1 left over.
               14 sticks – 4 triangles and 2 left over.
               15 sticks – 5 triangles and 0 left over.

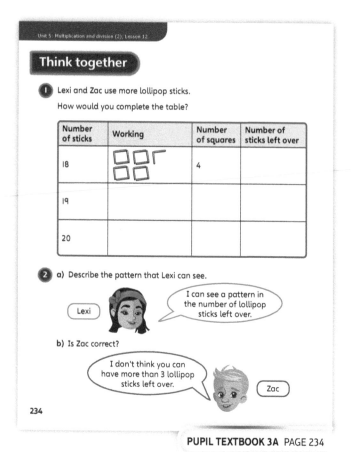

PUPIL TEXTBOOK 3A PAGE 234

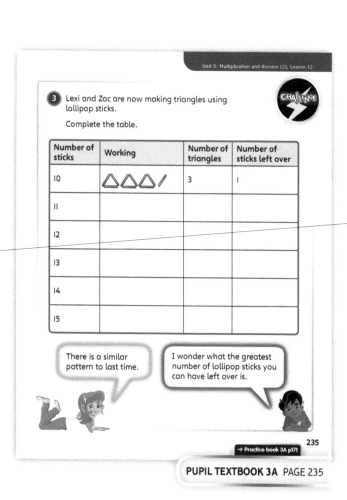

PUPIL TEXTBOOK 3A PAGE 235

Practice

WAYS OF WORKING Independent thinking

IN FOCUS Question ❶ builds on the **Textbook** by recapping making squares and triangles from lollipop sticks. Children must record their working out as well as the number of whole shapes made and the remainder. The word 'remainder' is now used instead of 'left over', to encourage correct mathematical language. In question ❷, children extend their thinking by making regular pentagons. They should see that they are now dividing by 5 and should be able to predict the pattern of whole pentagons and remainders.

STRENGTHEN Some children will be able to tackle this without having lollipop sticks or strips of paper to help them. Children who are still struggling to work out the wholes by drawing, can use lollipop sticks to do this, which will help them see the number of wholes and the remainder.

DEEPEN In question ❸, children work from the wholes and the remainder to work out the number of cubes that Max started with. They may need to draw the blocks to help them. Ask children to explain their reasoning. Give them other wholes and remainders and see if they can find a method to work out the original starting number of cubes. To further deepen understanding, ask children to predict the remainder if they had 45 cubes, or 82 cubes, by thinking of the numbers in the 4 times-table.

ASSESSMENT CHECKPOINT Children recognise that they have been doing division throughout and understand the terms 'whole' and 'remainder'. Children can see a pattern in the number of wholes and the remainder.

ANSWERS Answers for the **Practice** part of the lesson can be found in the *Power Maths* online subscription.

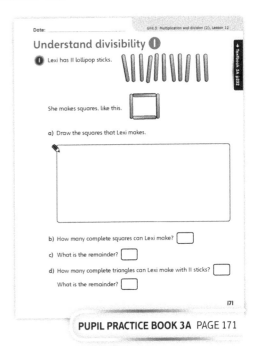

PUPIL PRACTICE BOOK 3A PAGE 171

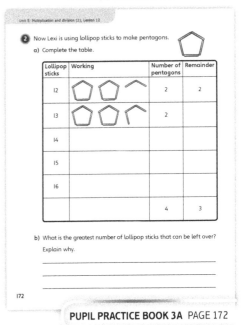

PUPIL PRACTICE BOOK 3A PAGE 172

Reflect

WAYS OF WORKING Pair work

IN FOCUS Children have been doing lots of repetitive practice and should realise that making squares is dividing by 4 and making triangles is dividing by 3. When dividing by 5, children should be able to communicate through words or diagrams that they are making pentagons. Children should be able to reason that the greatest remainder they can have is 4 because, if they had 5, they would be able to make another pentagon.

ASSESSMENT CHECKPOINT Check whether children know that, when dividing by 5, the greatest possible remainder is 4.

ANSWERS Answers for the **Reflect** part of the lesson can be found in the *Power Maths* online subscription.

PUPIL PRACTICE BOOK 3A PAGE 173

After the lesson

- Do children know that some situations will leave a remainder?
- Can they explain why the greatest possible remainder when dividing by 4 is 3, and so on?

Understand divisibility ②

Learning focus

In this lesson, children will continue to realise that some division problems leave a remainder. Children will learn to write the result of a division with a remainder more formally.

NATIONAL CURRICULUM LINKS

Year 3 Number – multiplication and division

Solve problems, including missing number problems, involving multiplication and division, including positive integer scaling problems and correspondence problems in which *n* objects are connected to *m* objects.

Write and calculate mathematical statements for multiplication and division using the multiplication tables that they know, including for two-digit numbers times one-digit numbers, using mental and progressing to formal written methods.

ASSESSING MASTERY

Children can understand that some divisions may leave a remainder and should start to recognise (using their knowledge of multiplication facts) when a division may lead to a remainder. Children understand that a remainder may result from both equal sharing and equal grouping and can find the result of a division with a remainder and write it in the form 'a remainder b'.

COMMON MISCONCEPTIONS

Children may lack understanding of what the remainder means. They need to understand that the remainder is the number left over at the end (because there are either not enough to group or not enough to share). Ask:
- *Do you have enough to make another whole? Why? Why not? What is that called?*

Children may mix up the whole and the remainder when writing out answers formally. To help overcome this, ask children to link the final answer with the context in the question. For example, 13 apples shared between 5 people; each person receives 2 apples and the remainder is 3 apples. Remind children that we always write the remainder at the end. Showing this as an array may also help. Ask:
- *Can every person have the same amount? How many do they each have? How many are left over?*

STRENGTHENING UNDERSTANDING

Ensure that children have access to concrete objects and demonstrate how to carry out both grouping and sharing procedures for division. Children should realise that the answer remains the same regardless of which type of division is used. Ask children to carry out the grouping or sharing step by step, explicitly linking it to the concrete objects. They may find putting objects into arrays helpful for understanding the whole and the remainder.

GOING DEEPER

Children work backwards and solve missing number problems such as: ☐ ÷ 4 = 5 remainder 3, and 15 ÷ ☐ = 7 remainder 1. Children may then tackle problems such as: ☐ ÷ ☐ = 2 remainder 3. Can they find several answers to this problem? Can they reason why this cannot be a division by 2 or 3?

KEY LANGUAGE

In lesson: left over, division, divide (÷), remainder, shared equally, array, group

Other language to be used by the teacher: whole, complete

STRUCTURES AND REPRESENTATIONS

Number lines, arrays

RESOURCES

Mandatory: counters or cubes

 In the eTextbook of this lesson, you will find interactive links to a selection of teaching tools.

Quick recap 🔁

Ask children to divide 16 by 3. What do they notice? How many wholes are there? What is the remainder? Ask other similar, simple remainder questions. Observe the strategies that children use.

Discover

Understand divisibility ②

WAYS OF WORKING Pair work

ASK

- Question ① a): *How many apples are there? How many should you put into a group? Do you think you will have any left? Can you form an array to help you? How many do you have left over once all the bags are full? What is the amount left over called?*
- Question ① b): *How can what you have done be written as a division? Is it grouping or sharing? What number are you dividing by what? How many whole lines or groups have you formed? What is the remainder?*

IN FOCUS This lesson formalises writing division with a remainder. In question ① a), children start with 22 counters (representing apples) and have to group them into 5s. Before doing the grouping, children should consider if they are likely to have any apples left over at the end. They may be able to reason that they will because it is not a multiple of 5. Children may find it easier to form an array made up of rows of 5 counters. This will help them see how many complete rows are formed and how many are left over. Question ① b) focuses on how children may record their answer. Children need to understand that they are carrying out a division, as they are grouping, and then they need to realise that the answer is not an exact whole and that there will be a remainder.

PRACTICAL TIPS Ensure that there are enough counters or cubes for children to be able to group the apples into 5s.

ANSWERS

Question ① a): There are 4 full bags and 2 apples left over.

Question ① b): 22 ÷ 5 = 4 remainder 2

Discover

① a) There are 22 apples
They are packed in bags of 5.
How many full bags are made?
How many apples are left over?

b) Write the calculation as a division.

☐ ÷ ☐ = ☐ remainder ☐

236

PUPIL TEXTBOOK 3A PAGE 236

Share

WAYS OF WORKING Whole class teacher led

ASK

- Question ① a): *Can you explain what Flo is doing in each step? Is this an example of grouping or sharing? Could you have predicted that there were going to be some left over at the end? If so, how did you know? How does the array help you find the number of groups and the remainder?*
- Question ① b): *Can you explain why it is a division? Why is this grouping and not sharing? Could we have done it as sharing? If so, what would the remainder have been?*

IN FOCUS In question ① a), talk through the approach to dividing 22 by 5. 5 apples are grouped in a bag and there are 17 left over, and so on, until there are 4 bags and 2 apples left over. Model this with cubes, counters or other objects. Also show the method where children form an array. The array helps children to keep track of the number of groups and ensures that they have the same number in each group. Arrays can be used for grouping and for sharing. Question ① b) shows how this can be written and recorded more formally. Children need to understand that this is a division, as they are grouping objects. Show that the answer would have been the same if they were sharing 22 apples between 5 people: each person would have 4 apples with 2 left over.

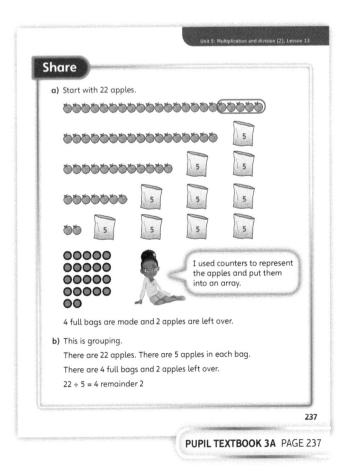

PUPIL TEXTBOOK 3A PAGE 237

Think together

WAYS OF WORKING Whole class teacher led (I do, We do, You do)

ASK

- Question ❶: *Is this equal grouping or sharing? How many pieces of fruit will go in each bowl? How do you know at the start that there are going to be some left over?*
- Question ❸: *Can you tell what the remainder is going to be at the start? Can you do any of these without using equipment? Did you use a sharing or a grouping method? What do you notice about numbers that do not have a remainder when you divide by 3? Where have you seen these numbers before?*

IN FOCUS
Question ❶ is a sharing example as opposed to a grouping example. Children should see that the remainder is the amount left over that cannot be shared equally. In question ❸, children can choose to use either a grouping or a sharing method to work out the whole and the remainder. Can children work out any of the remainders without doing the full calculation? Question ❸ b) asks children to find a number that can be divided by 3 with no remainder. Children may generalise that these are numbers in the 3 times-table.

STRENGTHEN
To support understanding of remainders, in question ❷, ask children to use concrete objects to carry out both grouping and sharing procedures. They should realise that the answer remains the same regardless of which type of division is used. Ask children to carry out the grouping or sharing step by step and, at the end, count the number of groups formed (or the number in each group for sharing) and the amount remaining. Show children how the answer can then be written. Children may find that making arrays helps them see the answers more clearly: the number of rows (or number in each row) will help them see the whole, and the number left over not in the array is the remainder.

DEEPEN
Build on question ❸ by asking: *When you divide by 10, what other numbers have a remainder that is also 1? What numbers have a remainder of 2 when you divide by 10? Can you spot a pattern?*

ASSESSMENT CHECKPOINT
Children use a grouping or sharing strategy to work out the answers to a simple division that includes a remainder. Children know how to record their answers in the form '☐ remainder ☐'. They should be able to talk about what each number in the answer means (such as the number of pieces of fruit in each bowl and the amount left over).

ANSWERS

Question ❶: There are 3 oranges in each bowl.
There is 1 orange left over.
$7 \div 2 = 3$ remainder 1

Question ❷: $14 \div 3 = 4$ remainder 2

Question ❸ a): $12 \div 5 = 2$ remainder 2
$17 \div 4 = 4$ remainder 1
$13 \div 8 = 1$ remainder 5
$51 \div 10 = 5$ remainder 1

Question ❸ b): Numbers in the 3 times-table have a remainder of 0.

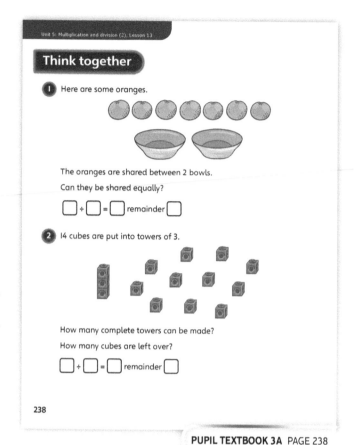

Think together

❶ Here are some oranges.

The oranges are shared between 2 bowls.
Can they be shared equally?

☐ ÷ ☐ = ☐ remainder ☐

❷ 14 cubes are put into towers of 3.

How many complete towers can be made?
How many cubes are left over?

☐ ÷ ☐ = ☐ remainder ☐

238

PUPIL TEXTBOOK 3A PAGE 238

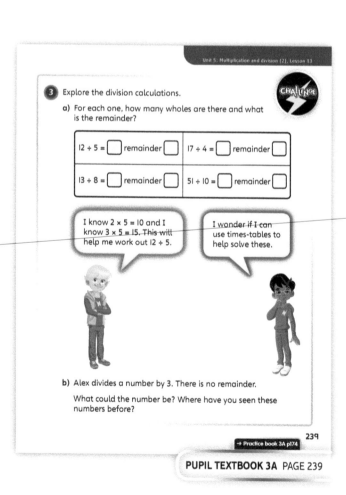

❸ Explore the division calculations.   CHALLENGE

a) For each one, how many wholes are there and what is the remainder?

$12 \div 5 = $ ☐ remainder ☐ $17 \div 4 = $ ☐ remainder ☐

$13 \div 8 = $ ☐ remainder ☐ $51 \div 10 = $ ☐ remainder ☐

I know $2 \times 5 = 10$ and I know $3 \times 5 = 15$. This will help me work out $12 \div 5$.

I wonder if I can use times-tables to help solve these.

b) Alex divides a number by 3. There is no remainder.
What could the number be? Where have you seen these numbers before?

→ Practice book 3A p174

239

PUPIL TEXTBOOK 3A PAGE 239

Practice

WAYS OF WORKING Independent thinking

IN FOCUS Question 4 keeps the amount children are dividing the same and changes the number they are dividing by. This helps children focus on the process. Question 5 provides abstract practice. Here, children need to work out which divisions will have a remainder: they should realise that the questions without a remainder are those where the number they are dividing is in the relevant times-table. You should ensure that children know what each number in a division and the answer mean.

STRENGTHEN To support understanding of remainders, children should use concrete objects and carry out both grouping and sharing procedures for division. In questions 4 and 5, children may need support to realise how many counters they need to take and the number of groups they need to make (or the number in each group). Children need to see the clear link between the numbers in the question and the objects.

DEEPEN To build on question 4, ask children to come up with a division where the remainder is 5. Ask: *What would you divide by?* Ask children to create their own problems similar to the ones in questions 6 and 7, where they have to work out the number being divided. You can also ask children to find examples, such as ☐ ÷ ☐ = 2 remainder 1.

THINK DIFFERENTLY Question 6 a) requires children to think about the numbers that will give a remainder of 1 when divided by 4. Children should notice that the numbers that have a remainder of 1 are all numbers that are 1 greater than a number in the 4 times-table. Question 6 b) asks children to reason that the remainder cannot be bigger than the number you are dividing by.

ASSESSMENT CHECKPOINT Children know how to find the answer to a division that includes a remainder and how to write their answers formally.

ANSWERS Answers for the **Practice** part of the lesson can be found in the *Power Maths* online subscription.

Reflect

WAYS OF WORKING Pair work

IN FOCUS This focuses on numbers that have a remainder of 0 when you divide by 3. Children should notice, through discussion, that numbers in the 3 times-table have no remainder when they divide by 3. They then attempt to find numbers that have a remainder of 1. They may use a trial and improvement strategy to realise that numbers that are 1 more than numbers in the 3 times-table have a remainder of 1.

ASSESSMENT CHECKPOINT Children know that numbers in the 3 times-table have no remainder when you divide by 3. They know that if a number is 1 more than a number in the 3 times-table, then it has a remainder of 1.

ANSWERS Answers for the **Reflect** part of the lesson can be found in the *Power Maths* online subscription.

After the lesson

- Can children complete a simple division that has a remainder, using either a grouping or a sharing strategy?
- Do children understand why sometimes a division has a remainder and can they predict when there will be a remainder?
- Can children record their answers for division formally, such as: 13 ÷ 5 = 2 remainder 3?

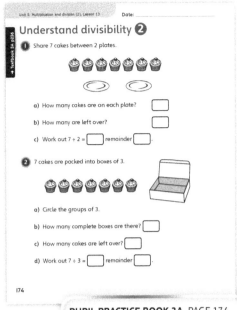

Unit 5: Multiplication and division (2), Lesson 13   Date:

Understand divisibility 2

1 Share 7 cakes between 2 plates.

a) How many cakes are on each plate? ☐

b) How many are left over? ☐

c) Work out 7 ÷ 2 = ☐ remainder ☐ .

2 7 cakes are packed into boxes of 3.

a) Circle the groups of 3.

b) How many complete boxes are there? ☐

c) How many cakes are left over? ☐

d) Work out 7 ÷ 3 = ☐ remainder ☐ .

174

PUPIL PRACTICE BOOK 3A PAGE 174

Unit 5: Multiplication and division (2), Lesson 13

3 These apples are shared between 4 children.

How many apples does each child get?
How many apples are left over?

9 ÷ 4 = ☐ remainder ☐

4 Use counters to work out the divisions.

a) 15 ÷ 2 = ☐ remainder ☐

b) 15 ÷ 3 = ☐ remainder ☐

c) 15 ÷ 4 = ☐ remainder ☐

d) 15 ÷ 5 = ☐ remainder ☐

e) 15 ÷ 6 = ☐ remainder ☐

5 Look at these calculations.
Circle the three that will have a remainder.
Discuss with a partner how you know.

13 ÷ 3   20 ÷ 5   19 ÷ 4   28 ÷ 10   30 ÷ 2   48 ÷ 8

175

PUPIL PRACTICE BOOK 3A PAGE 175

Unit 5: Multiplication and division (2), Lesson 13

6 a) When I divide a number by 4, the remainder is 1.
What could the number be?
Find three possible numbers.
_____

b) When Jamie divides her number by 5, the remainder is 7.
How do you know that Jamie has made a mistake?
_____
_____

7 Work out the missing number.   CHALLENGE
☐ ÷ 5 = 4 remainder 4

Reflect

What is the remainder when each of these numbers is divided by 3?
3, 6, 9, 12, 15, 18, 21, 24, 27, 30
How do you know?
What numbers will give a remainder of 1 when you divide by 3?
- _____
- _____
- _____
- _____

176

PUPIL PRACTICE BOOK 3A PAGE 176

End of unit check

Don't forget the unit assessment grid in your _Power Maths_ online subscription.

WAYS OF WORKING Group work adult led

IN FOCUS These questions cover the multiplication and division methods from this unit. They are designed to draw out particular misconceptions or misunderstandings.

Questions ❸ and ❹ look at what strategies children adopt. In question ❹, for example, do children just work out all the answers or do they reason correctly why b) must be correct (12 is double 6, and 4 is half of 8, so 12 × 4 must be the same)?

For question ❼, ask children to check that both sides are balanced once they have worked out the answer.

ANSWERS AND COMMENTARY Children who have mastered this unit will start to know their 3, 4 and 8 times-tables off by heart and be able to confidently apply the relevant multiplication facts to calculations. Children can explain when to multiply and when to divide. They will know that some divisions do not always give a whole answer and can have a remainder.

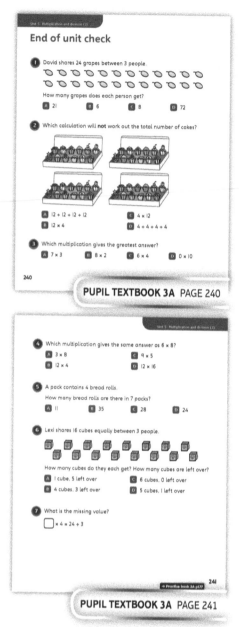

PUPIL TEXTBOOK 3A PAGE 240

PUPIL TEXTBOOK 3A PAGE 241

Q	A	WRONG ANSWERS AND MISCONCEPTIONS	STRENGTHENING UNDERSTANDING
1	C	A and D indicate children do not realise it is a division, B reflects a common misconception that 24 ÷ 3 = 6.	For question 1, use counters to represent the grapes and share them into 3 groups.
2	D	D shows children recognise there are four 12s being added together, not four 4s.	For question 4, children can rearrange an array to show why b) is the correct answer.
3	C	B shows that children think the greater the number in the calculation, the greater the answer.	In question 5, children can use a bar model to represent the situation and show it is a multiplication.
4	B	A, C or D shows children have not noticed both numbers have halved or doubled.	For question 6, children can use a sharing model (with counters).
5	C	The other answers show children have not understood they need to multiply or that they have multiplied incorrectly.	In question 7, both sides of the statement must be balanced.
6	D	A shows children have got the whole and the remainder confused. B or C shows they have not understood what a remainder is, or have divided incorrectly.	
7	2	Incorrect answers may suggest that children have misunderstood the problem and solved 24 ÷ 3 or created a multiplication with an answer of 24.	

My journal

WAYS OF WORKING Independent thinking

ANSWERS AND COMMENTARY

This activity draws together learning about multiplication and division facts.

Look for the different strategies that children use.

a) 30, 40, 50 … Look out for children who realise that this has to be a multiple of 10 above 25.

b) 24, 48 … Look out for children who realise that they need to focus on the 3 and 8 times-tables, as a number in the 8 times-table will definitely be in the 4 times-table.

c) 40 or 0. Children often overlook 0 in the times-tables. Remind them that 0 is in every times-table.

d) 0 or 60. Children may notice that 60 is 3 × 4 × 5 multiplied together because these numbers have no common factors except 1.

e) 0 or 120.

Power check

WAYS OF WORKING Independent thinking

ASK

- *How many multiplication facts do you think you know off by heart?*
- *Do you feel confident about deciding when a division is equal grouping and when it is equal sharing?*
- *Do you know how to solve word problems? Can you work out if it is multiplication or division?*

Power play

WAYS OF WORKING Independent thinking

IN FOCUS Use this **Power play** to test fluency with times-tables. Fluency and rapid recall of multiplication facts are essential for problem solving and free up valuable working memory; regular practice will help. Support children in understanding what multiplication facts mean; for example, 2 × 3 can be visualised as 2 groups of 3 or as a 2 × 3 array.

ANSWERS AND COMMENTARY The first set of wheels tests times-table fluency.

×3: 21, 6, 15, 18, 30, 36, 3, 0, 12, 9   ×5: 35, 2, 40, 12, 25, 9, 15, 6, 55

×4: 16, 24, 36, 48, 0, 4, 32, 12, 20, 44   ×8: 8, 1, 2, 3, 4, 0, 10, 11, 7, 5

The second set of wheels tests multiple times-tables. The inner number multiplied by the middle number gives the outer number.

Top left: 60, 20, 32, 8, 8, 18, 80, 48

Top right: inner: 4, 11, 12, 4, 8; middle: 8, 7; outer: 20, 21, 18

Bottom left: This wheel can have different answers as children need to find two numbers that make the outer number when multiplied together.

After the unit

- Do children know the 2, 3, 4, 5, 8 and 10 times-tables off by heart?
- Do children know how to use these times-tables to solve multiplication and division problems?
- Do they know that a division may have a remainder?

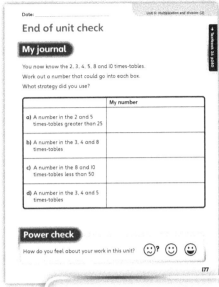

PUPIL PRACTICE BOOK 3A PAGE 177

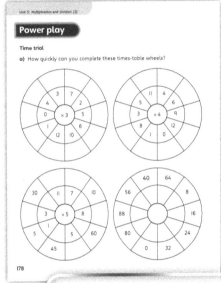

PUPIL PRACTICE BOOK 3A PAGE 178

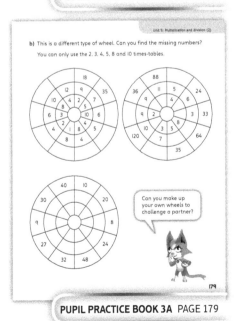

PUPIL PRACTICE BOOK 3A PAGE 179

Strengthen and **Deepen** activities for this unit can be found in the *Power Maths* online subscription.

279

Published by Pearson Education Limited, 80 Strand, London, WC2R 0RL.

www.pearsonschools.co.uk

Text © Pearson Education Limited 2018, 2022
Edited by Pearson and Florence Production Ltd
First edition edited by Pearson, Little Grey Cells Publishing Services and Haremi Ltd
Designed and typeset by Pearson and Florence Production Ltd
First edition designed and typeset by Kamae Design
Original illustrations © Pearson Education Limited 2018, 2022
Illustrated by Nigel Dobbyn, Virginia Fontanabona, Paul Moran, Nadene Naude and Emily Skinner
at Beehive Illustration, Kamae Design and Florence Production Ltd
Images: Bank of England: 245; The Royal Mint, 2017: 33, 239, 244
Cover design by Pearson Education Ltd
Back cover illustration © Diego Diaz and Nadene Naude at Beehive Illustration
Series editor: Tony Staneff; Lead author: Josh Lury
Authors (first edition): Tony Staneff and Josh Lury
Consultants (first edition): Professor Jian Liu and Professor Zhang Dan

The rights of Tony Staneff and Josh Lury to be identified as authors of this work have been
asserted by them in accordance with the Copyright, Designs and Patents Act 1988.

First published 2018
This edition first published 2022

26 25 24 23 22
10 9 8 7 6 5 4 3 2 1

British Library Cataloguing in Publication Data
A catalogue record for this book is available from the British Library

ISBN 978 1 292 45053 7

Printed in the UK by Ashford Press Ltd

For Power Maths online resources, go to:
www.activelearnprimary.co.uk

Note from the publisher
Pearson has robust editorial processes, including answer and fact checks, to ensure the
accuracy of the content in this publication, and every effort is made to ensure this publication
is free of errors. We are, however, only human, and occasionally errors do occur. Pearson is
not liable for any misunderstandings that arise as a result of errors in this publication, but it is
our priority to ensure that the content is accurate. If you spot an error, please do contact us at
resourcescorrections@pearson.com so we can make sure it is corrected.